Die Grundlehren der mathematischen Wissenschaften

in Einzeldarstellungen
mit besonderer Berücksichtigung
der Anwendungsgebiete

Band 158

Herausgegeben von
J. L. Doob · A. Grothendieck · E. Heinz · F. Hirzebruch
E. Hopf · W. Maak · S. Mac Lane · W. Magnus · J. K. Moser
M. M. Postnikov · F. K. Schmidt · D. S. Scott · K. Stein

Geschäftsführende Herausgeber
B. Eckmann und B. L. van der Waerden

Corneliu Constantinescu · Aurel Cornea

Potential Theory on Harmonic Spaces

Springer-Verlag Berlin Heidelberg New York 1972

Corneliu Constantinescu · Aurel Cornea
Mathematical Institute of the Roumanian Academy, Bucharest

AMS Subject Classifications (1970): 31C05, 31D05

ISBN 3-540-05916-4 Springer-Verlag Berlin Heidelberg New York
ISBN 0-387-05916-4 Springer-Verlag New York Heidelberg Berlin

 Library of Congress Catalog Card Number 72-86117. Printed in Germany. Typesetting, printing and binding: Universitätsdruckerei H. Stürtz AG, Würzburg.

Preface

There has been a considerable revival of interest in potential theory during the last 20 years. This is made evident by the appearance of new mathematical disciplines in that period which now-a-days are considered as parts of potential theory. Examples of such disciplines are: the theory of Choquet capacities, of Dirichlet spaces, of martingales and Markov processes, of integral representation in convex compact sets as well as the theory of harmonic spaces. All these theories have roots in classical potential theory.

The theory of harmonic spaces, sometimes also called axiomatic theory of harmonic functions, plays a particular role among the above mentioned theories. On the one hand, this theory has particularly close connections with classical potential theory. Its main notion is that of a harmonic function and its main aim is the generalization and unification of classical results and methods for application to an extended class of elliptic and parabolic second order partial differential equations. On the other hand, the theory of harmonic spaces is closely related to the theory of Markov processes. In fact, all important notions and results of the theory have a probabilistic interpretation.

Based on ideas of Brelot, Doob and Tautz, the theory of harmonic spaces has developed so rapidly in the last 15 years that up to now original papers, seminar and lecture notes were the only source for further studies in this field. In view of this situation, C. Constantinescu and A. Cornea who both had considerably influenced the development of the theory, undertook the difficult task to present our present-day knowledge of harmonic spaces as completely as possible.

The result of their effort is this book. It introduces the notion of a harmonic space in a somewhat refined form in order to proceed to the main examples of harmonic spaces without too many detours. It offers a wealth of material not only within the main text but also in the form of exercises.

With this monograph the authors do not only close a serious gap in the existing mathematical literature, the monograph certainly will also have a strong impact on future research in the field of potential theory.

Erlangen, July 1971 Heinz Bauer

Contents

Introduction

Potential theory is a very old area of mathematics. Its origins may be placed in the 18th century when J. Lagrange remarked in 1773 that the gravitational forces derive from a function (called a potential function by G. Green in 1828 and simply a potential by C. F. Gauss in 1840) and when P. S. Laplace showed in 1782 that in a mass free region this function satisfies the partial differential equation which today bears his name. The fundamental principles of this theory were elaborated during the last century and this theory constitutes today classical potential theory. In 1823 S.D. Poisson introduced his integral formula on the disc and on the ball in order to solve the first boundary-value problem for the Laplace equation (also called the Dirichlet problem); in 1828 G. Green invented the Green function and, with its aid, solved the Dirichlet problem for domains with sufficiently smooth boundary; in 1839 S. Earnshaw proved the minimum principle for the solution of the Laplace equation and in 1840 C. F. Gauss, in a celebrated paper, resolved the equilibrium problem, developed a capacity theory and gave a new solution for the Dirichlet problem. Needless to say, the rigour of these proofs left much to be desired. This prompted many mathematicians to come back to the problems: W. Thomson studied the Poisson integral formula and gave it the form known today; H. A. Schwarz, in 1870, gave the first rigorous proof of the behaviour of this integral at the boundary. The Dirichlet problem was also studied by L. Dirichlet and B. Riemann in 1853, but it was again H. A. Schwarz in 1870 who succeeded in giving the first rigorous proof for the two dimensional case, using the alternating method; the corresponding problem in three dimensions having to wait until 1887 for its solution via the balayage method of H. Poincaré. The proof in 1886 by A. Harnack of the inequality to which he lent his name must be considered an important contribution to potential theory. From his inequality he deduced the convergence property of monotone sequences. Thus by the end of the last century the three basic principles of potential theory, namely the Dirichlet problem, the minimum principle and the convergence property were established.

Gradually it became clear that these properties of the Laplace equation are also shared by other partial differential equations such as the heat equation or, more generally, linear elliptic or parabolic equations of second order. On the other hand, step by step, it was remarked that a large part of the results of potential theory could be obtained using only the above three principles. It seemed therefore quite natural to develop an axiomatic system which would unify these theories and extend potential theory to these partial differential equations. This axiomatic theory was constructed around the nineteen fifties, by G. Tautz, J. L. Doob, M. Brelot and H. Bauer. Their theory started with a linear sheaf of continuous real functions defined on a locally compact space (this sheaf playing the role of the sheaf of solutions of a partial differential equation) for which a convergence property, a minimum principle and the possibility of solving the Dirichlet problem for sufficiently many open sets was given. Such a construction had the advantage of being more elegant and more general and of giving better insight into the implications of the various results. It was used also as a guide for much research in the field of partial differential equations, and it drew attention to the key properties that had to be proved. Since it can be shown that under rather mild conditions there exist suitable Markov processes associated with the theory, it may be also used as a link between partial differential equations and Markov processes.

For the parabolic equations the proof that it is possible to solve the Dirichlet problem for a sufficiently large class of open sets is very difficult and even needs the development of a part of potential theory. We therefore considered it convenient to weaken the corresponding axiom, which can be done without giving up any interesting results of the theory. This leads to a more general axiomatic system but obliges us to take as the starting point the sheaf of hyperharmonic functions instead of the sheaf of harmonic functions.

The book is divided in three parts. In the first one (Chapter 1–3) harmonic spaces are presented, in the second, (Chapters 4–8) general problems are treated (balayage, natural and specific order, negligible sets) and in the last one (Chapters 9–11) three special problems (the axiom of domination and the axiom of polarity, Markov processes and the integral representation) are discussed. Two other special topics, namely duality and ideal boundaries are not treated here, since they are not yet sufficiently developed.

The connections between the chapters are roughly indicated by the following diagram:

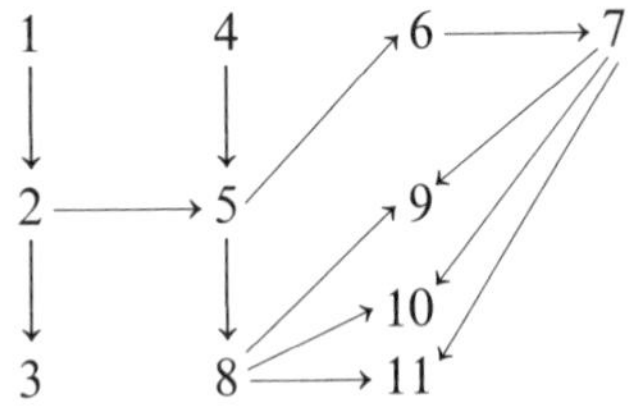

However, with the exception of the last three chapters which are independent from each other, it is assumed in each chapter that the definitions and the theorems of the preceding ones are known.

A difficult problem was to decide what attitude to take concerning the countable base of the underlying space. The interesting examples of the theory always have a countable base and the theory becomes simpler with this assumption. Nevertheless many of the theorems still hold in the uncountable case and therefore except for some sections we did not assume that the harmonic space had a countable base. The future evolution of the theory will show whether or not this decision was right.

Nearly all of the sections contain exercises. Their principal aim is to present supplementary results closely related to the material treated in the section. A few exercises give counterexamples, which indicate the limitations of the theory as well as the necessity of some of the hypotheses appearing in some of the theorems. There are also some exercises, which present to the reader important results (or sometimes, outlines of whole theories), that have been obtained in neighbouring fields. In the exercises, the references, besides their historical role, have the mission to indicate places were proofs may be found. We draw attention to the fact that the hypotheses assumed in a section are also supposed to hold in the exercises (unless the contrary is explicitly mentioned).

The historical notes and the bibliographical indications are certainly not complete. The emergence of a notion or of a result in mathematics is usually made in a long series of small steps. We tried to indicate those works where we felt that the principal contributions were made. Frequently we quoted the classical papers for theorems about harmonic spaces. It was done so when we thought that the essential difficulty was solved in the classical case. We apologise for the mistakes and the omissions. Besides the References (to which the reader is sent via the bibliographical indications) there exists also a list of papers, entitled Bibliography, which were not quoted in the book, but may present interest for the theory of harmonic spaces.

The book is intended for those wishing to do research within the theory itself and those wishing to use it in other fields (as for instance semi-elliptic partial differential equations or Markov processes). Elementary notions of general topology and integration theory on locally compact spaces are assumed; in these fields we quote N. Bourbaki. Some exercises may require more knowledge from other fields of analysis.

We express our gratitude to Prof. Heinz Bauer for his kindness in writing a preface to the book.

Terminology and Notation

The purpose of this section is to make precise the terminology and the notation for some of the notions which we use in the present book, especially for those terms and notations which either are not very usual or are used with different meanings in contemporary mathematical literature. Generally however, we have followed Bourbaki's terminology and notation.

If A, B are sets, $A \setminus B$ denotes the set of elements belonging to A and not belonging to B. If $R(x)$ is a proposition concerning x and A is a set, we denote by $\{x \in A \mid R(x)\}$, the set of elements x of A for which $R(x)$ holds. If A, B are sets and for any $x \in A$, $T(x)$ is an element of B, we shall denote by $x \mapsto T(x)$ the map of A into B which associates with any $x \in A$ the element $T(x) \in B$ and by $\{T(x) \mid x \in A\}$ the image of A through this map. Let A be a set. We say that a function is defined on A if its domain contains A. If f is a function defined on A (resp. if $\mathscr{F}$ is a set of functions defined on A) and if B is a subset of A we denote by $f|_B$ (resp. $\mathscr{F}|_B$) the restriction of f to B (resp. the set of restrictions to B of the functions of $\mathscr{F}$).

Let A be an ordered set and B a subset of A. We call the least upper bound or supremum (resp. the greatest lower bound or infimum) of B in A, if it exists, the smallest (resp. greatest) element of A which dominates every element of B (resp. is dominated by any element of B). The ordered set A is called inductive if any chain (i.e. linearily ordered subset) of A possesses a majorant in A. The ordered set A is called upper (resp. lower) directed if any finite non-empty subset possesses a majorant (resp. minorant). If A is upper directed we call the filter generated by the filter base

$$\{\{y \in A \mid y \geq x\} \mid x \in A\}$$

the section filter of A.

The ordered set A is called a lattice if any finite non-empty subset of A possesses a supremum and an infimum. The ordered set A is called upper complete (resp. conditionally complete) if any non-empty subset of A (resp. any non-empty subset of A possessing a majorant) possesses a least upper bound. A subset B of a lattice A is called a sublattice if the supremum and the infimum in A of any finite non-empty subset of B

belongs to B. If x, y are elements of an ordered set A we denote:

$$[x, y]:=\{z\in A \mid x\leq z\leq y\}, \quad [x, y[:=\{z\in A \mid x\leq z<y\},$$
$$]x, y]:=\{z\in A \mid x<z\leq y\}, \quad]x, y[:=\{z\in A \mid x<z<y\}.$$

Let X be a topological space. For any subset A of X we denote by ∂A the boundary of A and we call boundary (resp. adherent) point of A any point of ∂A (resp. $A\cup\partial A$). An open covering of X is a set $\mathfrak{W}$ of open sets of X such that

$$X=\bigcup_{W\in\mathfrak{W}} W.$$

A compact space will, in this book, always be Hausdorff (=separated). We say that the topological space X (resp. a subset of X) is σ-compact (resp. of type K_σ) if it is the union of a countable family of compact sets. A subset of a topological space is said to be meagre if it is the union of a countable family of nowhere dense sets.

We denote by $\mathbb{N}$, $\mathbb{Z}$, $\mathbb{R}$ and $\overline{\mathbb{R}}$ the set of natural numbers (0 included), the set of integers, the set of real numbers, and the extended real line respectively, each endowed with its natural order, algebraic and topological structure. We say that $\alpha\in\overline{\mathbb{R}}$ is positive (resp. strictly positive) if $\alpha\geq 0$ (resp. $\alpha>0$) and denote by $\mathbb{R}_+$ the set of positive real numbers. Throughout this book we shall use the conventions

$$0\cdot\infty=\infty\cdot 0=0\cdot(-\infty)=(-\infty)\cdot 0=\frac{1}{\infty}=0.$$

A real vector space is a vector space over the field $\mathbb{R}$. A subset A of a real vector space is called circled (resp. convex) if

$$x\in A, \quad \alpha\in\mathbb{R}, \quad |\alpha|\leq 1 \Rightarrow \alpha x\in A$$
$$(\text{resp. } x, y\in A, \quad \alpha, \beta\in\mathbb{R}_+, \quad \alpha+\beta=1 \Rightarrow \alpha x+\beta y\in A).$$

A vector lattice is a real vector space E endowed with an order relation such that for any $x, y, z\in E$ and for any positive real number α, we have

$$x\leq y\Rightarrow x+z\leq y+z, \qquad \alpha x\leq \alpha y$$

and such that E is a lattice with respect to the order relation. If E is a conditionally complete vector lattice, we call the subset A of E a band if

$$x, y\in A, \quad \alpha, \beta\in\mathbb{R} \Rightarrow \alpha x+\beta y\in A,$$
$$x\in E, \ y\in A, \quad 0\leq x\leq y \Rightarrow x\in A,$$

$B\subset A \Rightarrow$ the least upper bound of B in E belongs to A if it exists.

A real, numerical or lower finite numerical function is a function whose range of values is $\mathbb{R}$, $\overline{\mathbb{R}}$ or $\mathbb{R}\cup\{\infty\}$ respectively. We identify the elements of $\overline{\mathbb{R}}$ with the constant numerical functions. The set of numerical functions on a set A will be considered an ordered set by the natural order relation, i.e. $f \leq g$ if $f(x) \leq g(x)$ for any $x \in A$. A numerical function f on A will be called positive (resp. strictly positive) if $f(x) \geq 0$ (resp. $f(x) > 0$) for any $x \in A$. If $\mathscr{F}$ is a set of numerical functions on a set A then $\mathscr{F}_+$ denotes the set of positive functions of $\mathscr{F}$. For any numerical function f on A and for any subset B of A we denote by $\inf_B f$ (resp. $\sup_B f$) the infimum (resp. supremum) in $\overline{\mathbb{R}}$ of the set $f(B)$. If $(f_\iota)_{\iota \in I}$ is a family of numerical functions on A we denote by $\inf\limits_{\iota \in I} f_\iota$ (resp. $\sup\limits_{\iota \in I} f_\iota$) the infimum (resp. supremum) of this family in the set of all numerical functions on A; for two numerical functions f, g on A we denote by $\inf(f, g)$ (resp. $\sup(f, g)$) the infimum (resp. supremum) of f, g. Let f, g be two lower finite, numerical functions on a set A. We denote by $f+g$ the function on A equal to $f(x)+g(x)$ for every $x \in A$. If f is a numerical function on A and α is a real number, αf will denote the function $x \mapsto \alpha f(x)$ on A; moreover we write $-f$ instead of $-1f$. If $\mathscr{F}$ is a set of numerical functions on A we put

$$-\mathscr{F} := \{-f \mid f \in \mathscr{F}\}.$$

We say that a set $\mathscr{F}$ of numerical functions on a set A separates the points of A if for any $x, y \in A$, $x \neq y$, there exists $f, g \in \mathscr{F}$ such that

$$f(x)\, g(y) \neq f(y)\, g(x).$$

We say that a non-empty set of lower finite numerical functions on a set is a convex cone if for any $f, g \in \mathscr{F}$ and for any positive real numbers α, β the function $\alpha f + \beta g$ belongs to $\mathscr{F}$. Since by the above convention $0 \cdot \infty = 0$ the function 0 belongs to any convex cone of functions. A non-empty set $\mathscr{F}$ of real functions on a set is called a real vector space of real functions if for any $f, g \in \mathscr{F}$ and for any $\alpha, \beta \in \mathbb{R}$ we have $\alpha f + \beta g \in \mathscr{F}$.

Let X be a topological space. We shall say that an assertion holds locally on X if any point of X possesses a neighbourhood for which the given assertion holds. For any numerical function f on X we denote by $\hat{f}$ the greatest lower semi-continuous numerical function on X dominated by f; this function will be called the lower semi-continuous regularization of f. We set

$$\operatorname{Supp} f := \overline{\{x \in X \mid f(x) \neq 0\}}$$

and call this set the carrier of f. We denote by $\mathscr{C}(X)$ the set of continuous real functions on X endowed with its natural order and algebraic structures. Moreover if X is compact, $\mathscr{C}(X)$ will be endowed with the norm

$$f \mapsto \sup_X |f|.$$

If X is locally compact we denote by $\mathscr{K}(X)$ the subset of functions of $\mathscr{C}(X)$ with compact carrier. Let $\mathscr{F}$ be a real vector subspace of $\mathscr{C}(X)$. By the topology of pointwise convergence (resp. of compact convergence) on $\mathscr{F}$ we mean the topology having as a fundamental system of neighbourhoods of $f \in \mathscr{F}$, the sets of the form

$$\{g \in \mathscr{F} \mid x \in A \Rightarrow |g(x) - f(x)| < \varepsilon\},$$

were ε is an arbitrary strictly positive real number and A is an arbitrary finite (resp. compact) set of X.

Let X be a locally compact space. A numerical function on X is called a Borel (resp. Baire) function if it belongs to the smallest set of numerical functions on X containing the semi-continuous functions (resp. containing $\mathscr{K}(X)$) and closed with respect to the operation of taking the limits of monotone sequences. A Borel (resp. Baire) set is a set whose characteristic function is a Borel (resp. Baire) function. We call a positive linear map of $\mathscr{K}(X)$ into $\mathbb{R}$ a measure on X. Let Y be either a closed set or an open set of X; we shall identify the measures on X for which $X \setminus Y$ is of measure zero with the measures on Y for which the intersections with Y of compact sets of X are of finite measure. A linear map of $\mathscr{K}(X)$ into $\mathbb{R}$ which is the difference of two measures will be called a signed measure. The set of signed measures on a locally compact space will be considered a vector lattice with respect to the natural algebraic structure and with respect to the order relation $\mu \leq \nu$ if $\mu(f) \leq \nu(f)$ for any positive function of $\mathscr{K}(X)$. If μ is a signed measure on X we denote by $|\mu|$ the supremum in the set of signed measures on X of μ and $-\mu$. For any $x \in X$ we denote by ε_x the measure on X, $f \mapsto f(x)$. A filter $\mathfrak{F}$ on the set of measures on X is called vaguely convergent to a measure μ on X if

$$\lim_{\nu, \mathfrak{F}} \nu(f) = \mu(f)$$

for any $f \in \mathscr{K}(X)$. Let f be a lower semi-continuous, lower finite, numerical function on X which is positive outside a compact set and let $\mathscr{F}$ be the set of minorants of f belonging to $\mathscr{K}(X)$; for any measure μ on X we set

$$\int^{*} f \, d\mu := \sup_{g \in \mathscr{F}} \mu(g).$$

Let f be a numerical function on X and let $\mathscr{F}$ be the set of lower semi-continuous, lower finite, numerical functions on X which are positive outside a compact set of X and which dominate f; for any measure μ on X we set

$$\int^{*} f \, d\mu := \inf_{g \in \mathscr{F}} \int^{*} g \, d\mu, \qquad \int_{*} f \, d\mu := -\int^{*} (-f) \, d\mu.$$

Let X be a topological space. A sheaf of functions on X is a map $\mathscr{F}$ defined on the set of open sets of X such that:

a) for any open set U of X $\mathscr{F}(U)$ is a set of functions on U;

b) for any two open sets U, V of X such that $U \subset V$ the restriction of any function from $\mathscr{F}(V)$ to U belongs to $\mathscr{F}(U)$;

c) for any family $(U_\iota)_{\iota \in I}$ of open sets of X a function on $\bigcup_{\iota \in I} U_\iota$ belongs to $\mathscr{F}(\bigcup_{\iota \in I} U_\iota)$ if for any $\iota \in I$ its restriction to U_ι belongs to $\mathscr{F}(U_\iota)$.

We say that a real function defined on an open set of the n-dimensional space $\mathbb{R}^n$ is of class $\mathscr{C}^p$ $(p \in \mathbb{N})$ if it is p-times continuously differentiable.

PART ONE

Chapter 1

Harmonic Sheaves and Hyperharmonic Sheaves

Classical potential theory pointed out three main properties of harmonic and superharmonic functions, namely the Harnack convergence property, the resolutivity of the Dirichlet problem and the minimum principle and these in turn suggested the basic axioms for the recent axiomatic developements in potential theory. This chapter is devoted to an independent study of some aspects of each of these properties and to the introduction of some fundamental notions which are involved in them. They constitute the back-ground for the concept of a harmonic space which will be introduced and studied in the next chapter.

§ 1.1. Convergence Properties

A sheaf of functions $\mathscr{H}$ on a locally compact space is called a *harmonic sheaf* if for any open set U of X $\mathscr{H}(U)$ is a real vector space of real continuous functions on U. A function defined on a set containing an open set U is called an *$\mathscr{H}$-function on U* if its restriction to U belongs to $\mathscr{H}(U)$.

We shall say that a harmonic sheaf $\mathscr{H}$ on a locally compact space X possesses:

a) The Bauer convergence property if the limit function of any increasing sequence of $\mathscr{H}$-functions on any open set of X is an $\mathscr{H}$-function whenever it is locally bounded (H. Bauer 1960 [3]);

b) The Doob convergence property if the limit function of any increasing sequence of $\mathscr{H}$-functions on any open set of X is an $\mathscr{H}$-function whenever it is finite on a dense set (J. L. Doob 1956 [1]);

c) The Brelot convergence property if the limit function of any increasing sequence of $\mathscr{H}$-functions on any open connected set of X is an $\mathscr{H}$-function whenever it is finite at a point (M. Brelot 1957 [10]).

Each of these convergence properties was taken as an axiom in the axiomatic approach considered by H. Bauer, J. L. Doob and M. Brelot respectively and they are, in fact, the only essential distinction between them.

Obviously if X is locally connected, the Brelot convergence property implies the Doob convergence property and this in turn implies the Bauer convergence property. Neither of the remaining implications between these properties is true (see Exercises 3.1.7 and 11.1.9).

Proposition 1.1.1. *If a harmonic sheaf $\mathscr{H}$ possesses the Brelot convergence property then any positive $\mathscr{H}$-function on a connected open set vanishes identically if it vanishes at a point* (M. Brelot 1957 [10]).

Let h be a positive $\mathscr{H}$-function on a connected open set vanishing at a point. Then $(n\,h)_{n\in\mathbb{N}}$ is an increasing sequence of $\mathscr{H}$-functions whose limit is finite at a point. □

Proposition 1.1.2. *If a harmonic sheaf $\mathscr{H}$ possesses the Bauer (resp. Doob, resp. Brelot) convergence property then the supremum of any upper directed set of $\mathscr{H}$-functions on an open (resp. open, resp. open and connected) set of X is an $\mathscr{H}$-function whenever it is locally bounded (resp. finite on a dense set, resp. finite at a point)* (C. Constantinescu-A. Cornea 1963 [1], N. Boboc-P. Mustaţă 1967 [2], A. Cornea 1968 [2]).

Let us prove first the following lemma.

Lemma. *Let Y be a locally compact space and $\mathscr{F}$ be an upper directed set of numerical continuous functions on Y such that the supremum of any increasing sequence in $\mathscr{F}$ is continuous. Then the supremum of $\mathscr{F}$ is continuous* (A. Cornea 1968 [2]).

Denote by f the supremum of $\mathscr{F}$ and let us suppose that it is not continuous at a point $x\in Y$. Let α, β be two real numbers such that

$$f(x)<\alpha<\beta<\limsup_{y\to x} f(y).$$

We shall construct by induction an increasing sequence $(f_n)_{n\in\mathbb{N}}$ in $\mathscr{F}$, a decreasing sequence $(K_n)_{n\in\mathbb{N}}$ of compact neighbourhoods of x and a sequence $(x_n)_{n\in\mathbb{N}}$ in Y such that for any $n\in\mathbb{N}$

$$x_n\in K_n,\quad f_n(x_n)>\beta,\quad f_{n-1}<\alpha \quad \text{on } K_n.$$

We take as K_0 an arbitrary compact neighbourhood of x. By hypothesis, there exists a point x_0 of K_0 at which

$$f(x_0)>\beta.$$

Let f_0 be an element of $\mathscr{F}$ such that

$$f_0(x_0)>\beta.$$

Suppose now that the three sequences were constructed up to $n-1$. Let K_n be a compact neighbourhood of x such that $f_{n-1}<\alpha$ on K_n and x_n be a point of K_n at which $f(x_n)>\beta$. Since $\mathscr{F}$ is upper directed, there exists an element $f_n\in\mathscr{F}$ greater than f_{n-1} and such that $f_n(x_n)>\beta$.

Let g be supremum of the sequence $(f_n)_{n\in\mathbb{N}}$. By hypothesis g is continuous. Obviously

$$g\leq\alpha \quad \text{on} \quad \bigcap_{n\in\mathbb{N}} K_n.$$

Let y be a limit point of the sequence $(x_n)_{n\in\mathbb{N}}$ and note that for any $n\in\mathbb{N}$

$$g(x_n)\geq f_n(x_n)>\beta.$$

But this implies $g(y)\geq\beta$ which is a contradiction since

$$y\in\bigcap_{n\in\mathbb{N}} K_n. \quad \square$$

We prove now the proposition. Let $\mathscr{F}$ be an upper directed set of $\mathscr{H}$-functions on an open set U of X satisfying the hypotheses of the proposition and let h be its supremum. By the proceding lemma h is continuous. Let x be a point of U and K be a compact neighbourhood of x contined in U. By Dini's convergence theorem, there exists an increasing sequence $(h_n)_{n\in\mathbb{N}}$ in $\mathscr{F}$ converging to h on K. From the hypotheses of the proposition it follows that h is an $\mathscr{H}$-function on the interior of K, and therefore on a neighbourhood of x. Since $\mathscr{H}$ is a sheaf and x is an arbitrary point in U it follows that h is an $\mathscr{H}$-function. $\square$

We shall say that a harmonic sheaf $\mathscr{H}$ on a locally compact space X is *non-degenerate at a point* $x\in X$ if there exists an $\mathscr{H}$-function defined on a neighbourhood of x that does not vanish at x.

Theorem 1.1.1. *A locally compact space on which there exists a harmonic sheaf possessing the Bauer convergence property and non-degenerate at every point is locally connected* (N. Boboc-C. Constantinescu-A. Cornea 1965 [3]).

Let $\mathscr{H}$ be a harmonic sheaf on a locally compact space X possessing the Bauer convergence property and non-degenerate at every point of X. Choose $x\in X$ and U be an open neighbourhood of x on which there exists a strictly positive $\mathscr{H}$-function h. Let $\mathfrak{V}$ be the set of subsets of U which contain x and are both closed and open in U. For any $V\in\mathfrak{V}$ we denote by h_V the function on U equal to 0 on V and equal to h on $U\setminus V$. The family $(h_V)_{V\in\mathfrak{V}}$ is an upper directed family of $\mathscr{H}$-functions on U whose supremum h' is locally bounded. Hence h' is an $\mathscr{H}$-function and is therefore continuous. The set

$$\{y\in U|h'(y)=0\}=\{y\in U|h'(y)<h(y)\}$$

belongs to $\mathfrak{B}$ since it is open and closed in U and contains x. We deduce that $\mathfrak{B}$ contains a minimal element which must be therefore connected. □

Let X be a locally compact space and $\mathcal{H}$ be a harmonic sheaf on X. Let V be an open set of X; a family $\omega := (\omega_x)_{x\in V}$ of measures on ∂V will be called a *sweeping on* V. For any numerical function f defined on ∂V we shall denote by ωf the numerical function on V.

$$x \mapsto \int^{*} f \, d\omega_x.$$

The sweeping ω is called an *$\mathcal{H}$-sweeping* if :

a) V is relatively compact;

b) for any $f \in \mathscr{C}(\partial V)$ the function ωf is an $\mathcal{H}$-function;

c) for any $\mathcal{H}$-function h defined on an open neighbourhood of $\overline{V}$ we have $\omega h = h$ on V.

A sweeping system on X is a family $\Omega := ((\omega_{\iota x})_{x\in V_\iota})_{\iota\in I}$ such that $\{V_\iota | \iota \in I\}$ is a base for X of relatively compact sets and that for any $\iota \in I$ ω_ι is a sweeping on V_ι. The sweeping system Ω is called *$\mathcal{H}$-sweeping system* if for any $\iota \in I$ ω_ι is an $\mathcal{H}$-sweeping.

A relatively compact open set V of X is called *regular with respect to* $\mathcal{H}$ or *$\mathcal{H}$-regular* if any real continuous function f on ∂V possesses a unique continuous extention $\bar{f}$ to $\overline{V}$ which is an $\mathcal{H}$-function on V, positive if f is positive. For any $x \in V$ the map

$$f \mapsto \bar{f}(x) : \mathscr{C}(\partial V) \to \mathbb{R}$$

is a positive linear functional which defines therefore a measure ω_x^V on ∂V. The family $\omega^V := (\omega_x^V)_{x\in V}$ is obviously an $\mathcal{H}$-sweeping on V.

Proposition 1.1.3. *Let $\mathcal{H}$ be a harmonic sheaf on a locally compact space X, V be an $\mathcal{H}$-regular set and W be an open subset of V such that $\partial W \subset \partial V$. Then W is $\mathcal{H}$-regular and for any $x \in W$ we have $\omega_x^W = \omega_x^V$. Hence if X is locally connected, any component of a regular set is also regular. If $\mathcal{H}$ possesses the Brelot convergence property and if V is connected, then for any $x \in V$ the carrier of ω_x^V is equal to ∂V.* (M. Brelot 1958 [13]).

Let $f \in \mathscr{C}(\partial W)$. There exists $g \in \mathscr{C}(\partial V)$, positive if f is positive, such that $g|_{\partial W} = f$. The function $\bar{g}|_{\overline{W}}$ is a continuous extension of f to $\overline{W}$, is an $\mathcal{H}$-function on W, and is positive if f is positive. Let f', f'' be two continuous extentions of f to $\overline{W}$ which are $\mathcal{H}$-functions on W. The function on $\overline{V}$ which is equal to $f' - f''$ on $\overline{W}$ and 0 elsewhere is continuous, is an $\mathcal{H}$-function on V and vanishes on ∂V. Hence it vanishes identically and $f' = f''$. It follows that W is $\mathcal{H}$-regular and $\omega_x^W = \omega_x^V$ for any $x \in W$.

Suppose now that $\mathscr{H}$ possesses the Brelot convergence property and that V is connected. If $\partial V = \emptyset$ then the last assertion is trivial. Assume $\partial V \neq \emptyset$ and let $f \in \mathscr{C}(\partial V)$, $f \geq 0$, $f \neq 0$. If $\omega^V f(x) = 0$ for an $x \in V$ then $\omega^V f$ vanishes identically. (Proposition 1.1.1) We get for any $y \in \partial V$

$$f(y) = \lim_{x \to y} \omega^V f(x) = 0$$

which contradicts the hypothesis $f \neq 0$. We deduce that for any $x \in V$ the carrier of ω_x^V is equal to ∂V. □

Proposition 1.1.4. *Let $\mathscr{H}$ be a harmonic sheaf on a locally compact space, $(\omega_x)_{x \in V}$ be an $\mathscr{H}$-sweeping and f be a lower bounded numerical function on ∂V. If $\mathscr{H}$ possesses the Bauer convergence property then ωf is lower semi-continuous and for any $\mathscr{H}$-sweeping $(\omega'_x)_{x \in W}$ such that $\overline{W} \subset V$ we have $\omega' \omega f = \omega f$ on W. If f is bounded then ωf is an $\mathscr{H}$-function.*

Assume first that f is bounded and lower semi-continuous (resp. bounded) and let $\mathscr{F}$ be the set of real continuous (resp. bounded lower semi-continuous) functions on ∂V smaller (resp. greater) than f. Since $(\omega g)_{g \in \mathscr{F}}$ is an upper (resp. lower) directed family of $\mathscr{H}$-functions on V whose supremum (resp. infimum) is ωf, it follows by Proposition 1.1.2 that ωf is an $\mathscr{H}$-function.

When f is an arbitrary function then the proposition follows from

$$\omega f(x) = \int^* f \, d\omega_x = \lim_{n \to \infty} \int^* \inf(f, n) \, d\omega_x = \lim_{n \to \infty} (\omega \inf(f, n))(x)$$

for every $x \in V$. □

Proposition 1.1.5. *Let $\mathscr{H}$ be a harmonic sheaf on a locally compact space X. If there exists an $\mathscr{H}$-sweeping system on X, then the supremum of any upper directed family of $\mathscr{H}$-functions on an open set of X is an $\mathscr{H}$-function if it is continuous.*

Let $(h_\iota)_{\iota \in I}$ be an upper directed family of $\mathscr{H}$-functions on an open set U of X such that its supremum h is continuous. Let $(\omega_x)_{x \in V}$ be an $\mathscr{H}$-sweeping such that $\overline{V} \subset U$. Then for any $x \in V$

$$h(x) = \sup_{\iota \in I} h_\iota(x) = \sup_{\iota \in I} \int h_\iota \, d\omega_x = \int h \, d\omega_x .$$

Hence h is harmonic on V. □

Exercises

In all the exercises of this section X will denote a locally compact space and $\mathscr{H}$ a harmonic sheaf on X.

1.1.1. Let U be an open compact set of X. The following assertions are equivalent:

a) $\mathscr{H}(U)=\{0\}$;

b) there exists an $\mathscr{H}$-sweeping on U;

c) U is regular with respect to $\mathscr{H}$.

1.1.2. Let x be a point of X such that $\mathscr{H}$ is non-degenerated at x and let $((\omega_{\iota y})_{y\in V_\iota})_{\iota\in I}$ be an $\mathscr{H}$-sweeping system on X. Let $\mathfrak{V}$ be the set of neighbourhoods of x and for any $V\in\mathfrak{V}$ let

$$\mathfrak{M}_V:=\{\omega_{\iota y}|\iota\in I,\ y\in V_\iota\subset V\}.$$

Then $\{\mathfrak{M}_V|V\in\mathfrak{V}\}$ is a filter-base on the space of measures on X which converges vaguely to ε_x (M. Brelot 1960 [15], H. Bauer 1962 [5]).

1.1.3. Let f be a strictly positive continuous real function on X. We denote by $\mathscr{H}_f$ the harmonic sheaf on X $U\mapsto\left\{\frac{h}{f}\middle|h\in\mathscr{H}(U)\right\}$. If $(\omega_x)_{x\in V}$ is an $\mathscr{H}$-sweeping then $\left(\frac{f}{f(x)}\cdot\omega_x\right)_{x\in V}$ is an $\mathscr{H}_f$-sweeping. If $\mathscr{H}$ possesses the Bauer (resp. Doob, Brelot) convergence property then so does $\mathscr{H}_f$. The regular sets with respect to $\mathscr{H}$ and with respect to $\mathscr{H}_f$ coincide.

1.1.4. If there exists an $\mathscr{H}$-sweeping system on X then $\mathscr{H}(X)$ is complete with respect to the topology of compact convergence (H. Bauer 1962 [5]).

1.1.5. If $\mathscr{H}$ is non-degenerate at any point of X and possesses the Bauer convergence property, then $\mathscr{H}(X)$ is complete with respect to the topology of compact convergence.

1.1.6. Let V be a regular set with respect to $\mathscr{H}$, $\mathscr{F}$ be a uniformly bounded set of Borel functions on ∂V. For any $f\in\mathscr{F}$ we denote by $\bar{f}$ the function defined on $\bar{V}$ equal to f on ∂V and equal to $\omega^V f$ on V. If $\mathscr{F}$ is equicontinuous at a point $x\in\partial V$, then $\{\bar{f}|f\in\mathscr{F}\}$ is equicontinuous at x. If moreover $\mathscr{F}$ is equicontinuous on ∂V, then $\{\bar{f}|f\in\mathscr{F}\}$ is equicontinuous on $\bar{V}$ (G. Tautz 1943 [1]).

1.1.7. Let V be an open set of X such that ∂V is compact and metrisable and let $(\omega_x)_{x\in V}$ be a sweeping of V such that for any $f\in\mathscr{C}(\partial V)$, ωf is bounded and continuous. Then there exists a countable subset A of ∂V such that for any $f\in\mathscr{C}(\partial V)$ we have

$$\lim_{x\to y}\omega f(x)=f(y)$$

for all $y\in\partial V$ if this relation holds for all $y\in A$. (It is sufficient to assume in this exercise that X is a Hausdorff space) (G. Choquet 1968 [4]).

1.1.8. If for any open set U of X every uniformly bounded set of $\mathscr{H}$-functions on U is equicontinuous then $\mathscr{H}$ possesses the Bauer convergence property. (The converse is also true if there exists an $\mathscr{H}$-sweeping system on X: see Theorem 11.1.1).

1.1.9. Suppose that there exists an $\mathscr{H}$-sweeping system on X and that for any open set U of X and any compact set K of U, there exists a positive real continuous function δ on $K\times K$ vanishing on the diagonal $\{(x, y)\in K\times K \mid x=y\}$, and such that for any $h\in\mathscr{H}(U)$

$$\sup_{x,\, y\in K} \frac{1}{\delta(x, y)} |h(x)-h(y)| < \infty .$$

Then $\mathscr{H}$ possesses the Bauer convergence property. (Let U be an open σ-compact set of X and for any compact set K of U let p_K (resp. q_K) be the semi-norm on $\mathscr{H}(U)$ defined by

$$p_K(h)=\sup_{x\in K}|h(x)| \qquad \left(\text{resp. } q_K(h)=\sup_{x\in K}|h(x)|+\sup_{x,\, y\in K}\frac{1}{\delta(x, y)}|h(x)-h(y)|\right).$$

Then $\mathscr{H}(U)$ endowed with the family of semi-norms $(p_K)_K$ (resp. $(q_K)_K$) is a Frechet space (Exercise 1.1.4). Hence these associated locally convex topologies coincide (N. Bourbaki [3], Ch. I, §3, Corollaire 2 du Théorème 1). Let U' be an open set and let $\mathscr{F}$ be a set of locally uniformly bounded $\mathscr{H}$-functions on U'. In order to show that $\mathscr{F}$ is equicontinuous we may assume that U' is σ-compact. Then since the semi-norms p_K are bounded on $\mathscr{F}$ the same is true for the semi-norms q_K).

1.1.10. Let $\mathscr{H}$ be a harmonic sheaf on an open set X of $\mathbb{R}^n$ such that there exists an $\mathscr{H}$-sweeping system on X. If there exist a positive continuous real function δ on $\mathbb{R}_+$ vanishing at 0 such that for any open subset U of X, any compact subset K of U and any $h\in\mathscr{H}(U)$,

$$\sup_{x,\, y\in K}\frac{1}{\delta(|x-y|)}\,|h(x)-h(y)|<\infty ,$$

where

$$|x-y|=\left(\sum_{i=1}^{n}(x_i-y_i)^2\right)^{\frac{1}{2}},$$

then $\mathscr{H}$ possesses the Bauer convergence property. In particular, this is true if every $\mathscr{H}$-function is of class $\mathscr{C}^1$ (J.M. Bony 1967 [1]) (use Exercise 1.1.9).

1.1.11. If every set of locally uniformly bounded $\mathscr{H}$-functions on every open set of X is equicontinuous and if every point $x\in X$ possesses two open neighbourhoods U, V, $U\subset V$, such that $\mathscr{H}(V)$ separates the points of U, then X has locally a countable base (C. Constantinescu 1967 [5]).

1.1.12. If there exists an $\mathscr{H}$-sweeping system on X then the following assertions are equivalent:

a) $\mathscr{H}$ possesses the Doob convergence property;

b) for any $\mathscr{H}$-sweeping ω on an open set V of X and for any numerical function f on ∂V for which the upper integral with respect to ω_x is finite for x belonging to a dense subset of V, the function $\omega f \in \mathscr{H}(V)$;

c) there exists an $\mathscr{H}$-sweeping system $((\omega_{\iota x})_{x \in V_\iota})_{\iota \in I}$ on X such that for any $\iota \in I$ and for any lower semi-continuous real function f on ∂V_ι which is $\omega_{\iota x}$-integrable for x belonging to a dense set of V_ι, the function $\omega_\iota f \in \mathscr{H}(V_\iota)$. (H. Bauer 1963 [6].)

1.1.13. Suppose that X has a countable base and that $\mathscr{H}$ possesses the Doob convergence property. Then for any compact set K of X, there exists a finite subset A of X such that every positive $\mathscr{H}$-function on X vanishes on K if it vanishes on A (G. Mokobodzki 1966 [unpublished]). (Let $(x_n)_{n \in \mathbb{N}}$ be a dense sequence in X. If the assertion is not true, then there exists for any $n \in \mathbb{N}$ a positive $\mathscr{H}$-function, h_n, on X equal to zero at any point x_i $(i \leq n)$ and such that $\sup_K h_n \geq n$. The function $\sum_{n \in \mathbb{N}} h_n$ leads to a contradiction).

1.1.14. Assume that X is connected and that $\mathscr{H}$ possesses the Brelot convergence property. Then for any compact set K of X there exists a positive real number α such that for any positive $\mathscr{H}$-function on X and for $x, y \in K$, we have $h(x) \leq \alpha\, h(y)$.

1.1.15. Assume that there exists an $\mathscr{H}$-sweeping system on X. Then the following assertions are equivalent:

a) $\mathscr{H}$ possesses the Brelot convergence property;

b) for any open connected set U of X and any compact subset K of U there exists a positive real number α such that for any positive $\mathscr{H}$-function h on U and for any $x, y \in K$, we have $h(x) \leq \alpha\, h(y)$.

1.1.16. Let $((\omega_{\iota x})_{x \in V_\iota})_{\iota \in I}$ be an $\mathscr{H}$-sweeping system on X such that V_ι is connected for any $\iota \in I$. Then the following assertions are equivalent:

a) $\mathscr{H}$ possesses the Brelot convergence property;

b) for any $\mathscr{H}$-sweeping ω on a connected, open set V and for any numerical function f on ∂V for which there exists $x \in V$ such that the upper integral with respect to ω_x is finite, the function $\omega f \in \mathscr{H}(V)$.

c) there exists an $\mathscr{H}$-sweeping system $((\omega'_{\iota x})_{x \in V_\iota})_{\iota \in I}$ such that for any $\iota \in I$ and for any lower semi-continuous real function f on ∂V_ι which is $\omega'_{\iota x}$-integrable for an $x \in V_\iota$, the function $\omega'_\iota f \in \mathscr{H}(V_\iota)$.

d) for any $\mathscr{H}$-sweeping ω on a connected, open set V and for any compact set K of V, there exists a positive number α such that $\omega_x \leq \alpha\, \omega_y$ for any $x, y \in K$.

1.1.17. Assume that X is locally connected and that there exists an $\mathscr{H}$-sweeping system $\Omega := ((\omega_{\iota x})_{x \in V_\iota})_{\iota \in I}$ on X such that for any $x \in X$ there exists $I_x \subset I$ with the following properties: *a)* $\{V_\iota | \iota \in I_x\}$ is a fundamental system of neighbourhoods of x; *b)* for any $\iota \in I_x$, $\partial \overline{W}_\iota$ is contained in the carrier of $\omega_{\iota x}$, where W_ι denotes the component of V_ι containing x. If $\mathscr{H}$ possesses the Doob convergence property then $\mathscr{H}$ possesses the Brelot convergence property (C. Constantinescu 1965 [2], H. Bauer 1966 [9]). (Let $(h_n)_{n \in \mathbb{N}}$ be an increasing sequence of $\mathscr{H}$-functions on an open connected set U of X such that its limit function h is finite at a point of U but is not an $\mathscr{H}$-function. Then the set $A := \{x \in U | h(x) = \infty\}$ has a non-empty interior. Let W be a connected component of the interior of A and let $x \in U \cap \partial W$. There exists $\iota \in I_x$ such that $W \cap \partial \overline{W}_\iota \neq \varnothing$. Let f be the characteristic function of $W \cap \partial \overline{W}_\iota$. Since $\omega_\iota f$ is strictly positive at x and hence on a neighbourhood of x, we obtain the following contradiction

$$h = \lim_{n \to \infty} h_n = \lim_{n \to \infty} \omega_\iota h_n = \omega_\iota h = \infty .)$$

§ 1.2. Resolutive Sets

A sheaf of functions $\mathscr{U}$ on a locally compact space X is called a *hyperharmonic sheaf* if for any open set U of X, $\mathscr{U}(U)$ is a convex cone of lower semi-continuous, lower finite numerical functions on U. A function defined on a set containing an open set U is called a *$\mathscr{U}$-function on U* if its restriction to U belongs to $\mathscr{U}(U)$. The map $U \mapsto \mathscr{U}(U) \cap (-\mathscr{U}(U))$ is obviously a harmonic sheaf on X which will be denoted by $\mathscr{H}_{\mathscr{U}}$. An open set U of X will be called an *MP-set* (*with respect to* $\mathscr{U}$) (N. Boboc-C. Constantinescu-A. Cornea 1963 [2]) if every $\mathscr{U}$-function on U is positive, provided it is positive outside the intersection with U of a compact set of X and its lower limit at any boundary point of U is positive. An *MP*-set is in fact a set on which a kind of minimum principle is valid.

Proposition 1.2.1. *Let $\mathscr{U}$ be a hyperharmonic sheaf on a locally compact space X and let U, U' be open sets of X, $U' \subset U$ such that: a) U is an MP-set; b) if $u \in \mathscr{U}(U')$ then $\inf(u, 0) \in \mathscr{U}(U')$; c) for any $x \in U'$, there exists a $\mathscr{U}$-function v on U' and a compact set L of U' such that v is finite at x and $\inf_{U' \setminus L} v > 0$. Then U' is an MP-set.*

Let u be a $\mathscr{U}$-function on U', positive outside a compact set K of X and such that for any $y \in \partial U'$

$$\liminf_{x \to y} u(x) \geq 0.$$

Let $x \in K \cap U'$ and let v be a $\mathscr{U}$-function on U' and L be compact set of U' such that v is finite at x and $\inf_{U' \setminus L} v > 0$. Let further ε be a strictly positive real number. We set

$$K' := \{y \in (L \cup K) \cap U' \mid u(y) + \varepsilon v(y) \leq 0\}.$$

K' is a compact set of U'. Let w be the function on U equal to 0 on $U \setminus U'$ and equal to $\inf(u + \varepsilon v, 0)$ on U'. The function w is a $\mathscr{U}$-function on U' and equal to 0 on $U \setminus K'$. $\mathscr{U}$ being a sheaf and

$$U = (U \setminus K') \cup U'$$

it follows that w is a $\mathscr{U}$-function on U. Since U is an MP-set w is positive. Hence $u(x) + \varepsilon v(x) \geq 0$ and ε, x being arbitrary, it follows that u is positive. □

Let $\mathscr{U}$ be hyperharmonic sheaf on a locally compact space X. Let U be an MP-set of X, and f be a numerical function on ∂U. We denote by $\overline{\mathscr{U}}_f^U$ the set of $\mathscr{U}$-functions u on U which are lower bounded on U, positive outside the intersection with U of a compact set of X, and such that for any $y \in \partial U$

$$\liminf_{x \to y} u(x) \geq f(y).$$

We set $\underline{\mathscr{U}}_f^U := -\overline{\mathscr{U}}_{-f}^U$ and denote by $\overline{H}_f^U$ (resp. $\underline{H}_f^U$) the infimum (resp. supremum) of $\overline{\mathscr{U}}_f^U$ (resp. $\underline{\mathscr{U}}_f^U$). (If $\overline{\mathscr{U}}_f^U$ (resp. $\underline{\mathscr{U}}_f^U$) is empty then $\overline{H}_f^U$ (resp. $\underline{H}_f^U$) is identically $+\infty$ (resp. $-\infty$)). The following relations are immediate:

$$-\overline{H}_f^U = \underline{H}_{-f}^U, \qquad \underline{H}_f^U \leq \overline{H}_f^U,$$

$$f \leq g \Rightarrow \overline{H}_f^U \leq \overline{H}_g^U,$$

$$\alpha \in \mathbb{R}_+ \Rightarrow \overline{H}_{\alpha f} = \alpha \overline{H}_f,$$

$$\overline{H}_{f+g}^U \leq \overline{H}_f^U + \overline{H}_g^U,$$

where we made the convention $\infty - \infty = -\infty + \infty = \infty$.

A numerical function f on ∂U is called *resolutive* (*with respect to* $\mathscr{U}$) (for U) if the functions $\overline{H}_f^U$, $\underline{H}_f^U$ are $\mathscr{H}_{\mathscr{U}}$-functions and are equal; in this case we shall denote their common value by H_f^U. The finding of H_f^U is known in literature as the *Dirichlet problem* (*in the sense of Perron-Wiener-Brelot*). H_f^U itself is called sometimes the *solution of the* (*generalized*) *Dirichlet problem.* The name "resolutive" comes from the expression "resolutive for the Dirichlet problem" (see the historical note on the p. 21). The set of finite functions which are resolutive is a real vector space. An open set U of X is called *resolutive set* (*with respect to* $\mathscr{U}$) if every finite, continuous function with compact carrier on ∂U is resolutive.

Let V be a resolutive set of X. For any $x \in V$ the map

$$f \mapsto H_f^V(x) = \mathscr{K}(\partial V) \to \mathbb{R}$$

is a linear positive functional and defines therefore a measure on ∂V which will be denoted by μ_x^V and will be called the *harmonic measure on V at x* (*with respect to* $\mathscr{U}$). It is obvious that if V is relatively compact $(\mu_x^V)_{x \in V}$ is an $\mathscr{H}$-sweeping on V. The harmonic measure was introduced in classical potential theory by Ch. De la Vallée Poussin 1932 [1].

Proposition 1.2.2. *Let $\mathscr{U}$ be a hyperharmonic sheaf on a locally compact space X. Let V be a resolutive set and W be an open subset of V such that $\partial W \subset \partial V$. Then W is a resolutive set and for any $x \in W$, we have $\mu_x^W = \mu_x^V$. Hence if X is locally connected, any component of a resolutive set is also resolutive.*

It is easy to see that W is an MP-set. Let $f \in \mathscr{K}(\partial W)$. There exists $g \in \mathscr{K}(\partial V)$ such that $g|_{\partial W} = f$. For any $u \in \overline{\mathscr{U}}_g^V$ (resp. $u \in \underline{\mathscr{U}}_g^V$) we have $u|_W \in \overline{\mathscr{U}}_f^W$ (resp. $u|_W \in \underline{\mathscr{U}}_g^U$). Hence

$$H_g^V \le \underline{H}_f^W \le \overline{H}_f^W \le H_g^V$$

on W. □

Theorem 1.2.1. *Let $\mathscr{U}$ be a hyperharmonic sheaf on a locally compact space X, let V be a resolutive set of X, x be a point of V and f be a numerical function on ∂V. Then*

$$\underline{H}_f^V(x) \le \int_* f \, d\mu_x^V \le \int^* f \, d\mu_x^V \le \overline{H}_f^V(x).$$

If f is positive outside a compact set of ∂V, lower bounded and lower semi-continuous, then

$$\underline{H}_f^V(x) = \int^* f \, d\mu_x^V.$$

Let $u \in \overline{\mathscr{U}}_f^V$ and let u^* be the numerical function on ∂V defined by

$$u^*(y) = \liminf_{z \to y} u(z)$$

for any $y \in \partial V$. The function u^* is positive outside a compact set of ∂V, lower bounded, lower semi-continuous and majorizes f. We set $\mathscr{G} := \{g \in \mathscr{K}(\partial V) \mid g \le u^*\}$. For any $g \in \mathscr{G}$ we have

$$\int g \, d\mu_x^V = H_g^V(x) \le u(x).$$

Hence

$$\int^* f\,d\mu_x^V \le \int^* u^*\,d\mu_x^V = \sup_{g\in\mathscr{G}} \int g\,d\mu_x^V \le u(x).$$

Since u is arbitrary

$$\int^* f\,d\mu_x^V \le \overline{H}_f^V(x).$$

Similarly

$$\int_* f\,d\mu_x^V \ge \underline{H}_f^V(x).$$

Suppose now that f is positive outside a compact set of ∂V, lower bounded and lower semi-continuous. Let $g\in\mathscr{K}(\partial V)$, $g\le f$. Then

$$\int g\,d\mu_x^V = H_g^V(x) \le \underline{H}_f^V(x).$$

It follows that

$$\int^* f\,d\mu_x^V \le \underline{H}_f^V(x). \quad \square$$

Let X be a locally compact space, V an open set of X, ω a sweeping on V and $\mathfrak{F}$ a filter on V converging to a point y of ∂V. We shall say that $\mathfrak{F}$ is a *regular* (resp. *non-regular*) *filter for the sweeping* ω if ω converges (resp. does not converge) vaguely to ε_y along $\mathfrak{F}$. The filter $\mathfrak{F}$ is regular if and only if for any $f\in\mathscr{K}(\partial V)$

$$\lim_{x,\mathfrak{F}} \omega f(x) = f(y).$$

Let $\mathscr{U}$ be a hyperharmonic sheaf on X. A sweeping ω on a relatively compact open set V of X is called *quasi-regular* (*with respect to* $\mathscr{U}$) if:

a) for any $f\in\mathscr{C}(\partial V)$ the function ωf is a bounded $\mathscr{H}_{\mathscr{U}}$-function;

b) for any $x\in V$, there exists a positive hyperharmonic function on V, finite at x, which converges to ∞ along every ultrafilter which is non-regular for ω.

A relatively compact open set of X is called *quasi-regular* (*with respect to* $\mathscr{U}$) if there exists a quasi-regular sweeping on it.

Let V be a relatively compact open set of X. If V is regular with respect to $\mathscr{H}_{\mathscr{U}}$, then it is quasi-regular with respect to $\mathscr{U}$ and ω^V is a quasi-regular sweeping on V which possesses no non-regular filter. Conversely, if V is resolutive with respect to $\mathscr{U}$ and if μ^V possesses no non-regular filter, then V is regular with respect to $\mathscr{H}_{\mathscr{U}}$ and $\omega^V = \mu^V$.

Proposition 1.2.3. *Let $\mathscr{U}$ be a hyperharmonic sheaf on a locally compact space X. A quasi-regular set V contained in an MP-set is also an MP-set if, for any $u\in\mathscr{U}(V)$,* $\inf(u,0)\in\mathscr{U}(V)$.

Let ω be a quasi-regular sweeping on V. By the definition there exists for any $x\in V$, a positive $\mathscr{U}$-function u on V, finite at x and such that for any $y \in \partial V$

$$\liminf_{z \to y} (u(z) + \omega 1(z)) \geq 1.$$

Hence $\{y \in V \mid u(y) + \omega 1(y) \leq \frac{1}{2}\}$ is a compact set of V which is therefore an MP-set by Proposition 1.2.1. □

Theorem 1.2.2. *Let $\mathscr{U}$ be a hyperharmonic sheaf on a locally compact space X. Any quasi-regular MP-set is resolutive and the family of harmonic measures is the only quasi-regular sweeping on V. Moreover if g is a lower bounded lower semi-continuous real function on ∂V such that $\underline{H}_g^V$ is harmonic then g is resolutive.*

Let ω be a quasi-regular sweeping on V. For any $x \in V$ there exists a positive $\mathscr{U}$-function u_x on V, finite at x and converging to ∞ along every ultrafilter which is non-regular for ω.

For any $f \in \mathscr{C}(\partial V)$ and for any strictly positive real number ε we have

$$\omega f + \varepsilon u_x \in \overline{\mathscr{U}}_f^V, \qquad \omega f - \varepsilon u_x \in \underline{\mathscr{U}}_f^V.$$

$$\omega f(x) - \varepsilon u_x(x) \leq \underline{H}_f^V(x) \leq \overline{H}_f^V(x) \leq \omega f(x) + \varepsilon u_x(x).$$

Since ε and x are arbitrary, we obtain

$$\omega f \leq \underline{H}_f^V \leq \overline{H}_f^V \leq \omega f.$$

It follows that V is resolutive and $\omega_x = \mu_x^V$ for any $x \in V$.

From Theorem 1.2.1 we get for all $x \in V$

$$\underline{H}_g^V(x) = \int^* g \, d\mu_x^V.$$

Hence for all $x \in V$ and for any strictly positive real number ε, the function $\underline{H}_g^V + \varepsilon u_x$ belongs to $\overline{\mathscr{U}}_g^V$. It follows

$$\overline{H}_g^V(x) \leq \underline{H}_g^V(x) + \varepsilon u_x(x).$$

Again ε and x are arbitrary and therefore $\overline{H}_f^V = \underline{H}_f^V$. □

Historical note. Let U be a relatively compact, open set in $\mathbb{R}^n$ and let f be a continuous real function on ∂U. The classical Dirichlet problem (called also the first boundary problem) consisted of finding a (necessarily unique) harmonic function h on U (in the classical sense: i.e. h is a twice continuously differentiable real function, which is a solution of the Laplace equation $\sum_{i=1}^{n} \frac{\partial^2 h}{(\partial x_i)^2} = 0$) such that for any $y \in \partial U$

$$\lim_{x \to y} h(x) = f(y).$$

This problem does not always have a solution (S. Zaremba 1911 [1]). Because of this O. Perron 1923 [1] and R. Remak 1924 [1] were led to the formulation of the generalised Dirichlet problem. They defined the upper class of a bounded real function f on ∂U as the set of continuous real functions u on U such that u dominates its Poisson integral on any closed sphere contained in U and such that its lower limit at any boundary point dominates f. They then showed that the infimum, $\overline{H}_f^U$, of the upper class of f is harmonic. Similarly, they defined $\underline{H}_f^U$ and proved $\underline{H}_f^U \leq \overline{H}_f^U$. The equality of these functions for arbitrary U and arbitrary continuous f was proved by N. Wiener 1925 [2]. He also remarked that there exists a sweeping ω on U which gives the solution. Perron's method was extended by W. Sternberg 1929 [1] to the heat equation. A more general extention of Perron's method for the Laplace equation was given by M. Brelot 1940 [3]. He considered for arbitrary numerical functions f on ∂U, the Perron upper class, which was enlarged so as to contain the lower semi-continuous, superharmonic functions previously defined by F. Riesz 1925 [1]. M. Brelot also showed that $\overline{H}_f^U$ and $\underline{H}_f^U$ coincide if and only if f is integrable with respect to the sweeping ω.

Exercises

In all the exercises of this section, X will denote a locally compact space and $\mathscr{U}$ a hyperharmonic sheaf on X.

1.2.1. If U is an MP-set with respect to $\mathscr{U}$, then any open subset V of U such that $\partial V \subset \partial U$ is also an MP-set with respect to $\mathscr{U}$.

1.2.2. An open set of X with empty boundary is resolutive with respect to $\mathscr{U}$ if and only if it is an MP-set with respect to $\mathscr{U}$.

1.2.3. Let f be a strictly positive, continuous real function on X. We denote by $\mathscr{U}_f$, the hyperharmonic sheaf on X

$$U \mapsto \left\{\frac{u}{f} \,|\, u \in \mathscr{U}(U)\right\}.$$

Then the MP-sets with respect to $\mathscr{U}$ and with respect to $\mathscr{U}_f$ coincide. Let U be an open set of X and let g be a numerical function on ∂U. Then

$$\overline{H}_g'^U = \frac{1}{f}\overline{H}_{fg}^U, \qquad \underline{H}_g'^U = \frac{1}{f}\underline{H}_{fg}^U$$

where $\overline{H}_g'^U$, $\underline{H}_g'^U$ are defined as $\overline{H}_g^U$, $\underline{H}_g^U$ but with respect to $\mathscr{U}_f$ instead of $\mathscr{U}$. Hence g (resp. U) is resolutive with respect to $\mathscr{U}_f$ if and only if g (resp. U)

is resolutive with respect to $\mathscr{U}$. If U is resolutive, then for any $x \in U$

$$\mu_x'^U = \frac{f}{f(x)} \cdot \mu_x^U,$$

where $\mu_x'^U$ denotes the harmonic measure on U at x with respect to $\mathscr{U}_f$. U is quasi-regular with respect to $\mathscr{U}_f$ if and only if it is quasi-regular with respect to $\mathscr{U}$.

1.2.4. Let U be a relatively compact MP-set and let $\mathscr{S}$ be the set of continuous, real functions on $\overline{U}$ which are $\mathscr{U}$-functions on U. U is resolutive with respect to $\mathscr{U}$ if the following conditions are fulfilled: *a)* $\mathscr{H}_{\mathscr{U}}$ possesses the Bauer convergence property; *b)* there exists a bounded $\mathscr{U}$-function u_0 on U such that $\inf_U u_0 > 0$; *c)* $u, v \in \mathscr{S} \Rightarrow \inf(u, v) \in \mathscr{S}$; *d)* $\mathscr{S}|_{\partial U}$ separates the points of ∂U; *e)* any $x \in U$ has a neighbourhood $V \subset U$ such that for any $u \in \mathscr{S}$ there exists $v \in \mathscr{S}$ with the following properties: $v \leq u$ on U, $v = u$ outside a compact set of u, $v|_V \in \mathscr{H}_{\mathscr{U}}(V)$. (By *a)*, *c)* and *e)* the infimum of $\overline{\mathscr{U}}_u^U$ is an $\mathscr{H}_{\mathscr{U}}$-function for any $u \in \mathscr{S}$ (Proposition 1.1.2) and belongs therefore to $\underline{\mathscr{U}}_u^U$. It follows that $u|_{\partial U}$ is resolutive. The set $\{(u-v)|_{\partial U} \,|\, u, v \in \mathscr{S}\}$ is a real vector space and by *c)* a sublattice of $\mathscr{C}(\partial U)$. By *d)* it is dense in $\mathscr{C}(\partial U)$ (Stone's theorem). By *b)* and *a)* any function of $\mathscr{C}(\partial U)$ is resolutive.)

1.2.5. Let U be a resolutive set with respect to $\mathscr{U}$ such that the sum of any countable family of positive functions of $\mathscr{U}(U)$ belongs to $\mathscr{U}(U)$. *a)* For any $f \in \mathscr{K}(\partial U)$ and any $x \in U$, there exists a positive function $u \in \mathscr{U}(U)$ finite at x such that for every ultrafilter $\mathfrak{U}$ on V converging to a boundary point y of U, either

$$\lim_{\mathfrak{U}} H_f^U = f(y)$$

or

$$\lim_{\mathfrak{U}} u = \infty .$$

b) If U is relatively compact and ∂U is metrisable, then U is quasi-regular.

1.2.6. Let U be a resolutive set with respect to $\mathscr{U}$ and let $x \in U$. If, for every lower semi-continuous, lower finite, numerical function f on ∂U which is positive outside a compact set we have

$$\overline{H}_f^U(x) = \int^* f \, d\mu_x^U,$$

then the same equality holds for every numerical function f on ∂U (H. Bauer 1962 [5]).

1.2.7. Let V be an open set of X, ω be a sweeping on V and $\mathfrak{F}$ be a filter on V, regular for the sweeping ω and converging to a point $x \in \partial V$.

Let f be a bounded real function on ∂V with compact carrier. If f is continuous at x, then ωf converges to $f(x)$ along $\mathfrak{F}$. (The assertion is no longer true if f is not bounded (Exercise 3.2.14).)

1.2.8. Let V be a resolutive set with respect to $\mathscr{U}$ such that ∂V is compact and metrisable and H_1^V is bounded. Then there exists a countable set $A \subset \partial V$ such that for any $f \in \mathscr{C}(\partial V)$ we have

$$\lim_{x \to y} H_f^V(x) = f(y)$$

for every $y \in \partial V$ is this relation holds for every $y \in A$ (M.V. Keldych 1938 [1], 1941 [2]). (Use Exercise 1.1.7.)

§ 1.3. Minimum Principle

The minimum principle plays an important part in the theory of harmonic spaces. There are two important criteria from which this principle can be derived. The first one based on "ellipcity" was proved by M. Brelot 1957 [10]; the second one uses the "separation" and was proved by H. Bauer 1960 [3].

Let $\Omega := ((\omega_{\iota x})_{x \in V_\iota})_{\iota \in I}$ be a sweeping system on a locally compact space X. Let U be an open set of X and let u be a lower semi-continuous lower finite numerical function on U. u will be called *Ω-hyperharmonic* if for any $\iota \in I$ such that $\overline{V}_\iota \subset U$ we have $\omega_\iota u \leq u$ on V_ι. u will be called *Ω^*-hyperharmonic* if for any $x \in U$ and any neighbourhood V of x there exists $\iota \in I$ such that $x \in V_\iota$, $\overline{V}_\iota \subset U \cap V$ and $\omega_\iota u(x) \leq u(x)$. u will be called *locally Ω-hyperharmonic* if there exists an open covering $\mathfrak{W}$ of U such that for any $W \in \mathfrak{W}$, $u|_W$ is an Ω-hyperharmonic function on W. Obviously any Ω-hyperharmonic function is locally Ω-hyperharmonic, and any locally Ω-hyperharmonic function is Ω^*-hyperharmonic. The set of Ω-hyperharmonic (resp. locally Ω-hyperharmonic) functions on an open set U of X is a convex cone containing together with any two elements their greatest lower bound. The least upper bound of an upper directed family of Ω-hyperharmonic functions on an open set is an Ω-hyperharmonic function.

If for any open set U of X, $\mathscr{U}(U)$ denotes the set of locally Ω-hyperharmonic functions on U then $\mathscr{U}$ is a hyperharmonic sheaf on X which will be called *the hyperharmonic sheaf generated by Ω*.

Suppose X locally connected. A sweeping system $(((\omega_{\iota x})_{x \in V_\iota})_{\iota \in I})$ on X is called *elliptic at a point* $x \in X$ if there exists a neighbourhood U of x such that for any $\iota \in I$ for which $x \in V_\iota$ and $\overline{V}_\iota \subset U$, the carrier of $\omega_{\iota x}$ contains the boundary of the closure of the component of V containing x. A sweeping system on X is called *elliptic*, if it is elliptic at every point of X. The results related to ellipticity are not very numerous in this book and they constitute a marginal part of the theory which may be omitted by the non-interested reader.

Proposition 1.3.1. *If a harmonic sheaf* $\mathscr{H}$ *on a locally compact, locally connected space possesses the Brelot convergence property, then for any base* $\mathfrak{B}$ *of* X *of regular sets the sweeping system* $((\omega_x^V)_{x\in V})_{V\in\mathfrak{B}}$ *is elliptic* (M. Brelot 1958 [13]).

The proposition follows immediately from the last assertion of Proposition 1.1.3. □

Proposition 1.3.2. *Let* $\mathscr{H}$ *be a harmonic sheaf on a locally compact, locally connected space* X *and* $((\omega_{\iota x})_{x\in V_\iota})_{\iota\in I}$ *be an elliptic sweeping system on* X *such that for any* $\iota\in I$, $\omega_\iota 1\in\mathscr{H}(V_\iota)$. *Then* $\mathscr{H}$ *is non-degenerate at any non-isolated point of* X.

Let x be a non-isolated point of X and let U be an open connected set containing x with the property that for any $\iota\in I$ such that $x\in V_\iota$ and $\overline{V}_\iota\subset U$ the carrier of $\omega_{\iota x}$ contains the boundary of the closure of the component, W_ι, of V_ι containing x. Let now ι be an element of I such that $x\in V_\iota$, $\overline{V}_\iota\subset U$ and $U\setminus V_\iota\neq\emptyset$. Then $U\cap\overline{W}_\iota$ is non-empty and contained in the carrier of $\omega_{\iota x}$. We thus obtain $\omega_\iota 1(x)>0$. □

Proposition 1.3.3. *Let* Ω *be an elliptic sweeping system on a locally compact, locally connected and connected space* X. *Any positive* Ω^*-*hyperharmonic function on* X *vanishes identically if there exists a point of* X *at which it vanishes.*

Let u be a positive Ω^*-hyperharmonic function on X and let

$$U:=\{x\in X \mid u(x)>0\}.$$

Assume that U and $X\setminus U$ are not empty. Let V be a component of U and let y be a point of ∂V. Let Ω be the family $((\omega_{\iota x})_{x\in V_\iota})_{\iota\in I}$ and let W be a neighbourhood of y such that $V\setminus W\neq\emptyset$ and such that for any $\iota\in I$ for which $y\in V_\iota$ and $\overline{V}_\iota\subset W$, the carrier of $\omega_{\iota y}$ contains the boundary of $\overline{W}_\iota$ where W_ι is the component of V_ι containing x. Since V is connected and the sets $\overline{W}_\iota\cap V$, $V\setminus\overline{W}_\iota$ are non-empty, the set $V\cap\partial\overline{W}_\iota$ is non-empty and therefore $\omega_{\iota y}(V\cap\partial\overline{W}_\iota)>0$. Since u is strictly positive on V we get

$$\int^* u\,d\omega_{\iota y}\geq\int^*_{V\cap\partial\overline{W}_\iota} u\,d\omega_{\iota y}>0.$$

Now u being Ω^*-hyperharmonic implies there exists $\iota\in I$ such that $x\in V_\iota$, $\overline{V}_\iota\subset W$ and

$$\int^* u\,d\omega_{\iota y}\leq u(y)=0.$$

This is a contradiction. □

Theorem 1.3.1 (Brelot-Bauer). *Let Ω be a sweeping system on a locally compact space X and let U be an open set of X such that for any $x \in U$ and for any compact set K of X there exists a locally Ω-hyperharmonic function on U, finite at x, whose infimum on $K \cap U$ is strictly positive. Suppose further that one of the following conditions is fulfilled:*

a) the set of Ω-hyperharmonic functions on U separates the points of U;

b) U is locally connected, possesses no compact component and Ω is elliptic at every point of U.

Then every Ω^*-hyperharmonic function on U is positive if it is positive outside the intersection of U and a compact set of X and if its lower limit at any boundary point of U is positive. Hence if $\mathcal{U}$ is a hyperharmonic sheaf such that any $\mathcal{U}$-function is an Ω^*-hyperharmonic function, then U is an MP-set with respect to $\mathcal{U}$ (H. Bauer 1960 [3], 1962 [5] for *a)*, M. Brelot 1957 [10], 1958 [13] for *b)*).

We prove first the following lemmata:

Lemma 1. *Let Y be a compact space, x a point of Y and f, g two lower semi-continuous, lower finite numerical functions on Y such that $f(x)<0$, $g(x)<\infty$, and $g(y)>0$ for every $y \in Y$. Then there exists a strictly positive real number α such that $f+\alpha g \geq 0$ and*

$$\{y \in Y \mid f(y)+\alpha g(y)=0\} \neq \emptyset .$$

We set

$$\varphi := -\frac{\inf(f, 0)}{g} \qquad \left(\frac{1}{\infty}=0\right),$$

$$\alpha := \sup_Y \varphi .$$

Since φ is upper bounded and strictly positive at x, α is a strictly positive real number. Because φ is upper semi-continuous it achieves its supremum at a point of Y. It follows that α has the required properties. □

Lemma 2. *Let Y be a compact space and $\mathcal{F}$ be a set of lower semi-continuous, lower finite numerical functions on Y which separates the points of Y. If there exist $x \in Y$ and $f, g \in \mathcal{F}$ such that $f(x)<0$, $g(x)<\infty$, and $g(y)>0$ for every $y \in Y$, then there exists $x_0 \in Y$ such that $f(x_0)<0$ and such that ε_{x_0} is the only measure μ on Y satisfying the relation*

$$\int^* u \, d\mu \leq u(x_0)$$

for any $u \in \mathcal{F}$ (H. Bauer 1958 [1], 1960 [2], 1961 [4], N. Boboc-A. Cornea 1967 [2]).

By the proceding lemma there exists a strictly positive real number α such that $f+\alpha g \geq 0$ and

$$K := \{y \in Y \mid f(y)+\alpha g(y)=0\} \neq \emptyset .$$

It follows immediately that f (resp. g) is finite and strictly negative (resp. finite and strictly positive) on K.

For any $y \in Y$ we denote by $\mathfrak{M}_y$ the set of measures μ on Y such that

$$\int^{*} u \, d\mu \leq u(y)$$

for any $u \in \mathscr{F}$. Let $\mathfrak{A}$ be the set of compact non-empty sets A of Y such that for any $y \in A$ and any $\mu \in \mathfrak{M}_y$ we have

$$\mu(Y \setminus A)=0 .$$

It is easy to see that $K \in \mathfrak{A}$ and that $\mathfrak{A}$ ordered by the converse inclusion relation is inductive. Hence, by Zorn's lemma, there exists a set $A \in \mathfrak{A}$, $A \subset K$, such that

$$A' \subset A, \; A' \in \mathfrak{A} \Rightarrow A' = A .$$

Let $u \in \mathscr{F}$ such that $u \not\equiv \infty$ on A. Then there exists a positive real number β such that $u+\beta f$ is strictly negative at a point of A. By Lemma 1 applied to A and to the restrictions to A of the functions $u+\beta f$ and g, we deduce that there exists a strictly positive real number γ such that $u+\beta f+\gamma g \geq 0$ on A and

$$A' := \{y \in A \mid u(y)+\beta f(y)+\gamma g(y)=0\} \neq \emptyset .$$

It is easy to see that $A' \in \mathfrak{A}$. Hence $A' = A$ and

$$u = -\beta f - \gamma g = (\alpha \beta - \gamma) g .$$

From the hypotheses that $\mathscr{F}$ separates the points of Y, it follows that A reduces to a point x_0 which therefore has the required properties. □

We now prove the theorem. Let u be an Ω^*-hyperharmonic function on U, positive outside the intersection of U and a compact set of X, and whose lower limit at any boundary point of U is positive. We set

$$A := \{x \in U \mid u(x) < 0\} .$$

Assume $A \neq \emptyset$. Let $x \in A$ and let v be a locally Ω-hyperharmonic function on U, finite at x, for which

$$\inf_{\bar{A} \cap U} v > 0 .$$

Let ε be a strictly positive real number such that

$$u(x) + \varepsilon v(x) < 0 .$$

We set

$$K := \{y \in A \mid u(y) + \varepsilon v(y) \leq 0\}.$$

K is a compact set. Let Y be a compact neighbourhood of K contained in $\{x \in U \mid v(x) > 0\}$.

Assume first that *a)* is fulfilled and let $\mathscr{F}$ be the set consisting of the restriction to Y of the Ω-hyperharmonic functions on U, and of the restrictions to Y of the functions $u + \varepsilon v$ and v. By Lemma 2 there exists a point $x_0 \in K$ such that ε_{x_0} is the only measure μ on Y satisfying the relation

$$\int^* f\, d\mu \leq f(x_0)$$

for any $f \in \mathscr{F}$. If Ω is the family $((\omega_{\iota x})_{x \in V_\iota})_{\iota \in I}$ then there exists $\iota \in I$ such that $x_0 \in V_\iota$, $\overline{V}_\iota \subset Y$,

$$\int^* u\, d\omega_{\iota x_0} \leq u(x_0), \qquad \int^* v\, d\omega_{\iota x_0} \leq v(x_0).$$

We have then, that

$$\int^* f\, d\omega_{\iota x_0} \leq f(x_0)$$

for any $f \in \mathscr{F}$ and this is a contradiction.

Assume now that *b)* is satisfied. By Lemma 1, there exists a strictly positive real number α such that $(u + \varepsilon v) + \alpha v \geq 0$ on Y and such that $(u + \varepsilon v) + \alpha v$ vanishes at a point $z \in Y$. We set

$$V := \{y \in U \mid v(y) > 0\},$$

and let W be the component of V containing z. Since W is not compact and since $Y \subset V$ it follows that $W \setminus Y$ is non-empty. On the other hand the function $u + (\varepsilon + \alpha) v$ is Ω^*-hyperharmonic, strictly positive on $W \setminus Y$, positive on W and vanishes at z. This contradicts Proposition 1.3.3. □

Exercises

1.3.1. Let Ω be a sweeping system on a locally compact space X and let $x \in X$. If there exists an Ω^*-hyperharmonic function defined on a neighbourhood of x and strictly negative at x, then x is not an isolated point of X.

1.3.2. Let $\Omega := ((\omega_{\iota x})_{x \in V_\iota})_{\iota \in I}$ be a sweeping system on a locally compact space X and let U be an open set of X. If there exists a positive Ω^*-hyperharmonic function u on U such that for any $y \in \partial U$ we have

$$\liminf_{x \to y} u(x) > 0$$

and for any $x \in U$ and for any neighbourhood V of x there exists $\iota \in I$ such that

$$x \in V_\iota, \ \overline{V}_\iota \subset U \cap V, \ \omega_{\iota x} u < u(x),$$

then U is an MP-set with respect to the hyperharmonic sheaf generated by Ω.

1.3.3. Let $\mathscr{H}$ be a harmonic sheaf on a locally connected, locally compact space X such that the regular sets with respect to $\mathscr{H}$ form a base of X. The following assertions are equivalent:

a) for any regular connected set V and for any $x \in V$, the carrier of ω_x^V is equal to ∂V;

b) there exists a base $\mathfrak{B}$ of X of regular sets such that the sweeping system $((\omega_x^V)_{x \in V})_{V \in \mathfrak{B}}$ is elliptic;

c) a positive harmonic function on an open connected set U vanishes identically if it vanishes at a point of U.

Chapter 2

Harmonic Spaces

§ 2.1. Definition of Harmonic Spaces

A locally compact space X endowed with a hyperharmonic sheaf $\mathscr{U}$ is called a *harmonic space* if the following axioms are satisfied:

Axiom of Positivity. *$\mathscr{H}_{\mathscr{U}}$ is non-degenerate at all points of X.*

Axiom of Convergence. *$\mathscr{H}_{\mathscr{U}}$ possesses the Bauer convergence property.*

Axiom of Resolutivity. *The resolutive sets (with respect to $\mathscr{U}$) form a base of X.*

Axiom of Completeness. *A lower semi-continuous, lower finite function u on an open set U of X belongs to $\mathscr{U}(U)$ if, for any relatively compact resolutive set V (with respect to $\mathscr{U}$) with $\overline{V} \subset U$, we have $\mu^V u \leq u$ on V, where μ^V is constructed with respect to $\mathscr{U}$.*

In the usual examples of harmonic spaces, the axiom of completeness is trivially fulfilled since it is used in order to define the hyperharmonic sheaf. The theory may be developed without the axiom of positivity since it can be shown that the set of points at which it is not satisfied is closed and totally disconnected, but in this case the theory becomes disagreeable; moreover in the usual examples, this axiom is also trivially satisfied. Equivalent formulations of the axiom of convergence and the axiom of resolutivity will be given later (Theorem 11.1.1 and Theorem 2.4.5).

The notations $\overline{\mathscr{U}}_f^U, \underline{\mathscr{U}}_f^U$, $\overline{H}_f^U$, $\underline{H}_f^U$, μ_x^V (resp. ω_x^V), introduced in § 1.2 (resp. § 1.1) will be always used on a harmonic space with respect to the sheaf $\mathscr{U}$ (resp. $\mathscr{H}_{\mathscr{U}}$).

The terminology introduced in the first chapter in connection with a harmonic or with a hyperharmonic sheaf will be used on a harmonic space in the following way. An $\mathscr{H}_{\mathscr{U}}$-function (resp. $\mathscr{U}$-function, will be called *harmonic* (resp. *hyperharmonic*) *function*. A function u will be called *hypoharmonic* if $-u$ is hyperharmonic. On a harmonic space we shall say *regular set* instead of "regular set with respect to $\mathscr{H}_{\mathscr{U}}$" and *MP-set* instead of "MP-set with respect to $\mathscr{U}$." Similarly *resolutive function, resolutive set, quasi-regular sweeping, quasi-regular set* will be used without the qualification "with respect to $\mathscr{U}$." If U is a resolutive set in a harmonic space and if $\mathfrak{F}$ is a filter on U converging to a boundary point of U, we shall say that $\mathfrak{F}$ is *regular* (resp. *non-regular*) if it is regular (resp. non-regular) for the sweeping μ^U. A point $x \in \partial U$ is called *regular boundary point* if any filter on U converging to x is regular. This last notion was introduced in classical potential theory by H. Lebesgue 1924 [3]. In these terms, a relatively compact resolutive set is regular if and only if every boundary point is a regular boundary point.

A harmonic space is locally connected (Theorem 1.1.1) and possesses therefore a base consisting of connected resolutive sets (Proposition 1.2.2). No point of a harmonic space is isolated. Indeed if x is such a point then $\{x\}$ is resolutive and any harmonic function on $\{x\}$ is positive. It follows that $\mathscr{H}(\{x\})=\{0\}$ and this contradicts the axiom of positivity. If the axiom of resolutivity is satisfied and if $\mathfrak{B}$ denotes the set of relatively compact resolutive sets, then the axiom of completeness is equivalent to the assertion that $\mathscr{U}$ is generated by the sweeping system $((\mu_x^V)_{x\in V})_{V\in\mathfrak{B}}$. From the axiom of completeness it follows immediately that the infimum of two hyperharmonic functions and the supremum of any upper directed set of hyperharmonic functions are also hyperharmonic functions.

If U is an open set of the harmonic space X, then U endowed with the restriction of $\mathscr{U}$ to U is also a harmonic space which will be called a *harmonic subspace of* X.

The following proposition may be used in order to show that the axioms of harmonic spaces are satisfied in special cases.

Proposition 2.1.1. *Let $\Omega := ((\omega_{\iota x})_{x\in V_\iota})_{\iota\in I}$ be a sweeping system on a locally compact space X and let $\mathscr{U}$ be the hyperharmonic sheaf generated by Ω. If for any $\iota\in I$, ω_ι is quasi-regular with respect to $\mathscr{U}$ and if there exists a covering of X by MP-sets with respect to $\mathscr{U}$, then the axiom of resolutivity and the axiom of completeness are satisfied.*

Choose $\iota\in I$ such that V_ι is contained in an MP-set. By Proposition 1.2.3, V_ι is an MP-set with respect to $\mathscr{U}$ and therefore resolutive with respect to $\mathscr{U}$ (Theorem 1.2.2). Hence the axiom of resolutivity is satisfied since the set of V_ι, $\iota\in I$, which are contained in an MP-set form a base for X.

Let u be a lower semi-continuous, lower finite function on an open set U of X such that for any relatively compact resolutive set V with $\overline{V}\subset U$, we have $\mu^V u \leq u$. Let $\mathfrak{W}$ be the set of MP-sets with respect to $\mathscr{U}$. The above considerations show that for any $W\in\mathfrak{W}$ and any $\iota\in I$ such that $\overline{V}_\iota \subset U\cap W$, we have $\omega_\iota u\leq u$ on V_ι. Hence the function $u|_{U\cap W}$ is

Ω-hyperharmonic and therefore belongs to $\mathscr{U}(U \cap W)$. But $\mathscr{U}$ being a sheaf and $\mathfrak{W}$ a covering of X imply that $u \in \mathscr{U}(U)$. Hence the axiom of completeness holds. □

Proposition 2.1.2. *Let U, U' be open sets of a harmonic space, $U' \subset U$, and let u, u' be hyperharmonic functions on U, U' respectively. If the function u^* on U equal to* $\inf(u, u')$ *on U' and u on $U \setminus U'$ is lower semi-continuous, then it is hyperharmonic* (M. Brelot 1960 [15], N. Boboc-C. Constantinescu-A. Cornea 1963 [2]).

Let V be a relatively compact resolutive set of U such that there exists a strictly positive harmonic function defined on a neighbourhood of $\overline{V}$. Let $f \in \mathscr{C}(\partial V)$, $f \leq u^*$ on ∂V. We have

$$u^* = u \geq H_f^V$$

on $V \setminus U'$ and thus the lower limit of $(u^* - H_f^V)|_{U' \cap V}$ at any boundary point of $U' \cap V$ is positive. Since $U' \cap V$ is an MP-set (Proposition 1.2.1), we deduce that $H_f^V \leq u^*$ on $V \cap U'$. □

Corollary 2.1.1. *On a harmonic space, any open subset of an MP-set is also an MP-set.*

Let U be an MP-set and U' be an open subset of U. Assume u is a hyperharmonic function on U' which is positive outside the intersection of U' and a compact set of X and whose lower limit at any boundary point of U' is positive. By the proposition, the function on U which is equal to $\inf(u, 0)$ on U' and 0 on $U \setminus U'$ is hyperharmonic. Since U is an MP-set, this function is positive. Hence u is positive. □

Proposition 2.1.3. *Let u be a hyperharmonic function on a harmonic space, U be a relatively compact resolutive set and v be a hyperharmonic function on U greater than $\mu^U u$. If the function u^* on the harmonic space which is equal to* $\inf(u, v)$ *on U,*

$$y \mapsto \inf\left(u(y), \liminf_{x \to y} v(x)\right)$$

on ∂U, and u elsewhere is lower finite, then it is hyperharmonic.

It is obvious that u^* is lower semi-continuous. Let V be a relatively compact resolutive set, let $f \in \mathscr{C}(\partial V)$ with $f \leq u^*$ on ∂V, and let $\underline{u} \in \underline{\mathscr{U}}_f^V$. We extend $\underline{u}$ to $\overline{V}$ by setting

$$\underline{u}(y) := \limsup_{x \to y} u(x)$$

for any $y \in \partial V$. Since $u \geq f$ on ∂V and since V is an MP-set, we obtain $\underline{u} \leq u$ on $\overline{V}$. But u is lower semi-continuous and $\underline{u}$ upper semi-continuous and hence there exists $g \in \mathscr{C}(\partial U)$ such that $g \leq u$ on ∂U and $\underline{u} \leq g$ on $\overline{V} \cap \partial U$.

Let $\bar{u} \in \bar{\mathcal{U}}_g^U$. The function on $U \cap V$ equal to $v + \bar{u} - \underline{u} - \mu^U g$ is hyperharmonic and its lower limit at any boundary point of $U \cap V$ is positive. Since $U \cap V$ is an MP-set (Corollary 2.1.1) we obtain

$$v + \bar{u} - \underline{u} - \mu^U g \geq 0$$

on $U \cap V$. Since $\bar{u}$ was arbitrary, we deduce that $v - \underline{u} \geq 0$ on $U \cap V$. Furthermore, $\underline{u}$ being arbitrary, we obtain $v \geq \mu^V f$ on $U \cap V$. Hence for any $x \in V \cap \partial U$

$$\mu^V f(x) = \lim_{\substack{y \to x \\ y \in U}} \mu^V f(y) \leq \liminf_{y \to x} v(y).$$

The relation $\mu^V f \leq u$ on V is trivial and thus $\mu^V f \leq u^*$ on V. But again f is arbitrary and thus we have that $\mu^V u^* \leq u^*$ on V. □

Let u be a hyperharmonic function on a harmonic space X and let V be a relatively compact resolutive set. We denote by u_V the function on X which is equal to u on $X \setminus \bar{V}$, $\mu^V u$ on V and

$$y \mapsto \inf\left(u(y), \liminf_{x \to y} \mu^V u(x)\right)$$

on ∂V. If $\mu^V 1$ is bounded (e.g. if there exists a strictly positive harmonic function defined on a neighbourhood of $\bar{V}$), then u_V is lower finite.

Corollary 2.1.2. *If u_V is lower finite, then it is a hyperharmonic function less than u.* □

Theorem 2.1.1. *Let $\mathcal{U}, \mathcal{U}'$ be two hyperharmonic sheaves on a locally compact space X and assume that X endowed with either of the sheaves $\mathcal{U}$, $\mathcal{U}'$ is a harmonic space. If $\mathcal{U}(U) \subset \mathcal{U}'(U)$ for every open set U of X, then $\mathcal{U} = \mathcal{U}'$.*

Let U be an open set of X and $u \in \mathcal{U}'(U)$. Let W be an MP-set with respect to $\mathcal{U}'$, and let V be a relatively compact resolutive set with respect to $\mathcal{U}$ with $\bar{V} \subset U \cap W$. Choose $\underline{u} \in \underline{\mathcal{U}}_u^V$. Then $u|_V - \underline{u}$ belongs to $\mathcal{U}'(V)$ and its lower limit at any boundary point of V is positive. Since V is an MP-set with respect to $\mathcal{U}'$ (Corollary 2.1.1), it follows that $u - \underline{u}$ is positive on V. Hence

$$\mu^V u = \underline{H}_u^V \leq u$$

on V (Theorem 1.2.1). □

Theorem 2.1.2. *For any open set U of $\mathbb{R}$ let $\mathcal{U}(U)$ be the set of lower semi-continuous, lower finite, numerical functions u on U which are decreasing on every connected component of U. Then $\mathcal{U}$ is a hyperharmonic sheaf and $\mathbb{R}$ endowed with $\mathcal{U}$ is a harmonic space for which the sheaf of harmonic functions possesses the Brelot convergence property.*

For any open set U of $\mathbb{R}$, $\mathcal{H}_{\mathcal{U}}(U)$ is the set of real functions on U which are constant on connected components of U. Obviously $\mathcal{H}_{\mathcal{U}}$ possesses the Brelot convergence property and therefore the axiom of convergence is satisfied. The axiom of positivity is trivially satisfied.

Let $\alpha, \beta \in \mathbb{R}$, $\alpha < \beta$. It is obvious that $]\alpha, \beta[$ is resolutive and that for any $x \in]\alpha, \beta[$, $\mu_x^{]\alpha, \beta[} = \varepsilon_\beta$. It follows immediately that the axiom of resolutivity and the axiom of completeness are satisfied. □

Remark. This example shows that on a harmonic space, the sheaf of hyperharmonic functions is not determined by the sheaf of harmonic functions. Indeed, if in the above definition of $\mathcal{U}(U)$, we replace decreasing by increasing, we get another harmonic space with the same sheaf of harmonic functions.

Historical remark. The axiomatic systems considered in literature up to now started with a locally compact space X on which was given (or was constructed with the aid of a sweeping system) a harmonic sheaf $\mathcal{H}$. The axioms differed according to each author but all of them assumed that the regular sets with respect to $\mathcal{H}$ formed a base $\mathfrak{B}_r$ of X. Let $\mathcal{H}^*$ be the hyperharmonic sheaf generated by the sweeping system $((\omega_x^V)_{x \in V})_{V \in \mathfrak{B}_r}$. The first approach was made by G. Trautz 1943 [1], in which be took X equal to an open set of $\mathbb{R}^2$. He assumed that any uniformly bounded set of harmonic functions defined on an open set of X is equicontinuous (it will be shown in Theorem 11.1.1 that this property is equivalent to the Bauer convergence property) and that any hyperharmonic function which attains a strictly negative minimum is constant. From this last axiom, it follows that any open set of X is an MP-set with respect to $\mathcal{H}^*$. In this framework he studied the Dirichlet problem and proved the resolutivity of relatively compact open sets with "sufficiently smooth" boundaries. Tautz then applied these results to some elliptic equations. In 1949 [2] he came back to this axiomatic system but allowed X to be a locally compact metric space and in 1952 [3], he gave a criterion in order that a sheaf $\mathcal{H}$ on an open set of $\mathbb{R}^n$ ($n \in \mathbb{N}$, $n \neq 0$) be the sheaf of solutions of an elliptic partial differential equation.

Another approach was made by J.L. Doob 1956 [1]. For him, the space has a countable base, the constant functions are harmonic, $\mathcal{H}$ possesses the Doob convergence property and a complicated axiom which is equivalent to the existence of an open covering with MP-sets with respect to $\mathcal{H}^*$ is assumed. He showed that the sheaf of solutions of the heat equation satisfies his axioms. The problems studied in this context were the Dirichlet problem and Fatou's theorem.

Later M. Brelot 1957 [10] introduced a more restrictive system of axioms, requiring that $\mathcal{H}$ possesses the Brelot convergence property. He showed that in this case there exists a covering of the space with MP-sets. Subsequently, a long list of results was obtained by him and his Paris school (especially R.-M. Hervé) within the framework of this axiomatic system which achieved quickly the status of an elaborate theory.

Another attempt of axiomatic theory was developed for n-dimensional manifolds by N. Boboc and N. Radu 1958 [1]. The interest of this axiomatic system is in the fact that among other axioms, it assumed that $\mathscr{H}$ possesses the Bauer convergence property.

In 1960 H. Bauer [3] proposed a new axiomatic system by assuming that the constants are harmonic functions, that $\mathscr{H}$ possesses the Bauer convergence property and that the set of harmonic functions separates the points of the space. Later 1962 [5], Bauer weakened these axioms by replacing the assumption that the constants are harmonic with the assumption that there exists a strictly positive harmonic function on X and that the set of hyperharmonic functions divided by this harmonic function separates the points of X. He proved that any open set is an MP-set with respect to $\mathscr{H}^*$ and constructed a theory parallel to that of Brelot, but obviously more general.

In 1963 N. Boboc-C. Constantinescu-A. Cornea [2] considered a somewhat more general axiomatic theory which in fact was a combination of Doob's and Bauer's systems; namely, it assumed that $\mathscr{H}$ possesses the Bauer convergence property and that there exists a covering of the space by MP-sets. In 1965 [3], they added to this system the condition that $\mathscr{H}$ is non-degenerate at any point of the space. It will be proved in this book (Theorem 3.1.1) that this last axiomatic theory is nothing else but a localized version of Bauer's axiomatic theory.

The existence of a base of regular sets is somewhat difficult to verify for some important examples, as for instance, the heat equation or more generally, for parabolic equations. This led J. Köhn 1968 [1] to construct a new axiomatic system in which the base of regular sets was replaced by a base of "semi-regular" sets. A pair (V, V^r) is called semi-regular if V is an open, relatively compact set and V^r is a compact non-empty subset of ∂V such that every $f \in \mathscr{C}(V^r)$ possesses uniquely a continuous extention, $\bar{f}$, to $\bar{V}$ whose restriction to V is harmonic and such that $f \geq 0 \Rightarrow \bar{f} \geq 0$. It was this axiomatic approach which suggested to the authors the axioms proposed above. In the next chapter it will be shown that these axioms, although more general than the older ones, are equivalent to them if for example, the basic sweeping system is elliptic or if the points are polar.

Exercises

2.1.1. If in exercise 1.2.3, X endowed with $\mathscr{U}$ is a harmonic space, then so is X endowed with $\mathscr{U}_f$.

2.1.2. Let X be a locally compact space and $\mathscr{U}$ be a hyperharmonic sheaf. For any set $\mathscr{F}$ of locally uniformly lower bounded hyperharmonic functions on an open set U of X, we denote by $\wedge\mathscr{F}$ the function on U

$$x \mapsto \liminf_{y\to x} \big(\inf_{u\in\mathscr{F}} u(y)\big).$$

X endowed with $\mathscr{U}$ is a harmonic space if $\mathscr{U}$ possesses the following properties:

a) for any point x of X, there exists a hyperharmonic function defined on a neighbourhood of x and strictly negative at x;

b) for any open set U, for any set $\mathscr{F}$ of locally uniformly lower bounded of $\mathscr{U}$-functions on U, and any $\mathscr{U}$-function u on U, we have

$$\wedge(u+\mathscr{F}) = u + \wedge\mathscr{F};$$

c) the axiom of resolutivity and the axiom of completeness are satisfied.

2.1.3. Let X be an open set of $\mathbb{R}^n$ ($n\in\mathbb{N}, n\neq 0$), let a_{ij}, b_i, c ($i,j=1,2,\ldots,n$) be locally bounded real functions on X such that the matrix $(a_{ij})_{ij}$ is positively definite, and for any compact set K in X, $\inf_K \left(\sum_{i=1}^{n} a_{ii}\right) > 0$. Let L be the differential operator

$$\sum_{i,j=1}^{n} a_{ij} \frac{\partial^2}{\partial x_i\, \partial x_j} + \sum_{i=1}^{n} b_i \frac{\partial}{\partial x_i} + c.$$

Assume further that $\Omega := \big((\omega_{\iota x})_{x\in V_\iota}\big)_{\iota\in I}$ is a sweeping system on X such that for any $\iota\in I$ we have; *a)* $\omega_\iota 1$ is bounded and strictly positive; *b)* for any bounded Borel function f on ∂V_ι, $\omega_\iota f$ is continuous; *c)* there exists a dense set $\mathscr{C}_\iota$ of $\mathscr{C}(\partial V_\iota)$ such that for any $f\in\mathscr{C}_\iota$, $\omega_\iota f$ is of class $\mathscr{C}^2$ and $L\omega_\iota f = 0$; *d)* there exists a positive real function u_ι of class $\mathscr{C}^2$ on V_ι such that $Lu_\iota \le 0$ and u_ι converges to ∞ along any ultrafilter which is non-regular for ω_ι. Then X endowed with the hyperharmonic sheaf generated by the sweeping system Ω is a harmonic space. (Use *a)* and *c)* and Exercise 2.1.1 in order to reduce the problem locally to the case where $c=0$. Let $\mathscr{V}$ be the sheaf of functions u of class $\mathscr{C}^2$ for which $Lu<0$. Since the matrix $(a_{ij})_{ij}$ is positive definite, no $\mathscr{V}$-function has a local minimum. By *a)*, *c)* and *d)*, it follows that any $\mathscr{V}$-function is Ω-hyperharmonic. Use this fact and the fact that for any compact set K of X, $\inf_K \left(\sum_{i=1}^{n} a_{ii}\right) > 0$, in order to show that the set of Ω-hyperharmonic functions separates the points of sufficiently small open sets. By the Brelot-Bauer proposition, there exists a covering of X with MP-sets with respect to $\mathscr{U}$. Show that there exists a base consisting of quasi-regular sets. Use Proposition 2.1.1 and Exercise 1.1.10.)

2.1.4. Let X be a harmonic space, $\mathscr{U}$ its structural sheaf, Y a locally compact space and f a continuous open map of Y into X. Assume that every $y \in Y$ has a neighbourhood U such that the map $U \mapsto f(U)$ induced by f is a homeomorphism. For any open set U of Y, we set

$$\mathscr{V}_0(U) = \{u \circ f^{-1} \mid u \in \mathscr{U}(f(U))\}.$$

Then $\mathscr{V}_0$ is a presheaf. If $\mathscr{V}$ is the sheaf generated by $\mathscr{V}_0$, then Y endowed with $\mathscr{V}$ is a harmonic space.

2.1.5. Let Y be a locally compact space and for any $y \in Y$, let u_y be a positive hyperharmonic function on a harmonic space X. If the function on $X \times Y$, $(x, y) \mapsto u_y(x)$ is lower semi-continuous, then for any measure μ on Y, the function on X

$$x \mapsto \int^{*} u_y(x)\, d\mu(y)$$

is a hyperharmonic function.

§ 2.2. Superharmonic Functions and Potentials

A hyperharmonic function u on a harmonic space is called *superharmonic* if, for any relatively compact, resolutive set V, the function $\mu^V u$ is harmonic. This notion was introduced into classical potential theory by F. Riesz 1925 [1]. Every superharmonic function is finite on a dense set. The sheaf of harmonic functions possesses the Doob convergence property if, and only if, every hyperharmonic function on an open set which is finite on a dense set, is superharmonic. The locally bounded hyperharmonic functions and the hyperharmonic minorants of a superharmonic function are superharmonic functions (Proposition 1.1.4). Hence the infimum of two superharmonic functions is also superharmonic. A function which is locally superharmonic is superharmonic. If u is a superharmonic function and if V is a relatively compact resolutive set such that u_V is lower finite, then u_V is a superharmonic function which is harmonic on V. The set of superharmonic functions on a harmonic space is a convex cone. A hypoharmonic function u is called *subharmonic* if $-u$ is superharmonic. If u is a hyperharmonic function on X which possesses a subharmonic minorant, then for any relatively compact, resolutive set V, the function u_V is lower finite and therefore is hyperharmonic. A positive superharmonic function for which any positive harmonic minorant vanishes identically is called a *potential* (M. Brelot 1958 [11]). A positive hyperharmonic function dominated by a potential is also a potential.

A lower directed set $\mathscr{V}$ of hyperharmonic functions on a harmonic space X is called a *Perron set* if it possesses a subharmonic minorant and if every point of X has a relatively compact, resolutive neighbourhood

V such that: *a)* for any $v \in \mathscr{V}$ we have $v_V \in \mathscr{V}$; *b)* there exists $v \in \mathscr{V}$ such that $\mu^V v$ is harmonic. For any superharmonic function u on a harmonic space X which possesses a subharmonic minorant and for any covering $\mathfrak{W}$ of X by relatively compact, resolutive sets the set of functions of the form $u_{V_1, V_2, \ldots, V_n}$ where $(V_i)_{1 \leq i \leq n}$ is a finite sequence in $\mathfrak{W}$ will be called *the Perron set generated by u and* $\mathfrak{W}$.

Theorem 2.2.1 (O. Perron). *The infimum of a Perron set $\mathscr{V}$ is a harmonic function. For every compact set K and strictly positive real number ε, there exists a $u \in \mathscr{V}$ such that $u < \inf_{v \in \mathscr{V}} v + \varepsilon$ on K* (O. Perron 1923 [1]).

The proposition is an immediate consequence of the definition of a Perron set and of the convergence axiom. □

Theorem 2.2.2 (F. Riesz). *Every superharmonic function u on a harmonic space which has a subharmonic minorant may be written uniquely as the sum of a potential and a harmonic function. This harmonic function is the greatest hypoharmonic minorant of u and is the infimum of any Perron set generated by u* (F. Riesz 1925 [1]).

By Perron's theorem, the infimum of a Perron set generated by u is a harmonic minorant, h, of u. If v is a hypoharmonic minorant of u, then we may show inductively that for any sequence $(V_i)_{1 \leq i \leq n}$ of relatively compact, resolutive sets, we have $v \leq u_{V_1, V_2, \ldots, V_n}$ and this shows that h is the greatest hypoharmonic minorant of u. We set $p := u - h$. Obviously p is a positive superharmonic function. If h_0 is a positive harmonic minorant of p, then $h + h_0$ is a harmonic minorant of u and therefore smaller than h. We deduce that $h_0 = 0$. Hence p is a potential. Let $u := p' + h'$ where p' is a potential and h' is a harmonic function. Then h' is a harmonic minorant of u and therefore $h' \leq h$. Since $h - h'$ is a positive harmonic minorant of p' one obtains $h = h'$ and $p = p'$. □

Corollary 2.2.1. *The real vector space of the differences of positive harmonic functions on a harmonic space is a conditionally complete vector lattice with respect to the natural order* (M. Brelot 1957 [10], H. Bauer 1960 [3]).

Let h', h'' each be difference of positive harmonic functions. Then $\inf(h', h'')$ is a superharmonic function which has a harmonic minorant. By the theorem, it possesses a greatest harmonic minorant. It follows that the above space is a vector lattice. From the convergence axiom, we deduce immediately that it is conditionally complete. □

Proposition 2.2.1. *Let u be a positive superharmonic function on a harmonic space X. The following assertions are equivalent:*

a) u is a potential;

b) (resp. b′)) the infimum of any (resp. a) Perron set generated by u is identically 0;

c) every hyperharmonic function v on X for which u+v is positive, is positive;

d) every hypoharmonic minorant of u is negative.

$a \Rightarrow b$ follows from the theorem of Riesz.

$b \Rightarrow b'$ is trivial.

$b' \Rightarrow c$. Let v be a hyperharmonic function on X such that $u+v$ is positive. Let $\mathscr{V}$ be a Perron set generated by u whose infimum is 0. From

$$u_{V_1, V_2, \ldots, V_n} + v \geq (u+v)_{V_1, V_2, \ldots, V_n} \geq 0$$

it follows that $u'+v \geq 0$ for all $u' \in \mathscr{V}$. Hence

$$v(x) = v(x) + \inf_{u' \in \mathscr{V}} u'(x) = \inf_{u' \in \mathscr{V}} (u'+v)(x) \geq 0$$

for any $x \in X$.

$c \Rightarrow d$. Let v be a hypoharmonic minorant of u. Then $u-v$ is positive and we have

$$-v \geq 0, \qquad v \leq 0.$$

$d \Rightarrow a$ is trivial. □

Corollary 2.2.2. *If p is a potential on a harmonic space X, continuous at any point where it vanishes, then the restriction of p to* $\{x \in X \mid p(x) > 0\}$ *is also a potential.*

We set $U := \{x \in X \mid p(x) > 0\}$. Let h be a positive harmonic minorant of $p|_U$ and let h^* be the function on X which is equal to h on U and to 0 on $X \setminus U$. Since h^* is continuous, it is hypoharmonic (Proposition 2.1.2) and being a minorant of p, it vanishes identically (by $a \Rightarrow d$). □

Proposition 2.2.2. *If the sum of a family of potentials on a harmonic space is a superharmonic function, then it is also a potential.*

Let $(p_\iota)_{\iota \in I}$ be a family of potentials on a harmonic space such that

$$u := \sum_{\iota \in I} p_\iota$$

is superharmonic and let h be a positive harmonic minorant of u. For any finite subset J of I, $h - \sum_{\iota \in I \setminus J} p_\iota$ is a hypoharmonic minorant of the potential $\sum_{\iota \in J} p_\iota$ and is therefore negative. It follows that h vanishes on the dense set where u is finite and hence is identically 0. □

Let f be a numerical function on a harmonic space X. We denote by Rf the infimum of the set of hyperharmonic functions on X which dominate f. This function was considered for the first time in classical potential theory by M. Brelot 1945 [8]. We have

$$f \leq g \Rightarrow Rf \leq Rg,$$
$$R(f+g) \leq Rf + Rg, \qquad R(\alpha f) = \alpha Rf,$$

for any positive real number α.

Proposition 2.2.3. *For any lower semi-continuous, lower finite, numerical function f on a harmonic space X, Rf is a hyperharmonic function. If f has a superharmonic majorant, then Rf is superharmonic and is finite and continuous at any point where f is finite and continuous. Moreover Rf is harmonic on any open set where f is either subharmonic or continuous and strictly smaller than Rf* (M. Brelot 1945 [8], C. Constantinescu-A. Cornea 1963 [2], N. Boboc-C. Constantinescu-A. Cornea 1965 [4]).

Let u be the function on X

$$x \mapsto \liminf_{y \to x} Rf(y).$$

Obviously u is lower semi-continuous and dominates f, hence it does not take the value $-\infty$. Let V be a resolutive set and let v be a hyperharmonic majorant of f. Then $\mu^V u \leq \mu^V v \leq v$ on V. Since v is arbitrary, we obtain $\mu^V u \leq Rf$ on V. $\mu^V u$ being lower semi-continuous (Proposition 1.1.4) implies $\mu^V u \leq u$. Hence u is hyperharmonic and $Rf = u$.

Assume now that f has a superharmonic majorant and let U be an open set such that either f is subharmonic on U or f is continuous and $f < Rf$ on U and choose $x \in U$. There exists then, a relatively compact, resolutive set V containing x and a harmonic function defined on an open neighbourhood of $\overline{V}$ such that $f \leq h \leq Rf$ on $\overline{V}$. Then $(Rf)_V$ is greater than f. Hence $Rf \leq (Rf)_V$ and we have $Rf = (Rf)_V$. Since Rf is superharmonic it follows that it is harmonic on V. In particular we deduce that Rf is finite everywhere on U.

Let x be a point of X at which f is finite and continuous and let h be a harmonic function defined on an open neighbourhood of x and equal to 1 at x. Let ε be a strictly positive real number and V be a relatively compact, resolutive set such that $x \in V$, h is defined on $\overline{V}$ and

$$f \leq (f(x)+\varepsilon)\, h, \qquad Rf \geq (Rf(x)-\varepsilon)\, h$$

on $\overline{V}$. By Proposition 2.1.3 the function u^* on X equal to

$$\inf(Rf, \mu^V Rf + 2\varepsilon h)$$

on V, equal to Rf on $X \setminus \overline{V}$ and equal to

$$y \mapsto \inf(Rf(y), \liminf_{z \to y} (\mu^V Rf(z) + 2\varepsilon h(z)))$$

on ∂V is hyperharmonic. Since

$$\mu^V Rf + 2\varepsilon h \geq (f(x)+\varepsilon)\, h \geq f$$

on V we deduce $f \leq u^*$. Hence

$$Rf \leq u^*, \qquad Rf \leq \mu^V Rf + 2\varepsilon h$$

on V. We obtain

$$\limsup_{y \to x} Rf(y) \leq \mu^V Rf(x) + 2\varepsilon \leq Rf(x) + 2\varepsilon.$$

Since ε was arbitrary, Rf is continuous. □

We call an *Evans function* of a potential p on a harmonic space X any positive hyperharmonic function u on X such that for all $\alpha \in \mathbb{R}_+$ the set $\{x \in X \mid u(x) < \alpha\, p(x)\}$ is relatively compact. If u is an Evans function of p, then for any strictly positive real number ε, εu is also an Evans function of p. If $\mathfrak{U}$ is an ultrafilter on X converging to the Alexandroff point of X, then either p converges to 0 along $\mathfrak{U}$ or u converges to ∞ along $\mathfrak{U}$.

Proposition 2.2.4. *Let p be a potential on a harmonic space X. For any $x \in X$, there exists an Evans function of p which is finite at x. If X is σ-compact, then there exists an Evans function u of p such that : a) u is a potential; b) u is finite whenever p is finite; c) u is continuous at any point where p is continuous* (N. Boboc-C. Constantinescu-A. Cornea 1962 [1]).

Let $\mathscr{F}$ be the set of $f \in \mathscr{K}_+(X)$ such that $f \leq 1$. By the preceding proposition, $f \in \mathscr{F}$ implies $R((1-f)\,p)$ is harmonic on the interior of the set $\{x \in X \mid f(x) = 1\}$. Since the family $(R((1-f)\,p))_{f \in \mathscr{F}}$ is lower directed, its infimum is a harmonic function by the convergence axiom. Being a positive minorant of a potential, it vanishes identically.

Let $x \in X$. There exists a sequence $(f_n)_{n \in \mathbb{N}}$ in $\mathscr{F}$ such that

$$\sum_{n \in \mathbb{N}} R((1-f_n)\,p)(x) < \infty.$$

The function

$$u := \sum_{n \in \mathbb{N}} R((1-f_n)\,p)$$

is an Evans function of p since for any $\alpha \in \mathbb{R}_+$

$$\{x \in X \mid u(x) < \alpha\, p(x)\} \subset \bigcup_{n < \alpha + 1} \operatorname{Supp} f_n.$$

Assume now that X is σ-compact. Let $(U_n)_{n \in \mathbb{N}}$ be an increasing sequence of open, relatively compact sets of X such that

$$\bigcup_{n \in \mathbb{N}} U_n = X.$$

There exists for any $n \in \mathbb{N}$, a function $f_n \in \mathscr{K}_+(X)$ such that

$$R((1-f_n)\,p) \leq \frac{1}{2^n}$$

on U_n. Then

$$u := \sum_{n \in \mathbb{N}} R((1-f_n)\,p)$$

is a superharmonic function and therefore a potential (Proposition 2.2.2). Properties *b)* and *c)* follow immediately from the preceding proposition. □

Exercises

2.2.1. Let u be a positive hyperharmonic function on a harmonic space X. We consider the following properties:

a) If $\mathscr{V}$ is the set of functions of the form $u_{V_1, V_2, \ldots, V_n}$, where $(V_i)_{1 \leq i \leq n}$ is a finite sequence of resolutive sets of X, then 0 is the greatest lower bound of $\mathscr{V}$.

b) Any hyperharmonic function v on X is positive if $u+v$ is positive.

c) Any harmonic minorant of u is negative.

d) Any positive harmonic minorant of u is identically 0. Prove

$$a \Rightarrow b \Rightarrow c \Rightarrow d.$$

(We do not know if any of the implications $d \Rightarrow c$, $c \Rightarrow b$, $b \Rightarrow a$ is true.)

2.2.2. Let U be an open MP-set of a harmonic space X for which there exists a base $\mathfrak{B}$ of X such that for any $V \in \mathfrak{B}$, the set $U \cap V$ is regular. Then, for any potential p on U which is harmonic outside a compact set of U and for any $y \in \partial U$, we have

$$\lim_{\substack{x \to y \\ x \in U}} p(x) = 0.$$

§ 2.3. $\mathfrak{S}$-Harmonic Spaces and $\mathfrak{P}$-Harmonic Spaces

A harmonic space X will be called $\mathfrak{S}$-*harmonic space* (resp. $\mathfrak{P}$-*harmonic space*) if for any point $x \in X$, there exists a positive superharmonic function (resp. a potential) on X which is strictly positive at x. An open set in a harmonic space will be called $\mathfrak{S}$-*set* (resp. $\mathfrak{P}$-*set*) if it is an $\mathfrak{S}$-harmonic space (resp. $\mathfrak{P}$-harmonic space) as harmonic subspace. Obviously every $\mathfrak{P}$-harmonic space is an $\mathfrak{S}$-harmonic space and every open set of an $\mathfrak{S}$-harmonic space is an $\mathfrak{S}$-set. It will be proved (Corollary 2.3.3) that every open set of a $\mathfrak{P}$-harmonic space is a $\mathfrak{P}$-set. If K is a compact set of an $\mathfrak{S}$-harmonic space (resp. $\mathfrak{P}$-harmonic space), then there exists a positive superharmonic function (resp. potential) u on X such that $\inf_K u > 0$. Indeed for any $x \in K$, there exists a positive superharmonic function (resp. potential) u_x on X which is strictly positive at x. Then

$$K \subset \bigcup_{x \in K} \{y \in X \mid u_x(y) > 0\}.$$

Since K is compact and the functions u_x are lower semi-continuous, there exists a finite set $A \subset K$ such that

$$K \subset \bigcup_{x \in A} \{y \in X \mid u_x(y) > 0\}.$$

The function $u := \sum_{x \in A} u_x$ is a positive superharmonic function (resp. potential) on X which is strictly positive on K.

Proposition 2.3.1. *Let f be a positive, finite, continuous real function with compact carrier on a harmonic space X. If X is an $\mathfrak{S}$-harmonic space (resp. a $\mathfrak{P}$-harmonic space), then Rf is a positive superharmonic function (resp. a potential) on X, continuous on X and harmonic outside the carrier of f.*

This proposition follows immediately from the above considerations and Proposition 2.2.3. ▯

Corollary 2.3.1. *Let X be an $\mathfrak{S}$-harmonic space (resp. $\mathfrak{P}$-harmonic space), u a positive hyperharmonic function on X and let $\mathscr{V}$ be the set of positive superharmonic functions (resp. potentials) on X which are continuous on X, harmonic outside a compact set and dominated by u. Then $\mathscr{V}$ is upper directed and u is the supremum of $\mathscr{V}$. If X has a countable base, then there exists an increasing sequence in $\mathscr{V}$ converging to u* (H. Bauer 1963 [6]).

If $v, w \in \mathscr{V}$ then $f := \sup(v, w)$ is finite and continuous everywhere and is subharmonic outside a compact set. But f is dominated by $v + w$ and thus we deduce that $Rf \in \mathscr{V}$. Hence $\mathscr{V}$ is upper directed.

Let $(f_\iota)_{\iota \in I}$ be a family in $\mathscr{K}_+(X)$ such that

$$u = \sup_{\iota \in I} f_\iota.$$

By the proposition we have that for any $\iota \in I$, $Rf_\iota \in \mathscr{V}$ and hence

$$u = \sup_{\iota \in I} f_\iota \le \sup_{\iota \in I} Rf_\iota \le \sup_{v \in \mathscr{V}} v \le u.$$

If X has a countable, base, we may take in the above proof an increasing sequence as family $(f_\iota)_{\iota \in I}$. ▯

Remark. F. Riesz 1930 [2] proved in classical potential theory that any positive superharmonic function is the limit of an increasing sequence of positive superharmonic functions of class $\mathscr{C}^2$.

Proposition 2.3.2. *Let X be a harmonic space. The following assertions are equivalent:*

a) X is a $\mathfrak{P}$-harmonic space;

b) the set $\mathscr{P}_c$ of finite, continuous potentials on X such that any $p \in \mathscr{P}_c$ is harmonic outside a compact set, separates the points of X;

c) the set of positive superharmonic functions on X separates the points of X;

d) for any relatively compact, resolutive set V and for any $x \in V$, there exists a positive, finite, continuous superharmonic function u on X such that $\mu^V u(x) < u(x)$.

(R.-M. Hervé 1962 [4], H. Bauer 1962 [5], N. Boboc-C. Constantinescu-A. Cornea 1965 [4].)

a ⇒ *b*. Let $x, y \in X$ and let p be a potential on X, which is strictly positive at x and y. By Corollary 2.3.1, we may suppose that p is finite at x and y. Let $\mathfrak{B}$ be the set of relatively compact, resolutive sets such that for any $V \in \mathfrak{B}$, we have $\{x, y\} \not\subset \overline{V}$. By the theorem of Riesz, the infimum of the Perron set generated by p and $\mathfrak{B}$ is identically 0. Hence there exists a finite sequence $(V_i)_{1 \le i \le n}$ in $\mathfrak{B}$ such that either

$$p_{V_1, V_2, \ldots, V_n}(x) < p_{V_1, V_2, \ldots, V_{n-1}}(x) \quad \textit{and} \quad p_{V_1, V_2, \ldots, V_{n-1}}(y) = p(y)$$

or

$$p_{V_1, V_2, \ldots, V_n}(y) < p_{V_1, V_2, \ldots, V_{n-1}}(y) \quad \textit{and} \quad p_{V_1, V_2, \ldots, V_{n-1}}(x) = p(x).$$

If the first alternative holds, then $x \in \overline{V}_n$ and we have $y \notin \overline{V}_n$ and

$$p_{V_1, V_2, \ldots, V_{n-1}}(y) = p_{V_1, V_2, \ldots, V_n}(y).$$

If the second alternative holds, we similarly deduce that

$$p_{V_1, V_2, \ldots, V_{n-1}}(x) = p_{V_1, V_2, \ldots, V_n}(x).$$

Hence

$$p(x)\, p_{V_1, V_2, \ldots, V_n}(y) \neq p(y)\, p_{V_1, V_2, \ldots, V_n}(x).$$

The assertion follows now from this relation and Corollary 2.3.1.

b ⇒ *c* is trivial.

c ⇒ *d*. If the carrier of μ_x^V is empty, then any superharmonic function on X which is strictly positive at x may be taken in place of the function u. Suppose now that y belongs to the carrier of μ_x^V and let u', u'' be positive superharmonic functions on X such that

$$u'(x)\, u''(y) < u'(y)\, u''(x).$$

Now u' (resp. u'') is strictly positive at x, since otherwise it would also vanish at y and this would contradict the above inequality. Using

Corollary 2.3.1, we may suppose that u' and u'' are finite and continuous. Multiplying u' by a real number we may suppose that u' and u'' are equal at x. Then $u''(y) < u'(y)$ and we obtain

$$\mu^V \inf(u', u'')(x) < \mu^V u'(x) \leq u'(x) = \inf(u', u'')(x).$$

$d \Rightarrow a$ follows from the theorem of Riesz. □

Corollary 2.3.2. *Every $\mathfrak{P}$-set of an $\mathfrak{S}$-harmonic space is an MP-set.*

Let U be a $\mathfrak{P}$-set of an $\mathfrak{S}$-harmonic space X. Since for any compact set K of X, there exists a positive superharmonic function on X which is strictly positive on K, we deduce from Brelot-Bauer theorem *a)* (using $a \Rightarrow c$ and the sweeping system $((\mu_x^V)_{x \in V})_{V \in \mathfrak{V}}$, where $\mathfrak{V}$ is the set of relatively compact, resolutive sets) that U is an MP-set. □

Corollary 2.3.3. *Every open set of a $\mathfrak{P}$-harmonic space is a $\mathfrak{P}$-set and an MP-set.*

The corollary follows immediately from $a \Leftrightarrow c$ and from the preceding corollary. □

Theorem 2.3.1. *Let X be a $\mathfrak{P}$-harmonic space and let f be a positive continuous real function on X with compact carrier K. For any strictly positive real number ε there exist two finite continuous potentials p, q on X such that:*

a) p, q are harmonic on $X \setminus K$;

b) $0 \leq p - q \leq f \leq p - q + \varepsilon$.

(R.-M. Hervé 1959 [1].)

We set

$$L := \left\{ x \in X \mid f(x) \geq \frac{\varepsilon}{3} \right\}$$

and denote by $\mathscr{F}$ the set of continuous real functions g on L for which there exist two finite continuous potentials p', q' on X such that $g = p' - q'$ on L. Obviously $\mathscr{F}$ is a real vector subspace of $\mathscr{C}(L)$. If p', q' are finite continuous potentials on X then so is $\inf(p', q')$ and from

$$\sup(p' - q', 0) = p' - \inf(p', q')$$

we see that $\mathscr{F}$ is a sublattice of $\mathscr{C}(L)$. By Stone's theorem (N. Bourbaki [2], §4, Corollaire de la Proposition 2) there exist two finite continuous potentials u, v on X such that

$$\left| \left(f - \frac{2\varepsilon}{3} \right) - (u - v) \right| < \frac{\varepsilon}{3}$$

on L. Since $u < v$ on the boundary of L, the function u' on X which is equal to u on L and equal to $\inf(u, v)$ on $X \setminus L$ is a positive, continuous hyperharmonic function; being dominated by u, it is a potential. We set

$$v' := \inf(u', v)$$

and note that u' and v' are equal on $X \setminus L$ and that

$$0 \le u' - v' \le f \le u' - v' + \varepsilon.$$

Let g be a function of $\mathcal{K}(X)$, whose carrier is contained in K, $0 \le g \le 1$, and equal to 1 on a neighbourhood of L. We set

$$p := R(g\,u'), \qquad q := R(g\,v').$$

p and q are finite, continuous potentials on X, harmonic on $X \setminus K$ (Proposition 2.3.1). We have

$$g\,u' \le p \le u', \qquad g\,v' \le q \le v'$$

and therefore

$$p = u', \qquad q = v'$$

on a neighbourhood U of L. Hence $p = q$ on $U \setminus L$. The function p' on X which is equal to q on $X \setminus L$ and to p on U is a hyperharmonic function on X, greater than $g\,u'$. It follows that $p \le p'$. Since $q \le p$ we get that $q = p$ on $X \setminus L$. The potentials p, q satisfy the required conditions. □

Remark. This result can be obtained immediately in classical potential theory by using the density of the set of functions of class $\mathscr{C}^\infty$ (M. Brelot 1945 [8]). The present proof of R.-M. Hervé is inspired by a proof in classical potential theory given by J. Deny.

Theorem 2.3.2. *Let X be a $\mathfrak{P}$-harmonic space and U be an open set of X with compact boundary. For any superharmonic function u' defined on an open neighbourhood U' of $\overline{U}$, there exists a superharmonic function u on X and a finite, continuous potential p on X such that:*

a) $u = u' + p$ on U;

b) p is harmonic on $U \cup (X \setminus \overline{U}')$;

c) $u = p$ on $X \setminus \overline{U}'$.

If $\overline{U}'$ is compact, then u is necessarily a potential (R.-M. Hervé 1959 [1]).

Let U_1, U_2 be two open sets of X such that

$$\overline{U} \subset U_1 \subset \overline{U}_1 \subset U_2 \subset \overline{U}_2 \subset U'$$

and such that $\overline{U}_2 \setminus U_1$ is compact. Let W be a relatively compact, open neighbourhood of $\overline{U}_2 \setminus U_1$, $\overline{W} \subset U' \setminus \overline{U}$ and let f be the function on U' which is equal to u' on $U' \setminus \overline{W}$ and to $\inf_{\overline{W}} u'$ on $\overline{W}$. By Proposition 2.2.3, Rf constructed on the harmonic space U' is a superharmonic function and it is finite and continuous on W. Let ε be a strictly positive real number and let g be a continuous real function on X whose carrier lies in W

and such that

$$g = Rf + \varepsilon \quad \text{on} \quad \partial U_1, \qquad g = Rf - \varepsilon \quad \text{on} \quad \partial U_2.$$

By Theorem 2.3.1 there exist two finite, continuous potentials p, q on X such that p, q are harmonic and equal on $X \setminus \overline{W}$ and

$$|g - (q - p)| < \varepsilon$$

on X. We denote by u the function on X which is equal to q on $X \setminus U_2$, equal to $\inf(q, Rf + p)$ on $U_2 \setminus U_1$ and equal to $Rf + p$ on U_1. Since

$$Rf + p < q \quad \text{on} \quad \partial U_1, \qquad q < Rf + p \quad \text{on} \quad \partial U_2,$$
$$Rf = u' \quad \text{on} \quad U_1 \setminus (\overline{W} \cup \overline{U}),$$

the function u is superharmonic. *a)* follows from the fact that Rf is equal to u' on U. *b)* and *c)* are trivial.

In order to prove the last assertion we remark first that X is an MP-set (Corollary 2.3.3). Hence u is positive. Let h be a harmonic minorant of u. Then $p - h$ is a hyperharmonic function positive outside a compact set. We deduce $h \leq p$ and therefore $h = 0$. $\square$

Proposition 2.3.3. *Let X be an $\mathfrak{S}$-harmonic space. Any relatively compact, open set of X whose closure has an MP-neighbourhood is a $\mathfrak{P}$-set.*

Let U be a relatively compact, open set of X and U' be an MP-set of X such that $\overline{U} \subset U'$. Suppose that U is not a $\mathfrak{P}$-set. Then there exists $x \in U$ such that every potential on U vanishes at x.

Let $\mathscr{F}$ be the set of positive, bounded harmonic functions on U which are equal to 1 at x. We show first that $\mathscr{F}$ is non-empty. By Proposition 2.3.1, there exists a positive, finite, continuous superharmonic function u on X which is strictly positive at x. By the theorem of Riesz, $u := p + h$ on U where p is a potential on U and h is a positive bounded harmonic function on U. Since p vanishes at x, $h(x)$ is strictly positive and $\frac{1}{h(x)} h \in \mathscr{F}$. Let $h', h'' \in \mathscr{F}$ and let h''' be the greatest harmonic minorant of h' and h'' (Corollary 2.2.1). Since $\inf(h', h'') - h'''$ is a potential on U, it vanishes at x and we obtain that $h''' \in \mathscr{F}$. Hence $\mathscr{F}$ is lower directed and its infimum h_0 belongs to $\mathscr{F}$.

Let v be the function on U' which is equal to h_0 on U, equal to 0 on $U' \setminus \overline{U}$ and equal to

$$y \mapsto \limsup_{z \to y} h_0(z)$$

on ∂U. We want to show that v is subharmonic. It is obvious that v is upper semi-continuous, bounded and is subharmonic on U. Let V be a relatively compact, resolutive set such that $x \notin \bar{V}$ and $\bar{V} \subset U'$ and let $\bar{u} \in \bar{\mathscr{U}}_v^V$. By Proposition 2.1.2, the function w on U which is equal to h_0 on $U \setminus V$ and to $\inf(\bar{u}, h_0)$ on $U \cap V$ is superharmonic. Let h' be its greatest harmonic minorant (Riesz' theorem). Since $w - h'$ is a potential on U it vanishes at x and we get that $h' \in \mathscr{F}$. Hence

$$h_0 \leq h' \leq w \leq \bar{u}$$

on $U \cap V$. Since $\bar{u}$ is arbitrary we deduce $h_0 \leq \bar{H}_v^V = \mu^V v$ on $U \cap V$ (Theorem 1.2.1). Now $\mu^V v$ is continuous (Proposition 1.1.4) and thus $v \leq \mu^V v$; this shows that v is subharmonic.

Since U' is an MP-set and since v is 0 outside a compact set, v is negative everywhere and this contradicts the relation $h_0(x) = 1$. □

Theorem 2.3.3. *Any harmonic space possesses a covering by* 𝔓*-sets.*

Let x be a point of a harmonic space. By the axiom of positivity and the axiom of resolutivity, there exists an MP-neighbourhood U of x on which there is a strictly positive harmonic function. By the preceding proposition, any relatively compact, open neighbourhood of x whose closure is contained in U is a 𝔓-set. □

Corollary 2.3.4. *Let u be a lower semi-continuous, lower finite, numerical function on a harmonic space X. If, for every point $x \in X$ and every neighbourhood U of x, there exists a relatively compact, resolutive set V such that $x \in V \subset U$ and $\mu^V u(x) \leq u(x)$, then u is hyperharmonic. Hence if* 𝔅 *is a base of X of relatively compact, resolutive sets, then the sweeping system $((\mu_x^V)_{x \in V})_{V \in \mathfrak{B}}$ generates $\mathscr{U}$* (G. Tautz 1943 [1], M. Brelot 1959 [14], H. Bauer 1962 [5]).

Let W be a 𝔓-set, 𝔅 the set of relatively compact resolutive sets V such that $\bar{V} \subset W$ and Ω, the sweeping system $((\mu_x^V)_{x \in V})_{V \in \mathfrak{B}}$. Assume further that $V \in \mathfrak{B}$ and $\underline{u} \in \underline{\mathscr{U}}_u^V$. Then $u - \underline{u}$ is Ω^*-hyperharmonic on V and its lower limit at any boundary point of V is positive. By Proposition 2.3.2, $a \Rightarrow c$, the set of Ω-hyperharmonic functions on V separates the points of V. Hence by Brelot-Bauer theorem, $u - \underline{u}$ is positive. We have

$$\mu^V u = \underline{H}_u^V \leq u$$

on V (Theorem 1.2.1). It follows that u is hyperharmonic on W. The assertion follows now from the theorem. □

Exercises

2.3.1. If in Exercise 1.2.3, X endowed with $\mathscr{U}$ is an 𝔖- (resp. 𝔓)-harmonic space, then so is X endowed with $\mathscr{U}_f$.

2.3.2. The harmonic space defined in Theorem 2.1.2 is a $\mathfrak{P}$-harmonic space.

2.3.3. On a compact $\mathfrak{P}$-harmonic space, every harmonic function is identically 0 and every hyperharmonic function is positive (C. Constantinescu-A. Cornea 1963 [1]).

§ 2.4. Resolutive Sets on Harmonic Spaces

Proposition 2.4.1. *Let U be an MP-set of a harmonic space X and let f be a numerical function on ∂U.*

a) If $\overline{\mathcal{U}}_f^U$ contains a superharmonic function and $\underline{\mathcal{U}}_f^U$ contains a subharmonic function, then $\overline{H}_f^U$ and $\underline{H}_f^U$ are harmonic functions; moreover if U is resolutive, then $\mu^U f$ is harmonic.

b) If X is an $\mathfrak{S}$-harmonic space and if f is bounded and has a compact carrier, then $\overline{\mathcal{U}}_f^U$ contains a superharmonic function and $\underline{\mathcal{U}}_f^U$ contains a subharmonic function and for any $y \in \partial U$

$$-\infty < \liminf_{x \to y} \underline{H}_f^U(x) \le \limsup_{x \to y} \overline{H}_f^U(x) < \infty .$$

a) The first assertion follows from Perron's theorem. The second assertion follows from the first one with the aid of Theorem 1.2.1.

b) Let g be a positive, continuous, real function on X with compact carrier such that $|f| \le g$ on ∂U. By Proposition 2.3.1, Rg is a finite, continuous superharmonic function. The assertion follows now from

$$Rg|_U \in \overline{\mathcal{U}}_f^U, \qquad -Rg|_U \in \underline{\mathcal{U}}_f^U. \quad \square$$

Proposition 2.4.2. *Let U be an MP-set of a harmonic space and $(f_n)_{n \in \mathbb{N}}$ be an increasing sequence of numerical functions on ∂U such that for any $n \in \mathbb{N}$, $\overline{H}_{f_n}^U$ is harmonic. Then*

$$\overline{H}^U_{\lim\limits_{n\to\infty} f_n} = \lim_{n \to \infty} \overline{H}_{f_n}^U .$$

Let $x \in U$. For any $n \in \mathbb{N}$ let $u_n \in \overline{\mathcal{U}}_{f_n}^U$ with

$$u_n(x) < \overline{H}_{f_n}^U(x) + \frac{1}{2^n} .$$

Then for any $m \in \mathbb{N}$ the function

$$\lim_{n \to \infty} \overline{H}_{f_n}^U + \sum_{n=m}^{\infty} (u_n - \overline{H}_{f_n}^U)$$

belongs to $\overline{\mathcal{U}}^U_{\lim\limits_{n\to\infty} f_n}$. Hence

$$\overline{H}^U_{\lim\limits_{n\to\infty} f_n}(x) \le \lim_{n \to \infty} \overline{H}_{f_n}^U(x) + \frac{1}{2^{m-1}} .$$

Since m and x are arbitrary, we have

$$\overline{H}^U_{\lim\limits_{n\to\infty} f_n} \leq \lim_{n\to\infty} \overline{H}^U_{f_n}.$$

The converse inequality is trivial. □

Theorem 2.4.1. *Let U be an MP-set of a harmonic space, $\mathscr{L}$ be the set of resolutive functions on ∂U and $\mathscr{L}_0 := \{f \in \mathscr{L} \mid |f| \in \mathscr{L}\}$.*

a) If $f \in \mathscr{L}$, then $f \in \mathscr{L}_0$ if and only if H^U_f is a difference of two positive harmonic functions on U.

b) If $f \in \mathscr{L}$, U is resolutive and $\overline{\mathscr{U}}^U_f$ contains a superharmonic function, then $f \in \mathscr{L}_0$.

c) If $f, g \in \mathscr{L}_0$, then $\sup(f, g)$, $\inf(f, g) \in \mathscr{L}_0$ and $H^U_{\sup(f,g)}$ (resp. $H^U_{\inf(f,g)}$) is the least harmonic majorant (resp. greatest harmonic minorant) of H^U_f and H^U_g.

d) If $f, g \in \mathscr{L}$ (resp. $\mathscr{L}_0$), then for all real numbers α, β, the numerical function f' on ∂U which is equal to $\alpha f + \beta g$, whenever the expression has a meaning, and defined arbitrarily elsewhere, belongs to $\mathscr{L}$ (resp. $\mathscr{L}_0$) and

$$H^U_{f'} = \alpha H^U_f + \beta H^U_g.$$

e) If $(f_n)_{n\in\mathbb{N}}$ is an increasing or decreasing sequence in $\mathscr{L}$ (resp. $\mathscr{L}_0$) such that $\lim\limits_{n\to\infty} H^U_{f_n}$ is harmonic, then $\lim\limits_{n\to\infty} f_n \in \mathscr{L}$ (resp. $\mathscr{L}_0$) and

$$H^U_{\lim\limits_{n\to\infty} f_n} = \lim_{n\to\infty} H^U_{f_n}.$$

Choose $f, g \in \mathscr{L}$ such that H^U_f, H^U_g are differences of positive harmonic functions on U and let h be the least harmonic majorant of H^U_f and H^U_g (Corollary 2.2.1). Let $u \in \overline{\mathscr{U}}^U_f$, $v \in \overline{\mathscr{U}}^U_g$. Then

$$h + (u - H^U_f) + (v - H^U_g) \in \overline{\mathscr{U}}^U_{\sup(f,g)}.$$

But u and v are arbitrary and thus

$$h \leq \underline{H}^U_{\sup(f,g)} \leq \overline{H}^U_{\sup(f,g)} \leq h.$$

Hence, $\sup(f, g) \in \mathscr{L}$ and $h = H^U_{\sup(f,g)}$.

From these considerations we deduce immediately *a)* and *c)*.

b) Let u be a superharmonic function in $\overline{\mathscr{U}}^U_f$. Since u is positive outside a compact set of X and lower bounded on U, there exists a negative continuous, real function g on ∂U, with compact carrier and satisfying

$$g(y) \leq \liminf_{x\to y} u(x)$$

for every $y \in \partial U$. Then $u \in \overline{\mathscr{U}}^U_g$ and therefore $H^U_g \leq u$. If h denotes the greatest harmonic minorant of u, then $h - H^U_g$ is positive and dominates H^U_f. Hence H^U_f is a difference of positive harmonic functions and by *a)* we deduce that $f \in \mathscr{L}_0$.

d) follows from the general properties of $\overline{H}$ and $\underline{H}$ and from *a)*. *e)* is an immediate consequence of the preceding proposition. □

Corollary 2.4.1. *Let U be a resolutive set of a harmonic space. Let f be a numerical function on ∂U such that f is μ_x^U-integrable for any $x \in U$ and such that the function $\mu^U |f|$ is harmonic. Then f is resolutive if one of the following conditions is fulfilled:*

a) f is a Baire function;

b) f vanishes outside a σ-compact set and every positive upper semi-continuous real function with compact carrier is resolutive;

c) $|f|$ is dominated by a resolutive, lower semi-continuous function and every positive, lower semi-continuous function g with $\mu^U g$ harmonic, is resolutive;

d) U is quasi-regular;

e) ∂U has a countable base.

We may assume that f is positive.

a) Let g be a positive, real continuous function on ∂U with compact carrier and let $\mathscr{F}$ be the set of numerical functions, g', on ∂U with $\inf(g, \sup(g', 0))$ resolutive. Obviously $\mathscr{F}$ contains every real, continuous function. By the theorem, part *e)*, it contains the limit of every increasing or decreasing sequence in $\mathscr{F}$ and hence contains every Baire function. It follows that $\inf(g, f)$ is resolutive. Since f' vanishes outside a σ-compact set, there exists an increasing sequence $(g_n)_{n \in \mathbb{N}}$ of positive, real, continuous functions on ∂U with compact carrier such that

$$f = \lim_{n \to \infty} \inf(g_n, f).$$

The assertion follows now from the above considerations by using again part *e)* of the theorem.

b) & *c)*. Assume first that f possesses a lower semi-continuous majorant g such that any positive, lower semi-continuous minorant of g is resolutive. Then any positive, bounded, upper semi-continuous function with compact carrier and dominated by g is resolutive. Let $\mathscr{F}'$ be the set of positive, upper semi-continuous minorants of f with compact carrier and $\mathscr{F}''$ be the set of lower semi-continuous majorants of f dominated by g. Then

$$\mu^U f = \sup_{f' \in \mathscr{F}'} \mu^U f' = \sup_{f' \in \mathscr{F}'} H_{f'}^U \leq \underline{H}_f^U \leq \overline{H}_f^U \leq \inf_{f'' \in \mathscr{F}''} H_{f''}^U = \inf_{f'' \in \mathscr{F}''} \mu^U f'' = \mu^U f.$$

Hence f is resolutive.

c) follows immediately from these considerations. For *b)*, we remark first that any positive, lower semi-continuous, real function on ∂U dominated by a real, continuous function with compact carrier is resolutive. Since f vanishes outside a σ-compact set, there exists an increasing sequence $(g_n)_{n\in\mathbb{N}}$ of positive, real, continuous functions on ∂U with compact carrier such that

$$f=\lim_{n\to\infty}\inf(g_n,f).$$

By the above considerations the function $\inf(g_n,f)$ is resolutive for any $n\in\mathbb{N}$. The assertion $b)$, now follows from part $e)$ of the theorem.

$d)$ follows from Theorem 1.2.2 and $b)$ or $c)$.

$e)$ follows from $a)$ and $b)$. □

Remark. $e)$ was proved in classical potential theory by M. Brelot 1940 [3].

Proposition 2.4.3. *Any finite, continuous potential on a harmonic space is resolutive for any MP-set.*

Let U be an MP-set of a harmonic space X, p be a finite, continuous potential on X. Let x be a point of U and u be an Evans function of p which is finite at x (Proposition 2.2.4). For any strictly positive real number ε, we have $\overline{H}_p^U-\varepsilon u\in\underline{\mathscr{U}}_p^U$ (Proposition 2.4.1 a)). Hence

$$\overline{H}_p^U(x)-\varepsilon u(x)\le\underline{H}_p^U(x).$$

The assertion follows now from the fact that ε and x are arbitrary. □

Theorem 2.4.2. *Any open set of a $\mathfrak{P}$-harmonic space is resolutive* (N. Wiener 1925 [2], M. Brelot 1959 [14], H. Bauer 1960 [3]).

Let U be an open set of a $\mathfrak{P}$-harmonic space X and let f be a continuous, real function on ∂U with compact carrier. Obviously U is an MP-set (Corollary 2.3.3). Let W be a relatively compact neighbourhood of the carrier of f and p_0 be a potential on X which is greater than 1 on W. If ε is a strictly positive real number, Theorem 2.3.1 implies that there exist two finite, continuous potentials p,q on X such that $p=q$ outside W and $|f-(p-q)|<\varepsilon$ on ∂U. Then $\varepsilon p_0\in\overline{\mathscr{U}}_{|f-(p-q)|}^U$ and since p and q are resolutive (Proposition 2.4.3) we obtain

$$-\varepsilon p_0+H_{p-q}^U\le\underline{H}_f^U\le\overline{H}_f^U\le\varepsilon p_0+H_{p-q}^U,\qquad \overline{H}_f^U-\underline{H}_f^U\le 2\varepsilon p_0$$

on U. The assertion follows from the fact that ε was arbitrary. □

Corollary 2.4.2. *Any point of a harmonic space possesses a neighbourhood W such that any open subset of W is resolutive.*

The assertion follows from the theorem using Theorem 2.3.3. □

Proposition 2.4.4. *Let U be an MP-set of a harmonic space and f be a numerical function on ∂U such that $\overline{H}_f^U$ is harmonic. Let U' be an open* *subset of U and f' be the function on $\partial U'$ which is equal to f on $\partial U \cap \partial U'$ and to $\overline{H}_f^U$ on $U \cap \partial U'$. Then $\overline{H}_f^U = \overline{H}_{f'}^{U'}$, on U'* (M. Brelot 1959 [14]).

Let $u \in \overline{\mathscr{U}}_f^U$ and $u' \in \overline{\mathscr{U}}_{f'}^{U'}$. The function on U which is equal to $\overline{H}_f^U + (u - \overline{H}_f^U)$ on $U \setminus U'$ and to $\inf(\overline{H}_f^U, u') + (u - \overline{H}_f^U)$ on U' is hyperharmonic (Proposition 2.1.2) and thus belongs to $\overline{\mathscr{U}}_f^U$. Hence

$$\overline{H}_f^U \leq u' + (u - \overline{H}_f^U)$$

on U'. Since u and u' are arbitrary, we have $\overline{H}_f^U \leq \overline{H}_{f'}^{U'}$ on U'. The converse inequality is trivial. □

Let U be an open set of a harmonic space and $\mathfrak{F}$ be a filter on U converging to a boundary point y of U. A strictly positive hyperharmonic function defined on the intersection of U and an open neighbourhood of y, is called a *barrier of* $\mathfrak{F}$ if it converges to 0 along $\mathfrak{F}$.

The barrier was implicitely used by H. Poincaré 1890 [1]. The first explicit definition, in a less general formulation, is due to H. Lebesgue 1912 [1]. The present definition for the case when $\mathfrak{F}$ is the trace on U of the filter of neighborhoods of y was given by G. Bouligand 1926 [1]. It was extended to arbitrary filters $\mathfrak{F}$ by N. Boboc-C. Constantinescu-A. Cornea 1962 [1].

Proposition 2.4.5 (G. Bouligand). *Let U be an MP-set of a harmonic space and $\mathfrak{F}$ be a filter on U converging to a boundary point y of U which possesses a barrier. Assume that there exists a fundamental system $\mathfrak{V}$ of resolutive neighbourhoods of y such that for any $V \in \mathfrak{V}$, the characteristic function of $U \cap \partial V$ is resolutive for V. If f is a numerical function on ∂U such that $\overline{H}_f^U$ is harmonic on U and*

$$\limsup_{x \to y} \overline{H}_f^U(x) < \infty,$$

then

$$\limsup_{\mathfrak{F}} \overline{H}_f^U \leq \limsup_{\substack{z \to y \\ z \in \partial U}} f(z)$$

(G. Bouligand 1926 [1], M. Brelot 1960 [15], N. Boboc-C. Constantinescu-A. Cornea 1963 [2]).

We may assume

$$\alpha := \limsup_{\substack{z \to y \\ z \in \partial U}} f(z) < \infty.$$

Let U' be an open neighbourhood of y and u be a barrier of $\mathfrak{F}$ defined on $U \cap U'$. We may assume that $\overline{H}_f^U$ is upper bounded on $U \cap U'$. Let ε be a strictly positive real number and h be a strictly positive harmonic function which is defined on an open neighbourhood of y and equal to 1 at y. Choose $V \in \mathfrak{V}$ such that $\overline{V}$ is compact and contained in U', h is defined on $\overline{V}$ and $f \leq (\alpha + \varepsilon) h$ on $\overline{V} \cap \partial U$, and let g be a continuous, real function on ∂V, $0 \leq g \leq 1$, whose carrier is contained in $U \cap \partial V$ and

such that
$$\mu_y^V(U \cap \partial V) < \mu^V g(y) + \varepsilon.$$

Since the characteristic function g' of $U \cap \partial V$ is resolutive, $g' - g$ is resolutive (Theorem 2.4.1 *d)*) and
$$H_{g'-g}^V(y) = \mu^V(g'-g)(y) < \varepsilon$$

(Theorem 1.2.1). Let f' be the function on $\partial(U \cap V)$ which is equal to f on $\overline{U \cap V} \cap \partial U$ and to $\overline{H}_f^U$ on $U \cap \partial V$. Choose any $\bar{u} \in \overline{\mathscr{U}}_{g'-g}^V$. Since u is strictly positive, there exists a positive real number β such that the function on $U \cap V$ which is equal to $(\alpha + \varepsilon)\, h + \beta\, u + \gamma\, \bar{u}$ belongs to $\overline{\mathscr{U}}_{f'}^{U \cap V}$ where
$$\gamma := |\alpha| \sup_V h + \sup_{U \cap U'} \overline{H}_f^U.$$

Hence
$$\overline{H}_{f'}^{U \cap V} \leq (\alpha + \varepsilon)\, h + \beta\, u + \gamma\, \bar{u}$$

on $U \cap V$. But $\bar{u}$ being arbitrary implies that
$$\overline{H}_{f'}^{U \cap V} \leq (\alpha + \varepsilon)\, h + \beta\, u + \gamma\, H_{g'-g}^V$$

on $U \cap V$. By Proposition 2.4.4
$$\limsup_{\mathfrak{F}} \overline{H}_f^U = \limsup_{\mathfrak{F}} \overline{H}_{f'}^{U \cap V} \leq \alpha + \varepsilon + \gamma\, \varepsilon.$$

The assertion follows now from the fact that ε is arbitrary. □

Remark 1. It will follow from Theorem 2.4.3 that any harmonic space has a base consisting of quasi-regular sets. With the aid of Theorem 1.2.2, we have that the existence of $\mathfrak{B}$ in the above proposition is always assured.

Remark 2. The hypothesis
$$\limsup_{x \to y} \overline{H}_f^U(x) < \infty$$

may not be dropped (Exercise 3.2.14).

Proposition 2.4.6. *Let U be an MP-set in a harmonic space. Assume that there exists a strictly positive potential on U and that for any $y \in \partial U$, there exists a fundamental system $\mathfrak{B}$ of resolutive neighbourhoods of y such that for any $V \in \mathfrak{B}$, the characteristic function of $U \cap \partial V$ is resolutive for V. If f is a positive, upper semi-continuous function on ∂U with $\overline{H}_f^U$ harmonic and*
$$\limsup_{x \to y} \overline{H}_f^U(x) < \infty$$

for every $y \in \partial U$, then f is resolutive (N. Boboc-C. Constantinescu-A. Cornea 1963 [2]).

Let p be a strictly positive potential on U and $\mathscr{V}$ be the set of positive, hyperharmonic functions, v, on U for which there exists a compact set K of X such that $v \geq \overline{H}_f^U$ on $U \setminus K$. Since $\mathscr{V}$ is a Perron set, its infimum h is harmonic. Let $\mathfrak{F}$ be a filter on U which converges to a boundary point y of U, and on which, p converges to 0. Let V be a relatively compact, open neighbourhood of y and f' be the function on $\partial(U \cap V)$ which is equal to $\overline{H}_f^U$ on $U \cap \partial V$ and to 0 elsewhere. Choose any $\bar{u} \in \overline{\mathscr{U}}_{f'}^{U \cap V}$. The function on U which is equal to $\inf(\bar{u}, \overline{H}_f^U)$ on $U \cap V$ and to $\overline{H}_f^U$ on $U \setminus V$ is hyperharmonic (Proposition 2.1.2) and belongs therefore to $\mathscr{V}$. We have that $h \leq \bar{u}$ on $U \cap V$. Since $\bar{u}$ was arbitrary, we get $h \leq \overline{H}_{f'}^{U \cap V}$ and the preceding proposition implies that

$$\limsup_{\mathfrak{F}} h \leq \limsup_{\mathfrak{F}} \overline{H}_{f'}^{U \cap V} = 0.$$

We show now that h vanishes identically. Choose $x \in U$ and let u be an Evans function of p which is finite at x (Proposition 2.2.4). Choose $\bar{u} \in \overline{\mathscr{U}}_f^U$ and ε, a strictly positive real number. Let $y \in \partial U$ and $\mathfrak{U}$ be an ultrafilter on U converging to y such that

$$\lim_{\mathfrak{U}} (\bar{u} - h + \varepsilon u) = \liminf_{z \to y} \big(\bar{u}(z) - h(z) + \varepsilon u(z)\big).$$

If p converges to 0 along $\mathfrak{U}$, then by the above considerations, h converges to 0 along $\mathfrak{U}$. If p does not converge to 0 along $\mathfrak{U}$, then the Evans function, u, of p converges to ∞. Hence in either case, $\bar{u} - h + \varepsilon u$ converges along $\mathfrak{U}$ to a value greater than $f(y)$. Since $\bar{u} - h + \varepsilon u$ is obviously positive, we have that it is in $\overline{\mathscr{U}}_f^U$, and

$$\overline{H}_f^U(x) \leq \bar{u}(x) - h(x) + \varepsilon u(x).$$

Now ε and $\bar{u}$ are arbitrary and thus

$$\overline{H}_f^U(x) \leq \overline{H}_f^U(x) - h(x), \qquad h(x) = 0.$$

Since x is arbitrary, h vanishes identically.

Let $x \in U$ and u be an Evans function of p which is finite at x. Let $v \in \mathscr{V}$ and ε be a strictly positive real number. Choose $y \in \partial U$ and $\mathfrak{U}$ be an ultrafilter on U converging to y such that

$$\lim_{\mathfrak{U}} (\overline{H}_f^U - v - \varepsilon u) = \limsup_{z \to y} \big(\overline{H}_f^U(z) - v(z) - \varepsilon u(z)\big).$$

If p converges to 0 along $\mathfrak{U}$, then $\overline{H}_f^U$ converges along U to a value smaller than $f(y)$ (Bouligand's Proposition). If p does not converge to 0 along $\mathfrak{U}$, then u converges to ∞ along $\mathfrak{U}$. Hence in either case, $\overline{H}_f^U - v - \varepsilon u$ converges along $\mathfrak{U}$ to a value smaller than $f(y)$. Since $\overline{H}_f^U - v - \varepsilon u$ is obviously negative outside the intersection of U and a compact set of X we obtain

$$\overline{H}_f^U - v - \varepsilon u \in \underline{\mathscr{U}}_f^U, \qquad \overline{H}_f^U(x) - v(x) - \varepsilon u(x) \leq \underline{H}_f^U(x).$$

Again, since v, ε and x are arbitrary, it follows that $\overline{H}_f^U \leq \underline{H}_f^U$. □

Theorem 2.4.3. *Any point of a harmonic space has a neighbourhood W such that every σ-compact, open subset of W is quasi-regular and resolutive.*

Let x be a point of a harmonic space. By Theorem 2.3.3, x possesses a $\mathfrak{P}$-neighbourhood, U. Let W be an open neighbourhood of x such that $\overline{W}$ is compact and contained in U. Let V be a σ-compact, open subset of W. By Corollary 2.3.3, V is an MP-set and $\mathfrak{P}$-set. Let $(K_n)_{n\in\mathbb{N}}$ be an increasing sequence of compact sets such that

$$V=\bigcup_{n\in\mathbb{N}} K_n.$$

For any $n\in\mathbb{N}$, there exists a finite, continuous potential p_n on V, which is strictly positive on K_n (Proposition 2.3.1). Then

$$p:=\sum_{n\in\mathbb{N}} \frac{1}{2^n \sup_{K_n} p_n} p_n$$

is a strictly positive potential on V.

Choose $y\in\partial V$ and let V' be a resolutive neighbourhood of y. Since V is σ-compact, the characteristic function of $V\cap\partial V'$ is resolutive for V' (Corollary 2.4.1 *a)*). From the preceding proposition, we deduce that V is resolutive. We shall show that μ^V is a quasi-regular sweeping on V. For any $f\in\mathscr{C}(\partial V)$, $\mu^V f$ is harmonic and bounded (Proposition 2.4.1). Let $x\in V$ and u be an Evans function of p which is finite at x. If $\mathfrak{U}$ is an ultrafilter on V which is non-regular for μ^V, then by the proposition of Bouligand, p does not converge to 0 along $\mathfrak{U}$ and thus u converges to ∞. □

Proposition 2.4.7. *Let U be a resolutive set of an $\mathfrak{S}$-harmonic space. Then any filter on U which converges to a boundary point of U is regular if it possesses a barrier. Hence, if all such filters possess barriers, and if U is relatively compact, then U is regular.*

Let $\mathfrak{F}$ be a filter on U converging to a boundary point of U and possessing a barrier and let f be a real, continuous function on U with compact carrier. By Proposition 2.4.1 *b)*

$$-\infty<\liminf_{x\to y} H_f^U(x)\leq\limsup_{x\to y} H_f^U(x)<\infty$$

for any $y\in\partial U$. With the aid of the Theorems 2.4.3 and 1.2.2 and of the proposition of Bouligand, we obtain

$$\lim_{\mathfrak{F}} \mu^U f=\lim_{\mathfrak{F}} H_f^U=f(y).$$

Since f is arbitrary, μ^U converges vaguely to ε_y along $\mathfrak{F}$. □

Theorem 2.4.4. *Let X be an $\mathfrak{S}$-harmonic space and U be an open set of X on which there exists a strictly positive potential. Then:*

a) Every positive upper semi-continuous real function on ∂U with compact carrier is resolutive; in particular U is resolutive.

b) A numerical function f on ∂U is resolutive if: α) f vanishes outside a σ-compact set; β) for every $x \in U$, f is μ_x^U-integrable; γ) $\mu^U |f|$ is harmonic.

c) If U is relatively compact, it is quasi-regular.

(N. Boboc-C. Constantinescu-A. Cornea 1963 [2].)

a) By Corollary 2.3.2, U is an MP-set. The assertion follows from Proposition 2.4.6, Theorem 2.4.3 and Proposition 2.4.1 *b)*.

b) follows immediately from *a)* and Corollary 2.4.1 *b)*.

c) We show that μ^U is a quasi-regular sweeping on U. By Proposition 2.4.1 *b)*, $f \in \mathscr{C}(\partial U)$ implies that $\mu^U f = H_f^U$ is harmonic and bounded. Let p be a strictly positive potential on U, $x \in U$ and u be an Evans function of p which is finite at x (Proposition 2.2.4). Let $\mathfrak{U}$ be a non-regular ultrafilter on U (for μ^U). By Proposition 2.4.7

$$\lim_{\mathfrak{U}} p > 0.$$

Hence

$$\lim_{\mathfrak{U}} u = \infty. \quad \square$$

Remark. We do not know whether the theorem still holds if we remove the hypothesis that X is an $\mathfrak{S}$-harmonic space.

Corollary 2.4.3. *Assume that X is a $\mathfrak{P}$-harmonic space. Any relatively compact, open set of X on which there exists a strictly positive, hyperharmonic function converging to 0 at every boundary point is regular.*

Let V be a relatively compact, open set of X and u be a strictly positive, hyperharmonic function on V converging to 0 at every boundary point of V. Since X is a $\mathfrak{P}$-harmonic space, there exists a superharmonic function on X which is strictly positive on $\overline{V}$. V being an MP-set (Corollary 2.3.3), the function $\inf(u, v|_V)$ is a potential on V because its hypoharmonic minorants are negative. The corollary follows now from the assertion *b)* of the theorem and from Proposition 2.4.7. $\square$

Theorem 2.4.5. *Let X be a locally compact space and $\mathscr{U}$ be a hyperharmonic sheaf on X such that the axiom of positivity, the axiom of convergence and the axiom of completeness are satisfied. Then the following assertions are equivalent:*

a) $\mathscr{U}$ satisfies the axiom of resolutivity;

b) there exists a sweeping system $\Omega := ((\omega_{\iota x})_{x \in V_\iota})_{\iota \in I}$ on X such that: α) for any $\iota \in I$, ω_ι is a quasi-regular sweeping with respect to $\mathscr{U}$; β) every

$\mathscr{U}$-function is locally Ω-hyperharmonic; γ) there exists an open covering $\mathfrak{W}$ of X such that for any $W \in \mathfrak{W}$, the set of Ω-hyperharmonic functions on W separates the points of W;

c) the quasi-regular sets with respect to $\mathscr{U}$ form a base of X and the MP-sets with respect to $\mathscr{U}$ form a covering of X.

$a \Rightarrow b$. We remark first that X endowed with $\mathscr{U}$ is a harmonic space. From Theorem 2.4.3, it follows that there exists a base $\mathfrak{B}$ of X consisting of quasi-regular resolutive sets. If we set $\Omega := ((\mu_x^V)_{x \in V})_{V \in \mathfrak{B}}$, then any hyperharmonic function is Ω-hyperharmonic. The existence of $\mathfrak{W}$ follows from Theorem 2.3.3 and Proposition 2.3.2.

$b \Rightarrow c$ follows by applying locally the Brelot-Bauer theorem.

$c \Rightarrow a$. By the axiom of completeness, the infimum of two $\mathscr{U}$-functions is a $\mathscr{U}$-function. The assertion follows now from Proposition 1.2.3 and Theorem 1.2.2. ▯

Exercises

2.4.1. Let U be an MP-set of a harmonic space, $y \in \partial U$, and $\mathfrak{F}$ be a filter on U converging to y. If, for every $x \in U$, there exists a positive, superharmonic function on U converging to 0 along $\mathfrak{F}$ and strictly positive at x, then for any numerical function f on ∂U such that

$$\limsup_{x \to y} \overline{H}_f^U(x) < \infty$$

we have

$$\limsup_{\mathfrak{F}} \overline{H}_f^U \leq \limsup_{z \to y} f(z).$$

2.4.2. Let U be an open set of a harmonic space such that for every $y \in \partial U$, there exists a positive superharmonic function u on U such that

$$\liminf_{x \to y} u(x) > 0$$

and

$$\limsup_{x \to z} u(x) < \infty$$

for every $z \in \partial U$.

a) If there exists a strictly positive potential on U, then all assertions of Theorem 2.4.4 remain valid for U.

b) U is resolutive if it is a $\mathfrak{P}$-set (for *b)* see N. Boboc-C. Constantinescu-A. Cornea 1963 [2]).

2.4.3. Let U, U' be open sets in a harmonic space, $U' \subset U$, and u, u' be hyperharmonic functions on U, U' respectively. If, for every $y \in U \cap \partial U'$, there exists a relatively compact neighbourhood W of y and a lower semi-continuous, lower finite, numerical function f on $\partial(W \cap U')$ such that: *a)* $\overline{W}$ is continued in a $\mathfrak{P}$-set; *b)* u' is lower bounded on $W \cap U'$; *c)* $f \leq u$

on $\partial(W \cap U')$; *d)* $f = u$ on $W \cap \partial U'$; *e)* $u' \geq \underline{H}_f^{W \cap U'}$ on $W \cap U'$; then the function u^* on U which is equal to u on $U \setminus \overline{U}'$, $\inf(u, u')$ on U, and

$$y \mapsto \inf\left(u(y), \liminf_{x \to y} u'(x)\right)$$

on $U \cap \partial U'$ is hyperharmonic. (The proof is similar to Corollary 2.1.2.)

2.4.4. Let V be an open set of a $\mathfrak{P}$-harmonic space, $\mathfrak{W}$ an upper directed (with respect to the inclusion relation) set of open subsets of V such that

$$V = \bigcup_{W \in \mathfrak{W}} W$$

and $\mathfrak{F}$, the section filter of $\mathfrak{W}$. Then, for any $x \in V$ and for any finite, continuous potential p, we have

$$\lim_{W, \mathfrak{F}} \mu^W p(x) = \mu^V p(x).$$

Hence μ_x^W converges vaguely to μ_x^V along $\mathfrak{F}$.

2.4.5. Let $\mathfrak{V}$ be a base of resolutive sets of an $\mathfrak{S}$-harmonic space. If the sweeping system $(\mu^V)_{V \in \mathfrak{V}}$ is elliptic, then any relatively compact, $\mathfrak{P}$-set is quasi-regular (use Theorem 2.4.4 and Proposition 1.3.3).

2.4.6. Let U, V be resolutive sets of an $\mathfrak{S}$-harmonic space such that U is relatively compact and contained in V. Let $x \in \partial U \cap \partial V$ and assume that there exists a neighbourhood W of x for which $U \cap W = V \cap W$. If x is a regular boundary point of U, then it is also a regular boundary point of V (M. Brelot 1960 [15]). (Let $f \in \mathscr{K}(\partial V)$ and let f_0 be the function on ∂U which is equal to f on $\partial U \cap \partial V$ and H_f^V on $V \cap \partial U$. Prove that f_0 is bounded by the use of Proposition 2.4.4 and Exercise 1.2.7.) (The converse is not true even if V is a regular $\mathfrak{P}$-set (see Exercise 6.3.10), but it is true if X is a $\mathfrak{P}$-harmonic space even without the assumption of the existence of W (see Corollary 6.3.2).)

2.4.7. Let U be a relatively compact, open set of a $\mathfrak{P}$-harmonic space X and let $\mathscr{F}$ be the set of finite, continuous real functions on $\overline{U}$ whose restrictions to U are hyperharmonic. A point $x \in \overline{U}$ is called extremal if ε_x is the only measure μ on $\overline{U}$ such that

$$\int u \, d\mu \leq u(x)$$

for all $u \in \mathscr{F}$. Prove that there exist extremal points and that any extremal point is a regular boundary point. Any $u \in \mathscr{F}$ is positive if it is positive at any extremal point (H. Bauer 1962 [5]). (The converse is also true under supplementary conditions; see Exercise 9.2.8.)

2.4.8. Let U be an MP-set of a harmonic space and let f be a numerical function on ∂U with $\overline{H}_f^U$ harmonic. Then, for any $x \in U$, there exists a positive, hyperharmonic function u on U, finite at x, such that for any strictly positive, real number ε, the function $\overline{H}_f^U + \varepsilon u \in \overline{\mathscr{U}}_f^U$. (Let $(u_n)_{n\in\mathbb{N}}$ be a sequence in $\overline{\mathscr{U}}_f^U$ such that

$$\sum_{n\in\mathbb{N}} (u_n(x) - \overline{H}_f^U(x)) < \infty.$$

Take

$$u := \sum_{n\in\mathbb{N}} (u_n - \overline{H}_f^U).)$$

2.4.9. Let U be an MP-set of a harmonic space X on which there exists a strictly positive potential and let f be a positive, numerical function on ∂U with $\overline{H}_f^U$ harmonic and such that for every $y \in \partial U$

$$\limsup_{\substack{x\to y\\ x\in U}} \overline{H}_f^U(x) < \infty.$$

Let $\mathfrak{B}$ be the set of relatively compact, open sets of X. For any $V\in\mathfrak{B}$, we denote by f_V (resp. f'_V), the function on $\partial(U\cap V)$ which is equal to f (resp. 0) on $\partial(U\cap V)\cap\partial U$ and 0 (resp. $\overline{H}_f^U$) on $U\cap\partial(U\cap V)$. Then

$$V, V' \in \mathfrak{B}, \qquad V\subset V' \Rightarrow \overline{H}_{f_V}^{U\cap V} \leq \overline{H}_{f_{V'}}^{U\cap V'}, \qquad \overline{H}_{f'_V}^{U\cap V} \geq \overline{H}_{f'_{V'}}^{U\cap V'}$$

and

$$\sup_{V\in\mathfrak{B}} \overline{H}^{U\cap V} = H_f^U, \qquad \inf_{V\in\mathfrak{B}} \overline{H}_{f'_V}^{U\cap V} = 0.$$

(Let u be an Evans function for the given potential. Prove that for every $y\in V\cap\partial U$ we have

$$\liminf_{\substack{x\to y\\ x\in U\cap V}} (u(x) - \overline{H}_{f'_V}^{U\cap V}(x)) \geq 0.$$

Set

$$h := \inf_{V\in\mathfrak{B}} \overline{H}_{f'_V}^{U\cap V}.$$

Then for every $v\in\overline{\mathscr{U}}_f^U$, we have that $u+v-h\in\overline{\mathscr{U}}_f^U$.)

2.4.10. Let U, V be open sets of an $\mathfrak{S}$-harmonic space such that $V\subset U$, U is an MP-set and there exists a strictly positive potential on U. Let f be a positive, bounded, real function with compact carrier on ∂V such that $\overline{H}_f^V = 0$. If we denote by g the function on ∂U which is equal to f on $\partial U \setminus \overline{U\setminus V}$ and 0 elsewhere, then $\overline{H}_g^U = 0$ (N. Wiener 1924 [1]). (Prove first that $\overline{H}_g^U = \overline{H}_{g'}^V$ on V, where g' is the function on ∂V which is equal to $\overline{H}_g^U$ on $U\cap\partial V$ and 0 elsewhere. Then deduce that for any Evans function u of the given potention, we have

$$\liminf_{\substack{x\to y\\ x\in U}} (u(x) - H_g^U(x)) \geq 0$$

for any $y\in\partial U\setminus\overline{U\setminus V}$. Prove then that for any $v\in\overline{\mathscr{U}}_g^U$ we have

$$u + v - \overline{H}_g^U \in \overline{\mathscr{U}}_g^U.)$$

Chapter 3

Bauer Spaces and Brelot Spaces

§ 3.1. Definitions and Fundamental Results

As it was pointed out in the historical remark of § 2.1, the various theories of harmonic spaces which were developed in the preceding years, started with a locally compact space, endowed with a harmonic sheaf satisfying some axioms but among them it was always assumed that the regular sets form a base of the space. The sheaf of hyperharmonic functions is then completely determined by the sweeping system constructed with the regular sets. The aim of this section is to show that the old theories may be included in the present one and to give criteria when the present theory is contained in the older ones.

Let $\mathscr{H}$ be a harmonic sheaf on a locally compact space X such that the set of regular sets with respect to $\mathscr{H}$ is a base of X. We denote by $\mathfrak{B}_r(\mathscr{H})=\mathfrak{B}_r$, the set of regular sets with respect to $\mathscr{H}$, by $\Omega_r(\mathscr{H})=\Omega_r$, the sweeping system $((\omega_x^V)_{x\in V})_{V\in\mathfrak{B}_r}$ on X, and by $\mathscr{H}^*$, the hyperharmonic sheaf on X generated by Ω_r. A base $\mathfrak{B}$ of X of regular sets with respect to $\mathscr{H}$ will be called *a strong base of regular* (*with respect to* $\mathscr{H}$) if the intersection of any two elements of $\mathfrak{B}$ is a regular set with respect to $\mathscr{H}$.

A locally compact space X endowed with a harmonic sheaf $\mathscr{H}$ will be called a *Brelot space* (M. Brelot 1957 [10]) if:

a) X has no isolated points and is locally connected;

b) the regular sets with respect to $\mathscr{H}$ form a base of X;

c) $\mathscr{H}$ has the Brelot convergence property.

A locally compact space X endowed with a harmonic sheaf $\mathscr{H}$ will be called a *Bauer space* if:

a) $\mathscr{H}$ is non-degenerate at any point of X;

b) there exists a strong base of X of regular sets with respect to $\mathscr{H}$.

c) $\mathscr{H}$ has the Bauer convergence property.

It will be shown later that every Brelot space is a Bauer space (Theorem 3.1.2) and that every Bauer space endowed with the hyperharmonic sheaf $\mathscr{H}^*$ is a harmonic space (Corollary 3.1.1). Because of the latter fact, we shall write "harmonic" (resp. "hyperharmonic") function on a Bauer space instead of "$\mathscr{H}$"-function (resp. "$\mathscr{H}^*$"-function). More generally, all of the notions introduced in this book for harmonic spaces will be considered as also defined for a Bauer space, by using in the definitions, the canonical harmonic space associated with the given Bauer space. Thus we say, $\mathfrak{S}$-Bauer (resp. *$\mathfrak{S}$-Brelot) space* and *$\mathfrak{P}$-Bauer* (resp. *$\mathfrak{P}$-Brelot) space* for a Bauer (resp. Brelot) space, if this space endowed with the sheaf of hyperharmonic functions is an $\mathfrak{S}$-harmonic space or a $\mathfrak{P}$-harmonic space.

The existence of a base of regular sets does not imply the existence of a strong base of regular sets even if a certain minimum principle holds (Exercises 3.1.1, 3.2.2, 3.3.8). Any connected $\mathfrak{S}$-Brelot space is σ-compact (Exercise 6.2.9). There exist connected $\mathfrak{P}$-Bauer spaces which are not σ-compact (and whose ground spaces are even one dimensional manifolds) (Exercise 3.1.9). Any Brelot and any Bauer space has a base consisting of regular, connected sets (Proposition 1.1.3). There exists $\mathfrak{P}$-Brelot spaces which are not metrisable (Exercise 3.1.8).

Proposition 3.1.1. *Let X be a Bauer space and $\mathfrak{V}$ be a strong base of X of regular sets. If we denote by Ω the sweeping system $((\omega_x^V)_{x\in V})_{V\in\mathfrak{V}}$, then the open sets, W, of X, for which the set of positive, finite, continuous, Ω-hyperharmonic functions on W separates the points of W, form a covering of X.*

For any finite, continuous, Ω-hyperharmonic function u defined on an open set U of X and for any $V\in\mathfrak{V}$ with $\overline{V}\subset U$, we denote by u_V, the function on U which is equal to $\omega^V u$ on V and u on $U\setminus V$. It is obvious that u_V is finite, continuous and smaller then u. We want to show that u_V is Ω-hyperharmonic. Let $V'\in\mathfrak{V}$, $\overline{V'}\subset U$. Obviously we have on $V'\setminus V$

$$u_V = u \geq \omega^{V'} u \geq \omega^{V'} u_V .$$

The function on $\overline{V\cap V'}$ which is equal to $u_V - \omega^{V'} u_V$ on $\overline{V}\cap V'$ and 0 elsewhere is continuous at any point of $\overline{V\cap V'}$, positive on $\partial(V\cap V')$ and harmonic on $V\cap V'$. Since by hypothesis, $V\cap V'$ is regular, this function is positive. Hence u_V is Ω-hyperharmonic.

Choose $x\in X$ and $V\in\mathfrak{V}$ such that $x\in V$ and there exists a strictly positive, harmonic function, h, defined on a neighbourhood of $\overline{V}$. Since $0<h(x)=\omega^V h(x)$, it follows that the carrier of ω_x^V is non-empty. Let x_0 be a point of the carrier of ω_x^V, and choose $V'\in\mathfrak{V}$ such that $x_0\in V'$, $x\notin V'$. Let f be a positive, continuous, real function on ∂V, strictly positive at x_0 and equal to 0 on $\partial V\setminus V'$. Let g be the function on $\partial(V\cap V')$ which is equal to $\omega^V f$ on $V\cap\partial V'$ and 0 elsewhere and p be the function on V which is equal to $\omega^{V\cap V'} g$ on $V\cap V'$ and $\omega^V f$ on $V\setminus V\cap V'$. Since g is continuous and $V\cap V'$ is regular, we easily deduce that $\omega^{V\cap V'} g\leq\omega^V f$ on

$V \cap V'$. p is obviously positive, finite, continuous, smaller than $\omega^V f$, strictly positive at x, and converges to 0 at every boundary point of V. We want to show that p is Ω-hyperharmonic. Let $V'' \in \mathfrak{B}$, $\overline{V''} \subset V$. We have on $V'' \setminus V'$,

$$p = \omega^V f = \omega^{V''} \omega^V f \geq \omega^{V''} p.$$

The function on $\overline{V' \cap V''}$ which is equal to $p - \omega^{V''} p$ on $V'' \cap \overline{V'}$ and 0 elsewhere is finite, continuous, harmonic on $V' \cap V''$ and positive on $\partial(V' \cap V'')$. Since $V' \cap V''$ is regular, we see that $\omega^{V''} p \leq p$ on $V' \cap V''$. Hence p is Ω-hyperharmonic.

Set $W := \{y \in V \mid p(y) > 0\}$. Obviously $x \in W$. We want to show that the set of positive, finite continuous, Ω-hyperharmonic functions on W separates the points of W. Let $y, z \in W$. For any finite sequence $\sigma := (V_i)_{1 \leq i \leq n}$ in $\mathfrak{B}$ such that for any $i = 1, \ldots, n$, $\{y, z\} \not\subset \overline{V_i} \subset V$, we denote by p_σ, the function $p_{V_1, V_2, \ldots, V_n}$. We denote by $\mathscr{F}$, the set of functions on V of the form p_σ. Since $\mathscr{F}$ is lower directed and for any $u \in \mathscr{F}$ and any $V'' \in \mathfrak{B}$, $\{y, z\} \not\subset \overline{V''} \subset V$, the function $u_{V''} \in \mathscr{F}$, we deduce that its infimum, h_0, is harmonic. Now h_0 is obviously positive and its limit at any boundary point of V is 0. Hence $h_0 = 0$, and there exists $V'' \in \mathfrak{B}$ for which $\{y, z\} \not\subset \overline{V''} \subset V$ and $u \in \mathscr{F}$ such that either

$$u_{V''}(y) < u(y), \quad u(z) > 0,$$

or

$$u_{V''}(z) < u(z), \quad u(y) > 0.$$

We get, in either case,

$$u(y)\, u_{V''}(z) \neq u(z)\, u_{V''}(y). \quad \square$$

Theorem 3.1.1. *Let $\mathscr{H}$ be a harmonic sheaf on a locally compact space X. Assume that $\mathscr{H}$ is non-degenerate at any point of X, has the Bauer convergence property and that the regular sets with respect to $\mathscr{H}$ form a base of X. Then the following assertions are equivalent:*

a) there exists a strong base of X of regular sets with respect to $\mathscr{H}$;

b) there exists an open covering $\mathfrak{W}$ of X, such that for any $W \in \mathfrak{W}$, the set of Ω_r-hyperharmonic functions on W separates the points of W;

c) the MP-sets with respect to $\mathscr{H}^*$ form a covering of X.

a ⇒ *b*. Let $\mathfrak{B}$ be a strong base of X of regular sets with respect to $\mathscr{H}$ and let Ω be the sweeping system $((\omega_x^V)_{x \in V})_{V \in \mathfrak{B}}$. By the preceding proposition, there exists an open covering $\mathfrak{W}$ of X such that for any $W \in \mathfrak{W}$, the set of Ω-hyperharmonic functions on W separates the points of W. Since $\mathscr{H}$ is non-degerate at any point of X, we may assume that on any $W \in \mathfrak{W}$ there exists a strictly positive $\mathscr{H}$-function. We want to show that for any $W \in \mathfrak{W}$, every Ω-hyperharmonic function u, on W, is Ω_r-hyperharmonic.

Let V be a regular set with $\overline{V} \subset W$, and let $f \in \mathscr{C}(\partial V)$, $f \leq u$ on ∂V. Then $u|_V - \omega^V f$ is an Ω-hyperharmonic function on V whose lower limit at any boundary point is positive. By the Brelot-Bauer theorem, applied to W, we deduce that $u|_V - \omega^V f$ is positive. Since f is arbitrary, we obtain that $\omega^V u \leq u$. Hence u is Ω_r-hyperharmonic and *b)* is proved.

$b \Rightarrow c$. Let $x \in X$ and U be an open neighbourhood of x such that there exists a $W \in \mathfrak{W}$ containing $\overline{U}$ and a strictly positive, harmonic function defined on a neighbourhood of $\overline{U}$. Again, the Brelot-Bauer theorem applied to W shows that U is an MP-set with respect to $\mathscr{H}^*$.

$c \Rightarrow a$. We remark first that X endowed with $\mathscr{H}^*$ is a harmonic space (Proposition 2.1.1). Let $\mathfrak{B}$ be the set of open, relatively compact sets V, for which there exists a strictly positive, superharmonic function on V which converges to 0 at every boundary point of V and such that $\overline{V}$ is contained in a $\mathfrak{P}$-set. We show first that $\mathfrak{B}$ is a base of X. Let $x \in X$ and let V be a neighbourhood of x. By Proposition 2.3.3 there exists a $\mathfrak{P}$-set W containing x. Let V' be a regular set such that $x \in V' \subset \overline{V'} \subset W \cap V$. This set being an MP-set (Corollary 2.3.3) implies that it is resolutive (Theorem 1.2.2). Hence, there exists a finite, continuous function $u \in \mathscr{H}^*(W)$ such that $\mu^{V'} u(x) < u(x)$. We set $V := \{y \in V' \mid \mu^{V'} u(y) < u(y)\}$. It is obvious that $(u - \mu^{V'} u)|_V$ is a positive, superharmonic function on V converging to 0 at every boundary point of V. Hence $V \in \mathfrak{B}$ and $\mathfrak{B}$ is a base of X.

We show now that $\mathfrak{B}$ is a strong base of regular sets. By Corollary 2.4.3, every element of $\mathfrak{B}$ is regular. $\mathfrak{B}$ is a strong base since the intersection of any two elements of $\mathfrak{B}$ belongs to $\mathfrak{B}$. □

Corollary 3.1.1. *If X endowed with $\mathscr{H}$ is a Bauer space, then X endowed with $\mathscr{H}^*$ is a harmonic space for which $\mathscr{H}_{\mathscr{H}^*} = \mathscr{H}$.*

The assertion follows immediately from $a \Rightarrow c$ and from Proposition 2.1.1. □

Corollary 3.1.2. *Let $\mathscr{U}$ be a hyperharmonic sheaf on a locally compact space. If X endowed with $\mathscr{U}$ is a harmonic space having a base of regular sets, then X endowed with $\mathscr{H}_{\mathscr{U}}$ is a Bauer space with $(\mathscr{H}_{\mathscr{U}})^* = \mathscr{U}$.*

It is obvious that $\mathscr{U}(U) \subset \mathscr{H}_{\mathscr{U}}^*(U)$ for any open set U of X. The reverse inclusion follows from Corollary 2.3.4. Hence there exists an open covering of X by MP-sets with respect to $\mathscr{H}_{\mathscr{U}}^*$. The corollary now follows from $c \Rightarrow a$. □

We give now two criteria (Corollary 3.1.3, Proposition 3.1.4) in order that a harmonic space have a base consisting of regular sets.

Proposition 3.1.2. *Let h be a strictly positive, harmonic function on a $\mathfrak{P}$-harmonic space X and let K be a compact set in X such that for any $x \in X \setminus K$, there exists a hyperharmonic function u on X, for which*

$$\frac{u(x)}{h(x)} < \inf_K \frac{u}{h}.$$

Then K possesses a fundamental system of regular neighbourhoods.

Let U be a relatively compact, open neighbourhood of K and x be a point of $U \setminus K$. Let u be the function, defined above, associated with x. Since u is lower bounded on $\bar{U}$, there exists a real number α such that $u - \alpha h$ is strictly positive on U. Obviously

$$\frac{u(x) - \alpha h(x)}{h(x)} < \inf_K \frac{u - \alpha h}{h}.$$

By Corollary 2.3.1, there exists a positive, finite, continuous superharmonic function v on U such that

$$\frac{v(x)}{h(x)} < \inf_K \frac{v}{h}.$$

Let W be an open neighbourhood of K such that $\bar{W} \subset U$. For any $x \in \partial W$, let v_x be a positive, finite, continuous, superharmonic function on U such that

$$\inf_K \frac{v_x}{h} = 1, \qquad \frac{v_x(x)}{h(x)} < 1.$$

Now the last inequality holds on a neighbourhood of x and since ∂W is compact, there exists a finite set $A \subset \partial W$ such that if we let

$$w := \inf_{y \in A} v_y,$$

we have $\dfrac{w}{h} < 1$ on ∂W. Let α be a real number such that

$$\sup_{\partial W} \frac{w}{h} < \alpha < 1,$$

and V be the set $\{y \in W \mid w(y) - \alpha h(y) > 0\}$. The function $w - \alpha h$ is a strictly positive, hyperharmonic function on V which converges to 0 at any boundary point of V. By Corollary 2.4.3, V is regular. The proof is now complete since $K \subset V \subset U$. $\square$

Corollary 3.1.3. *A harmonic space for which there exists an open covering $\mathfrak{W}$, such that for any $W \in \mathfrak{W}$ and any two points x, y of W, $x \neq y$, there exists a hyperharmonic function u on W and a strictly positive* *harmonic function h on W such that*

$$\frac{u(x)}{h(x)} < \frac{u(y)}{h(y)},$$

has a base of regular sets (G. Mokobodzki-D. Sibony 1966 [1], J. Köhn 1968 [1]).

Let $W \in \mathfrak{W}$ and let h_0 be a strictly positive, harmonic function on W. We want to show that for any two points x, y of W there exists a hyperharmonic function u on W such that

$$\frac{u(x)}{h_0(x)} < \frac{u(y)}{h_0(y)}.$$

Assume the contrary and let u, h be the functions associated, by the hypothesis, with x and y. Then

$$\frac{h(x)}{h_0(x)} \geq \frac{h(y)}{h_0(y)}, \qquad \frac{-h(x)}{h_0(x)} \geq \frac{-h(y)}{h_0(y)}.$$

We obtain

$$\frac{h(x)}{h_0(x)} = \frac{h(y)}{h_0(y)}, \qquad \frac{u(x)}{h_0(x)} = \frac{u(x)}{h(x)} \frac{h(x)}{h_0(x)} < \frac{u(y)}{h(y)} \frac{h(y)}{h_0(y)} = \frac{u(y)}{h_0(y)}$$

and this is the expected contradiction. The corollary follows immediately from the proposition. □

Remark. This criterion will be used later (Theorem 6.2.2) in order to show that any harmonic space for which the points are polar (see p. 142) has a base consisting of regular sets.

A harmonic space X is called *elliptic* if it possesses a base $\mathfrak{V}$ of relatively compact, resolutive sets such that the sweeping system $((\omega_x^V)_{x \in V})_{V \in \mathfrak{V}}$ is elliptic.

Proposition 3.1.3. *Let X be an elliptic* $\mathfrak{S}$*-harmonic space. Any open set of X which does not have compact components is an MP-set.*

Since for any compact set K of X, there exists a finite positive superharmonic function on X, strictly positive on K (Proposition 2.3.1), the assertion follows from the Brelot-Bauer theorem. □

Theorem 3.1.2. *Every Brelot space is an elliptic Bauer space.*

Let X be a Brelot space and $\mathscr{H}$ be its structural sheaf. By Proposition 1.3.1 and Proposition 1.3.2 $\mathscr{H}$ is non-degenerate at any point of X and the sweeping system Ω_r is elliptic. Let $\mathfrak{W}$ be the set of open, connected, non-compact sets W of X such that there exists a strictly positive, harmonic function defined on an neighbourhood of $\overline{W}$. By the Brelot-Bauer theorem, every $W \in \mathfrak{W}$ is an MP-set with respect to $\mathcal{H}^*$. Since $\mathfrak{W}$ is a covering of X, the assertion follows from Theorem 3.1.1 $c \Rightarrow a$. □

Remark. There exist elliptic Bauer spaces which are not Brelot spaces (see C. Constantinescu-A. Cornea 1969 [4]).

Proposition 3.1.4. *Let X be a harmonic space. Then the following assertions are equivalent:*

a) X is an elliptic harmonic space;

b) for any base $\mathfrak{V}$ of X of relatively compact, resolutive sets, the sweeping system $((\mu_x^V)_{x \in V})_{V \in \mathfrak{V}}$ is elliptic;

c) any positive, hyperharmonic function on a connected, open set is either finite on a dense set or identically ∞;

d) there exists a base of regular sets and any positive harmonic function on a connected open set is either strictly positive or identically 0;

e) any positive hyperharmonic function on a connected open set is strictly positive or identically 0.

$a \Rightarrow e$ follows from Proposition 1.3.3.

$e \Rightarrow d$. The second assertion of *d)* is trivially satisfied. Choose $x \in X$ and let U be a relatively compact, open set containing x with $\partial U = \partial \overline{U}$ and such that there exists a strictly positive, harmonic function h defined on an open neighbourhood U' of $\overline{U}$. Let $f \in \mathcal{K}_+(U)$, $f \leq h$ on U and $f(x) = h(x)$. Then the function Rf constructed on the harmonic subspace U is a finite, continuous, superharmonic function on U less than h on U, equal to $h(x)$ at x and harmonic on $U \setminus \operatorname{Supp} f$ (Proposition 2.2.3).

Let u be the function on $X \setminus \operatorname{Supp} f$ which is equal to Rf on $U \setminus \operatorname{Supp} f$, 0 on $X \setminus \overline{U}$, and

$$y \mapsto \limsup_{\substack{z \to y \\ z \in U}} Rf(z)$$

on ∂U. We want to show that u is subharmonic. It is obviously upper semi-continuous. Let V be a relatively compact, resolutive set of X such that $\overline{V} \subset X \setminus \operatorname{Supp} f$ and let $v \in \overline{\mathcal{U}}_u^V$. The function on U which is equal to $\inf(u, v)$ on $U \cap V$ and u on $U \setminus V$ is hyperharmonic (Proposition 2.1.2) and is greater than f. Hence it is greater than Rf and we get

$$Rf \leq v$$

on $U \cap V$. Since v is arbitrary, we must have

$$Rf \leq \overline{H}_u^V$$

on $U \cap V$. But $\overline{H}_u^V$ is a harmonic function on V (Proposition 2.4.1); it follows that

$$u \leq \overline{H}_u^V = \mu^V u$$

on V (Theorem 1.2.1). Hence u is subharmonic.

We set

$$A := \{y \in U \mid Rf(y) = h(y)\}.$$

Assume that $\bar{A} \cap \partial U$ is not empty and let y be a point of this set. If V is a connected, open neighbourhood of y contained in $U' \setminus \operatorname{Supp} f$, then $h - u$ is a positive, hyperharmonic function on V which is equal to 0 at y. By *e)*, it vanishes identically on V. But this is a contradiction, since $V \setminus \overline{U}$ is non-empty and $h - u$ is equal to h and therefore strictly positive on this set. Hence A is a compact set of U.

Let $y \in U \setminus A$. We have

$$\frac{Rf(y)}{h(y)} < 1 = \inf_A \frac{Rf}{h}.$$

By Proposition 3.1.2, there exists a regular set contained in U and containing A, and therefore also containing x.

d ⇒ *c*. Let u be a hyperharmonic function on a connected, open set U and let U' be a component of the interior of the set $\{x \in U \mid u(x) = \infty\}$. Let $x \in \partial U'$ and let V be a connected, regular MP-set, containing x, and such that $U' \cap \partial V$ is nonempty. Choose $f \in \mathscr{C}_+(\partial V)$ such that $f \not\equiv 0$ and $\operatorname{Supp} f \subset U'$. By *d)*, H_f^V is strictly positive on V. For any $\alpha \in \mathbb{R}_+$, $u|_V \in \overline{\mathscr{U}}_{\alpha f}^V$. Hence

$$\alpha H_f^V \leq u$$

on V. Since α is arbitrary, we get

$$u = \infty$$

on V and this is a contradiction since $y \in V \cap \partial U'$. Hence $\partial U'$ is empty and thus U' is either empty or equal to U.

c ⇒ *b*. Let V be a relatively compact, resolutive set. Let $x \in V$, W be the component of V containing x and $y \in \partial \overline{W}$. We want to show that y belongs to the carrier of μ_x^V. Assume the contrary. Then there exists $f \in \mathscr{C}_+(\partial V)$ such that

$$f(y) > 0, \quad H_f^V(x) = \mu_x^V(f) = 0.$$

Hence there exists a sequence $(u_n)_{n \in \mathbb{N}}$ in $\overline{\mathscr{U}}_f^V$ such that

$$\sum_{n \in \mathbb{N}} u_n(x) < \infty.$$

Let U be an open, connected set of X containing y and such that $f > 0$ on $U \cap \partial V$. The function u on $U \cup W$ which is equal to $\sum_{n \in \mathbb{N}} u_n$ on W and ∞

elsewhere is hyperharmonic (Proposition 2.1.2). $U \cup W$ being connected and u being finite at x, implies by $c)$, that u is finite on a dense set of $U \cup W$. This is impossible since u is infinite on the nonempty open set $U \setminus \overline{W}$.

$b \Rightarrow a$ is trivial. □

Exercises

3.1.1. Let $\mathscr{H}$ be a harmonic sheaf on a locally compact space X such that (this axiomatic system was proposed by M. Brelot 1962 [17]):

a) $\mathscr{H}$ is non-degenerate at any point of X;

b) there exists a base of X of regular sets with respect to $\mathscr{H}$;

c) $\mathscr{H}$ has the Bauer convergence property;

d) there exists an open covering $\mathfrak{W}$ of X such that for any $W \in \mathfrak{W}$, any $\mathscr{H}$-function on any open subset U of W is positive provided its lower limit at every boundary point of U is positive.

Then there exists a closed set F, of X, having only isolated points such that $X \setminus F$ endowed with the restriction of $\mathscr{H}$ is a Bauer space. (From Exercise 3.3.8 it will follow that F may be non-empty.) (The proof can be done in the following steps: $\alpha)$ Let X' be the union of all MP-sets with respect to $\mathscr{H}^*$ and $F = X \setminus X'$. Then F is closed and $X \setminus F$ endowed with the restriction of $\mathscr{H}$ is a Bauer space (Theorem 3.1.1 $c \Rightarrow a$). $\beta)$ If u is an Ω_r-hyperharmonic function on an open set U of X, and if V is a regular set with respect to $\mathscr{H}$ with $\overline{V} \subset U$, then the function u_V on U which is equal to u on $U \setminus V$ and $\omega^V u$ on V is also an Ω_r-hyperharmonic function. $\gamma)$ Let $\mathfrak{V}$ be a base of X of regular sets with respect to $\mathscr{H}$ and let Ω be the sweeping system $((\omega_x^V)_{x \in V})_{V \in \mathfrak{V}}$. Let u be a bounded, positive, Ω-hyperharmonic function on an open set U of X and let $\mathscr{V}$ be the set of functions on U of the form $u_{V_1, V_2, \dots, V_n}$, where $(V_i)_{1 \le i \le n}$ is a finite sequence in $\mathfrak{V}$ whose closures are contained in U. Then the infimum of $\mathscr{V}$ is the greatest $\mathscr{H}$-minorant of u. $\delta)$ Let $x \in X$, let U be an open neighbourhood of x and let p be a positive, Ω-hyperharmonic function on U, strictly positive at x and such that every positive $\mathscr{H}$-minorant of p vanishes identically. Then $x \notin F$ (Use $c)$ and the Brelot-Bauer proposition). $\varepsilon)$ Let $x \in F$ and let U be an open neighbourhood of x such that there exists a strictly positive $\mathscr{H}$-function defined on a neighbourhood of $\overline{U}$. Let $\mathscr{F}$ be the set of positive $\mathscr{H}$-functions on U which are equal to 1 at x and let h_U be its infimum. Then $\mathscr{F}$ is lower directed and h_U is harmonic. $\zeta)$ The function on X which is equal to $-h_U$ on U, 0 on $X \setminus \overline{U}$ and $y \mapsto \liminf_{z \to y}(-h_U(z))$ on ∂U is Ω-hyperharmonic; here $\mathfrak{V}$ is the set of regular sets V with respect to $\mathscr{H}$ such that either $x \notin \overline{V}$ or $\overline{V} \subset U$. $\eta)$ Let $y \in U \setminus \{x\}$ such that $h_U(y) > 0$ and let K

be a compact neighbourhood of y which is contained in $U\setminus\{x\}$. The function on U which is equal to $h_U - h_{U\setminus K}$ on $U\setminus K$, 0 on the interior of K, and $z \mapsto \liminf_{z'\to z}(h_U(z') - h_{U\setminus K}(z'))$ on ∂K is Ω-hyperharmonic where $\mathfrak{B}$ is the set of regular sets V with respect to $\mathscr{H}$ such that either $x \notin \overline{V}$ or $\overline{V} \subset U\setminus K$ and its greatest $\mathscr{H}$-minorant vanishes identically. Hence by *d)* $y \notin F$.)

3.1.2. In Exercise 1.1.3, X endowed with $\mathscr{H}$ and X endowed with $\mathscr{H}_f$ are simultaneously Bauer spaces, $\mathfrak{S}$-Bauer spaces, $\mathfrak{P}$-Bauer spaces, Brelot spaces, $\mathfrak{S}$-Brelot spaces and $\mathfrak{P}$-Brelot spaces.

3.1.3. An elliptic Bauer space for which the sheaf of harmonic functions possesses the Doob convergence property is a Brelot space (C. Constantinescu 1965 [2], H. Bauer 1966 [9]) (use Exercise 1.1.17).

3.1.4. Let α be a real number. For any open set U of $\mathbb{R}$, we denote by $\mathscr{H}(U)$, the set of real functions h, on U, of class $\mathscr{C}^2$ for which $\frac{d^2 h}{dx^2} + \alpha h = 0$. $\mathbb{R}$ endowed with the harmonic sheaf $\mathscr{H}$ is a Brelot space. If $\alpha > 0$, then $\mathbb{R}$ is not an $\mathfrak{S}$-Brelot space; if $\alpha = 0$, then $\mathbb{R}$ is an $\mathfrak{S}$-Brelot space but not a $\mathfrak{P}$-Brelot space; if $\alpha < 0$, then $\mathbb{R}$ is a $\mathfrak{P}$-Brelot space. If $\alpha \leq 0$, then any relatively compact open set of X is a regular MP-set. If $\alpha > 0$, then the interval $]a, b[$ is regular if and only if

$$b - a < \frac{\pi}{\sqrt{\alpha}}; \quad \text{if } \frac{\pi}{\sqrt{\alpha}} < b - a \neq \frac{n\pi}{\sqrt{\alpha}}$$

for any $n \in \mathbb{N}$, then any $f \in \mathscr{C}(\partial]a, b[)$ has a unique, continuous extension on $[a, b]$ whose restriction to a, b belongs to $\mathscr{H}(]a, b[)$ but is not positive if f does not vanish identically.

3.1.5. Let $\sim$ be the equivalence relation on $\mathbb{R}$

$$x \sim y :\Leftrightarrow \frac{1}{2\pi}|x - y| \in \mathbb{N},$$

let T be the quotient space $\mathbb{R}/\sim$ and let φ be the canonical map $\mathbb{R} \to T$. Let α be a real number. For any open set U of T we denote by $\mathscr{H}(U)$ the set of real functions h on U such that $h \circ \varphi$ is of class $\mathscr{C}^2$ on $\varphi^{-1}(U)$ and $\frac{d^2(h \circ \varphi)}{dx^2} + \alpha\, h \circ \varphi = 0$. T endowed with $\mathscr{H}$ is a Brelot space. If $\alpha > 0$, then T is not an $\mathfrak{S}$-Brelot space and if $\sqrt{\alpha} \notin \mathbb{N}$, every harmonic function on T vanishes identically; if $\alpha = 0$, then T is an $\mathfrak{S}$-Brelot space but not a $\mathfrak{P}$-Brelot space and every hyperharmonic function on T is constant; if $\alpha < 0$ then T is a $\mathfrak{P}$-Brelot space on which every harmonic function vanishes identically.

3.1.6. Let X be the interval $[0, 1[$ on the real axis. For any open set U of X, we denote by $\mathscr{H}(U)$ the set of continuous real functions h on U which are locally linear on $U \cap]0, 1[$ and such that if $0 \in U$, then h is constant on a neighbourhood of 0. Y endowed with the harmonic sheaf $\mathscr{H}$ is a $\mathfrak{P}$-Brelot space.

3.1.7. Let X be the interval $]-1, +1[$ on the real axis. For any open set U of X, we denote by $\mathscr{H}(U)$ the set of continuous real functions h on U which are locally linear on $U \setminus \{0\}$ and such that if $0 \in U$, there exists a strictly positive real number ε such that h is constant on $U \cap]-\varepsilon, 0[$. Then $\mathscr{H}$ is a harmonic sheaf possessing the Doob convergence property and X endowed with $\mathscr{H}$ is a Bauer space but not a Brelot space.

3.1.8. Let I be a set and X be $(]0, 1[\times I) \cup \{0\}$. We define a topology on X in the following way: a subset U of X is open if and only if: *a)* for any $\iota \in I$ the set $\{x \in]0, 1[\mid (x, \iota) \in U\}$ is open and *b)* if $0 \in U$ then there exists a finite subset J of I and a strictly positive real number ε such that $]0, 1[\times (I \setminus J) \subset U$ and $(]0, \varepsilon[\cup]1-\varepsilon, 1[) \times J \subset U$. For any open set U of X, we denote by $\mathscr{H}(U)$ the set of continuous, real functions h on U such that: *a)* for any $\iota \in I$, the function $x \mapsto h(x, \iota)$ is locally linear on $\{x \in]0, 1[\mid (x, \iota) \in U\}$ and *b)* if $0 \in U$, then there exists a finite subset J of I and a strictly positive real number ε such that $]0, 1[\times (I \setminus J) \subset U$, $(]0, \varepsilon[\cup]1-\varepsilon, 1[) \times J \subset U$ and

$$\sum_{\iota \in J} (h(\varepsilon, \iota) + h(1-\varepsilon, \iota) - 2h(0)) = 0.$$

Then X endowed with the harmonic sheaf $\mathscr{H}$ is a compact $\mathfrak{S}$-Brelot space which has no countable base if I is not countable.

3.1.9. Let ξ be an ordinal number and X be $[0, 1[\times [0, \xi[$. We introduce on X, the following topology: a subset U of X is open, if and only if, for any $\eta < \xi$ the following conditions are fulfilled: *a)* the set $\{x \in [0, 1[\mid (x, \eta) \in U\}$ is open; *b)* if $(0, \eta) \in U$ and if $\eta = \zeta + 1$ for an ordinal number ζ, then there exists a strictly positive real number ε such that $]1-\varepsilon, 1[\times \{\zeta\} \subset U$; *c)* if $(0, \eta) \in U$ and if η is a limit ordinal, then there exists an ordinal number $\zeta < \eta$ such that $[0, 1[\times [\zeta, \eta[\subset U$. For any open set U of X, we shall denote by $\mathscr{H}(U)$, the set of continuous, real functions h on U such that for any $\eta < \xi$ the following conditions are fulfilled: *a)* the function $x \mapsto h(x, \eta)$ is locally linear on $\{x \in]0, 1[\mid (x, \eta) \in U\}$; *b)* if $(0, \eta) \in U$, then there exists a strictly positive real number ε such that h is constant on $U \cap ([0, \varepsilon[\times \{\eta\})$. Then $\mathscr{H}$ is a harmonic sheaf possessing the Doob convergence property and X endowed with $\mathscr{H}$ is a $\mathfrak{P}$-Bauer space. If ξ is the first uncountable ordinal, we have an example of a connected Bauer space which has no countable base but is a one-dimensional manifold (and has therefore, locally, a countable base).

3.1.10. On a connected, elliptic harmonic space which is not a $\mathfrak{P}$-harmonic space, any two positive superharmonic functions are proportional and harmonic.

3.1.11. Any hyperharmonic function on a connected, elliptic harmonic space is either identically ∞ or finite on a dense set.

3.1.12. If X is an elliptic harmonic space, then there exists a base $\mathfrak{B}$ of X such that for any $V \in \mathfrak{B}$, there exists a strictly positive potential on V which converges to 0 at the boundary of V.

3.1.13. Any compact set of an elliptic $\mathfrak{P}$-harmonic space X is contained in a compact set K, such that every boundary point of $X \setminus K$ is regular (R.-M. Hervé 1962 [4]). (Use Exercise 2.2.2 and the preceding exercise.)

3.1.14. Any compact set of an elliptic $\mathfrak{P}$-harmonic space is contained in a regular set (R.-M. Hervé 1962 [4]). This result is not true for arbitrary $\mathfrak{P}$-Bauer spaces (H. Bauer 1966 [9]).

3.1.15. Let X be a locally compact space with a countable base such that any relatively compact, open set has a non-empty boundary and let $\mathscr{V}$ be a convex cone of lower semi-continuous, bounded, real functions on X, satisfying the following properties:

a) for any relatively compact, open set U of X and for any v

$$\inf_{\overline{U}} v = \inf_{\partial U} v;$$

b) any convex cone of lower semi-continuous, bounded, real functions on X with the property *a)* and containing $\mathscr{V}$ coincides with $\mathscr{V}$;

c) for any two points $x, y \in X$, there exists $v \in \mathscr{V}$ such that

$$v(x) < v(y);$$

d) for any $x \in X$, any open neighbourhood U of x, and any positive function $v \in \mathscr{V}$, the function

$$y \mapsto \inf w(y),$$

where w runs through the set of functions of $\mathscr{V}$ which are greater than v on $X \setminus \overline{U}$, is upper semi-continuous at x.

Then there exists a harmonic sheaf $\mathscr{H}$ such that X endowed with $\mathscr{H}$ is a $\mathfrak{P}$-Bauer space, the constant functions are harmonic and $\mathscr{V}$ coincides with the set of bounded hyperharmonic functions on X. (G. Mokobodzki-D. Sibony 1967 [2].)

3.1.16. Let X be an elliptic $\mathfrak{P}$-harmonic space, U be an open set of X and A be the set of regular points of U. Then $\partial \bar{U} \subset \bar{A}$ (R.-M. Hervé 1962 [4]).

3.1.17. Let X be an orientated one dimensional manifold and let $\mathscr{H}$ be a harmonic sheaf on X such that X endowed with $\mathscr{H}$ is a Bauer space. Let E be the set at which the sweeping system Ω_r is elliptic and A^+ (resp. A^-) be the set of points $x \in X$ for which there exists a neighbourhood U of x such that for any regular set V, $x \in V \subset U$, the carrier of ω_x^V lies on the right (resp. left) of x. Then:

a) E is an open dense set of X whose components have a countable base.

b) A^+, A^- are closed sets and $A^+ \cup A^- = X \setminus E$.

c) Let F_1, F_2 be two closed disjoint sets of X such that $X \setminus (F_1 \cup F_2)$ is an open dense set whose components have a countable base; then we may construct $\mathscr{H}$ such that $A^+ = F_1$, $A^- = F_2$.

(J. Král, J. Lukeš, I. Netuka 1970 [1].)

3.1.18. Let X be a Brelot space. The following assertions are equivalent: *a)* any point $x \in X$ possesses an open neighbourhood V such that for any open connected set $U \subset V$ and for any harmonic function h on U there exists a harmonic function h' on V equal to h on U; *b)* for any $x \in X$ there exists a finite family $(C_i)_{1 \le i \le n}$ $(n \ge 2)$ of arcs in X such that $\bigcup_{i=1}^{n} C_i$ is a neighbourhood of x in X and for any $1 \le i < j \le n$ we have $C_i \cap C_j = \{x\}$. (By arc in X it is meant a subspace of X homeomorphic with the segment $\{x \in \mathbb{R} \mid 0 \le x \le 1\}$ (J. Král, J. Lukeš 1971 [1]).)

§ 3.2. The Laplace Equation

The aim of this section is to show that classical potential theory, namely the theory based on the *Laplace equation* $\left(\text{i.e. the equation } \Delta h := \sum_{i=1}^{n} \frac{\partial^2 h}{(\partial x_i)^2} = 0\right)$ is a model for the theory of harmonic spaces. It is this model which guided, and is still guiding, a great deal of the research in this field.

Theorem 3.2.1. *Let n be a natural number different from* 0. *For any open set U of the topological space $\mathbb{R}^n$, we set*

$$\mathscr{H}(U) := \{h \in \mathscr{C}(U) \mid h \text{ is of class } \mathscr{C}^2 \text{ on } U \text{ and } \Delta h = 0\}.$$

Then $\mathscr{H}$ is a harmonic sheaf on $\mathbb{R}^n$ and $\mathbb{R}^n$ endowed with $\mathscr{H}$ is an $\mathfrak{S}$-Brelot space; it is a $\mathfrak{P}$-Brelot space if and only if $n \geq 3$.

The proof of this theorem is based on some well known classical results (i.e. the Poisson integral and Harnack inequality.)

For any point x of $\mathbb{R}^n$, we shall denote by x_i, its i-th coordinate, and we set

$$|x| := \left(\sum_{i=1}^{n} (x_i)^2 \right)^{\frac{1}{2}}.$$

If $a \in \mathbb{R}^n$ and r is a strictly positive real number, we set

$$U(a, r) := \{ y \in \mathbb{R}^n \,|\, |a - y| < r \}.$$

We denote by $\sigma_{a,r}$, the measure on $\partial U(a, r)$, invariant with respect to the rotations of $\partial U(a, r)$ and such that $\sigma_{a,r}(\partial U(a, r)) = 1$ (N. Bourbaki [6], Ch. VII, § 3, Exercise 8).

Lemma 1. *Let a be a point of $\mathbb{R}^n$, and r be a strictly positive real number. If f is a real function of class $\mathscr{C}^2$ on $U(a, r)$, whose lower limit at every boundary point of $U(a, r)$ is positive and such that $\Delta f \leq 0$, then f is positive.*

Suppose first $\Delta f < 0$ and assume $\inf_{U(a,r)} f < 0$. Let $x := (x_i)_{1 \leq i \leq n}$ be a point of $U(a, r)$ such that $f(x) = \inf_{U(a,r)} f$. For any i, $(1 \leq i \leq n)$, the function

$$t \mapsto f(x_1, \ldots, x_{i-1}, t, x_{i+1}, \ldots, x_n)$$

defined in a neighbourhood of $x_i \in \mathbb{R}$, is of class $\mathscr{C}^2$, and assumes its infimum at x_i. It follows that

$$\frac{\partial^2 f}{(\partial x_i)^2}(x) \geq 0,$$

and this contradicts the relation $\Delta f < 0$.

Suppose now $\Delta f \leq 0$ and assume again that $\inf_{U(a,r)} f < 0$. This leads to a contradiction applying the above considerations to the function

$$x \mapsto f(x) + \varepsilon (r^2 - |x - a|^2)$$

for a sufficiently small, strictly positive real number ε. □

Lemma 2. *Let a be a point of $\mathbb{R}^n$ and r be a strictly positive, real number. We denote by P_x, for any $x \in U(a, r)$, the function on $\partial U(a, r)$*

$$y \mapsto r^{n-2} \frac{r^2 - |x - a|^2}{|x - y|^n}$$

and by H_f for any numerical function f on $\partial U(a,r)$, the function on $U(a,r)$

$$x \mapsto \overset{*}{\int} P_x f \, d\sigma_{a,r}.$$

a) if f is lower bounded and $\sigma_{a,r}$-measurable, then H_f is either an $\mathscr{H}$-function or identically ∞;

b) if $f \in \mathscr{C}(\partial U(a,r))$, then for any $y \in \partial U(a,r)$, we have

$$\lim_{x \to y} H_f(x) = f(y);$$

c) if f is positive and $x', x'' \in U\left(a, \frac{r}{2}\right)$, then

$$H_f(x') \leq 3^{n+2} H_f(x'');$$

d) if g is a continuous function on $\overline{U(a,r)}$ and an $\mathscr{H}$-function on $U(a,r)$, then $g = H_f$ on $U(a,r)$ where $f = g|_{\partial U(a,r)}$.

a) If f is $\sigma_{a,r}$-integrable, then H_f is an $\mathscr{H}$-function. This follows from the fact that for any $y \in \partial U(a,r)$, the function on $U(a,r)$, $x \mapsto P_x(y)$, is an $\mathscr{H}$-function and from the fact that the operator Δ may be interchanged with the integral. If f is not $\sigma_{a,r}$-integrable, then H_f is identically ∞.

b) Let x', x'' be any two points of $U(a,r)$ such that $|x'-a| = |x''-a|$. There exists then, a rotation S of $\mathbb{R}^n$, with a as the centre, such that $x'' = Sx'$. Since $\sigma_{a,r}$ is invariant under S, we get $H_1(x') = H_1(x'')$. Hence, for any $r' < r$, the function H_1 is constant on $\partial U(a,r')$. Since H_1 and the constant functions are harmonic, we deduce from Lemma 1 that H_1 is constant. But $H_1(a) = 1$ which implies $H_1 = 1$.

Let $y \in \partial U(a,r)$ and assume first that there exists a strictly positive real number ε such that f vanishes on $U(y,\varepsilon) \cap \partial U(a,r)$. For any $x \in U(a,r) \cap U\left(y, \frac{\varepsilon}{2}\right)$ and for any $y' \in \partial U(a,r) \setminus U(y,\varepsilon)$, we have

$$P_x(y') \leq r^{n-2} \frac{r^2 - |x-a|^2}{\left(\frac{\varepsilon}{2}\right)^n}.$$

Hence

$$\left|\int P_x f \, d\sigma_{a,r}\right| \leq r^{n-2} \left(\frac{2}{\varepsilon}\right)^n \sup |f| (r^2 - |x-a|^2),$$

$$\lim_{x \to y} H_f(x) = 0.$$

Let now f be arbitrary and let η be a strictly positive real number. There exists $f', f'' \in \mathscr{C}(\partial U(a,r))$ such that $|f'| \leq \eta$, f'' vanishes on a neighbourhood of y and $f=f'+f''+f(y)$. We have

$$\limsup_{x\to y} |H_f(x)-f(y)| \leq \limsup_{x\to y} H_{|f'|}(x) + \limsup_{x\to y} |H_{f''}(x)| \leq \eta .$$

Since η is arbitrary, we obtain

$$\lim_{x\to y} H_f(x) = f(y).$$

c) $$H_f(x') = \int P_{x'} f \, d\sigma_{a,r} = \int \frac{P_{x'}}{P_{x''}} P_{x''} f \, d\sigma_{a,r} \leq \sup \frac{P_{x'}}{P_{x''}} H_f(x'')$$
$$\leq 3^{n+2} H_f(x'').$$

d) follows immediately from *a)* and *b)* using Lemma 1. □

Remark. The function P_x is called the *Poisson kernel* and H_f is called the *Poisson integral* (S.D. Poisson 1823 [1]); *b)* was rigorously proved for $n=2$ by H.A. Schwarz 1870 [1].

We prove now the theorem. From Lemma 2 *a)*, *b)*, *d)* it follows that every ball is regular with respect to $\mathscr{H}$; hence the regular sets form a base of $\mathbb{R}^n$. Let U be a connected open set of $\mathbb{R}^n$ and $(h_m)_{m\in\mathbb{N}}$ be an increasing sequence of $\mathscr{H}$-functions on U whose supremum h is finite at a point. From Lemma 2 *c)* it follows that the set where h is finite (resp. infinite) is open. Hence h is finite everywhere. Let $a\in U$ and let r be a strictly positive real number such that $\overline{U(a,r)}\subset U$. For any $x\in U(a,r)$, we have by Lemma 2 *d)*.

$$h(x) = \lim_{m\to\infty} h_m(x) = \lim_{m\to\infty} \int P_x h_m \, d\sigma_{a,r} = \int P_x h \, d\sigma_{a,r}.$$

Again by Lemma 2 *a)*, we deduce that h is an $\mathscr{H}$-function. Hence $\mathscr{H}$ possesses the Brelot convergence property (A. Harnack 1886 [1]). Since the constant functions are $\mathscr{H}$-functions, we deduce that $\mathbb{R}^n$ endowed with $\mathscr{H}$ is an $\mathfrak{S}$-Brelot space.

Assume $n\geq 3$ and let $y\in\mathbb{R}^n$. We denote by h_y, the function on $\mathbb{R}^n\setminus\{y\}$, $x\mapsto |x-y|^{2-n}$. A simple calculation shows that h_y is harmonic. For any positive real number α, the function $u_{y,\alpha}$ on $\mathbb{R}^n$, which is equal to α at y and $\inf(\alpha, h_y)$ on $\mathbb{R}^n\setminus\{y\}$, is a hyperharmonic function. Since the set of functions of the form $u_{y,\alpha}$ separates the points of $\mathbb{R}^n$, $\mathbb{R}^n$ is a $\mathfrak{P}$-Brelot space (Proposition 2.3.2 *c)* ⇒ *a)*).

Assume $n=1$ (resp. $n=2$) and let p be a potential on $\mathbb{R}^n$. Let r be a real number with $r>1$. We denote by h_r, the function on

$$\{x\in\mathbb{R}^n \mid 1<|x|<r\}, \qquad x\mapsto \frac{r-|x|}{r-1} \qquad \left(\text{resp. } x\mapsto \frac{\log\dfrac{|x|}{r}}{\log\dfrac{1}{r}}\right).$$

A simple calculations shows that h_r is harmonic. By Proposition 1.3.1 and the Brelot-Bauer theorem

$$p \geq (\inf_{|y|=1} p(y))\, h_r$$

on $\{x \in \mathbb{R}^n \mid 1 < |x| < r\}$. We obtain for any $x \in \mathbb{R}^n$, $|x| > 1$,

$$p(x) \geq (\inf_{|y|=1} p(y)) \lim_{r \to \infty} h_r(x) = \inf_{|y|=1} p(y).$$

Since the constant functions are harmonic, we get

$$\inf_{|y|=1} p(y) = 0$$

and by Proposition 1.3.1 and Proposition 1.3.3, p vanishes identically. □

Exercises

3.2.1. Any positive solution of the Laplace equation on $\mathbb{R}^n$ ($n \in \mathbb{N}$, $n \neq 0$) is constant.

3.2.2. Any Riemann surface endowed with the sheaf of harmonic functions is an $\mathfrak{S}$-Brelot space. Any connected Riemann surface is σ-compact and possesses therefore a countable base. An open set of a compact Riemann surface whose complement contains exactly one point (resp. two points) is a regular (resp. non-regular) MP-set. Hence there exist two regular MP-sets whose intersection is not regular.

3.2.3. Let X be the Alexandroff compactification of $\mathbb{R}^n$ ($n \in \mathbb{N}$, $n \geq 2$). For any open set U, we denote by $\mathscr{H}(U)$, the set of real, continuous functions h on U whose restriction to $U \cap \mathbb{R}^n$ are solutions of the Laplace equation and for which

$$\int_{\partial U(0,r)} \frac{\partial h}{\partial n}\, d\sigma_{0,r} = 0$$

for every $r \in \mathbb{R}$, such that $r > 0$ and $X \setminus U(0,r) \subset U$, where $\frac{\partial h}{\partial n}$ denotes the normal derivative. $\mathscr{H}$ is a harmonic sheaf on X and X endowed with $\mathscr{H}$ is an $\mathfrak{S}$-Brelot space (M. Brelot 1944 [5]). (A $\mathfrak{P}$-harmonic space which is locally isomorphic with the open sets of this space is called a Green space (M. Brelot-G. Choquet 1951 [1]).)

3.2.4. Let U be an open set of $\mathbb{R}^n$ ($n \in \mathbb{N}$, $n \neq 0$) and let f be a real function of class $\mathscr{C}^2$ on U. Then the following assertions are equivalent:

a) f is hyperharmonic for the Laplace equation;

b) for any $a \in U, r \in \mathbb{R}$ such that $r > 0$ and $\overline{U(a, r)} \subset U$ and any $x \in U(a, r)$ we have

$$\int P_x f \, d\sigma_{a,r} \leq f(x);$$

c) $\Delta f \leq 0$;

d) for any positive, real, continuous function g on U, of class $\mathscr{C}^\infty$ and with compact carrier, we have

$$\int_U f \, \Delta g \, d\tau \leq 0,$$

where τ is Lebesgue measure on $\mathbb{R}^n$.

(For $a \Leftrightarrow b$ use Lemma 2 of Theorem 3.2.1 and Corollary 2.3.4; for $b \Leftrightarrow c$ use Lemma 1 and 2 from Theorem 3.2.1; $c \Leftrightarrow d$ follows from Stokes' formula.)

3.2.5. Let U be an open, connected set of $\mathbb{R}^n$ ($n \in \mathbb{N}$, $n \neq 0$) and u be a lower semi-continuous function on U which does not take the value $-\infty$ and which is not identically ∞. Then u is a hyperharmonic function for the Laplace equation if and only if the following conditions are fulfilled:

a) if τ denotes Lebesgue measure on $\mathbb{R}^n$, then u is locally τ-integrable;

b) for any $x \in U$ we have

$$\lim_{\substack{r \to 0 \\ r > 0}} \frac{1}{\tau(U(0, r))} \int_{U(x, r)} u \, d\tau = u(x);$$

c) for any positive, real function f on U, of class $\mathscr{C}^\infty$ with compact carrier, we have

$$\int u \, \Delta f \, d\tau \leq 0.$$

(Prove first, that if u is a hyperharmonic function, then *a)* is satisfied. We set

$$k := \Big(\int_{U(0,1)} e^{\frac{1}{|x|^2 - 1}} \, d\tau(x) \Big)^{-1}.$$

For any strictly positive, real number ε, we denote by ρ_ε the real function on $[0, \infty[$ which is equal to $r \mapsto \frac{k}{\varepsilon^n} e^{\frac{\varepsilon^2}{r^2 - \varepsilon^2}}$ on $\{r \in \mathbb{R} \mid r < \varepsilon\}$ and 0 elsewhere. We denote by φ_ε (resp. u_ε), the real function on $\mathbb{R}^n$ (resp. $\{x \in U \mid \inf_{y \in \partial U} |x - y| > \varepsilon\}$)

$$x \mapsto \frac{k}{\varepsilon^n} \int_0^1 t \, e^{\frac{1}{t^2 - 1}} \, dt + \int_0^{|x|} \frac{1}{s^{n-1}} \left(\int_0^s t^{n-1} \frac{\partial \rho_\varepsilon}{\partial \varepsilon}(t) \, dt \right) ds$$

$$\Big(\text{resp. } x \mapsto \int \rho_\varepsilon(|x - y|) \, u(y) \, d\tau(y)\Big).$$

Then:

α) the functions on $\mathbb{R}^n$, $x \mapsto \rho_\varepsilon(|x|)$, and φ_ε are positive, of class $\mathscr{C}^\infty$ and vanish outside $U(0, \varepsilon)$;

β) the function u_ε is of class $\mathscr{C}^\infty$;

γ) $\Delta\varphi_\varepsilon(x) = \dfrac{\partial \rho_\varepsilon}{\partial \varepsilon}(|x|)$.

Suppose now that u satisfies the above conditions. Then by *c)*, *β)* and the preceding exercise, we deduce that u_ε is hyperharmonic. From *γ)* we deduce that for any $x \in U$, the function $\varepsilon \mapsto u_\varepsilon(x)$ is decreasing and by *b)*, it converges to $u(x)$. This shows that u is hyperharmonic.

Suppose now that u is hyperharmonic. *b)* is obvious. u_ε is hyperharmonic and, as $\varepsilon \downarrow 0$, increases to u. By the preceding exercise and *β)*, we deduce property *c)* for u_ε and therefore for u.)

3.2.6. Let n be a natural number different from 0 and let $(a_{ij})_{i,\,j=1,2,\ldots,n}$ be a strictly positive definite matrix of real numbers. We take X as the topological space $\mathbb{R}^n$ and $\mathscr{H}$ equal to the sheaf of functions on X defined by

$$\mathscr{H}(U) := \left\{ h \in \mathscr{C}(U) \,|\, h \text{ is of class } \mathscr{C}^2 \text{ on } U \text{ and } \sum_{i,\,j=1}^{n} a_{ij} \frac{\partial^2 h}{\partial x_i\, \partial x_j} = 0 \right\}.$$

Then X endowed with $\mathscr{H}$ is a Brelot space. (Use a linear map in order to transform the differential operator into the Laplace operator.)

3.2.7. Let X be an open set of $\mathbb{R}^n$ ($n \in \mathbb{N}$, $n \neq 0$) and let a_{ij}, b_i, c ($i, j = 1, 2, \ldots, n$) be continuous, real functions on X such that: *a)* $a_{ij} = a_{ji}$; *b)* (a_{ij}) is a strictly positive definite matrix at any point of X; *c)* a_{ij} are Dini continuous. For any open set U of X, let $\mathscr{H}(U)$ be the set of real functions h on U, of class $\mathscr{C}^2$, such that

$$\sum_{i,\,j=1}^{n} a_{ij} \frac{\partial^2 h}{\partial x_i\, \partial x_j} + \sum_{i=1}^{n} b_i \frac{\partial h}{\partial x_i} + c\,h = 0.$$

Then X endowed with the harmonic sheaf $\mathscr{H}$ is a Brelot space. A relatively compact, open set of X is regular if and only if it is regular with respect to the Laplace equation (M. Brelot 1958 [13], 1960 [15], R.-M. Hervé 1962 [4], N. Boboc-P. Mustaţă 1966 [1], 1968 [5]; J.-M. Bony 1967 [2]).

3.2.8. Let X be an open set of $\mathbb{R}^n$ ($n \in \mathbb{N}$, $n \neq 0$) and let a_{ij}, b_i, c ($i, j = 1, 2, \ldots, n$) be continuous, real functions on X such that $a_{ij} = a_{ji}$ and such that (a_{ij}) is a strictly positive definite matrix at any point of X. For any open set U of X, we denote by $\mathscr{H}(U)$, the set of functions h on U

which belong locally to the spaces B_2^p for $p>n$ and such that

$$\sum_{i,j=1}^{n} a_{ij} \frac{\partial^2 h}{\partial x_i \partial x_j} + \sum_{i=1}^{n} b_i \frac{\partial h}{\partial x_i} + c h = 0$$

almost everywhere. Then X endowed with the harmonic sheaf $\mathscr{H}$ is an elliptic Bauer space. If, at a point $x_0 \in X$, the functions a_{ij} are Dini continuous, then for any $\mathfrak{P}$-set U containing x_0, there exists a strictly positive potential on U, harmonic on $U \setminus \{x_0\}$, ∞ at x_0 (N. Boboc-P. Mustaţă 1967 [2]). (It is not known if, in this case, X is a Brelot space.)

3.2.9. Let X be a Riemannian space. For any open set U we denote by $\mathscr{H}(U)$ the set of real functions h of class $\mathscr{C}^2$ such that, in local coordinates, we have $\Delta h := \sum_i h_{;i;i} = 0$. X endowed with $\mathscr{H}$ is an $\mathfrak{S}$-Brelot space

3.2.10. Let τ be Lebesgue measure on $\mathbb{R}^n$, let X be an open set of $\mathbb{R}^n$ ($n \in \mathbb{N}$, $n \neq 0$) and let a_{ij} $(i, j = 1, 2, \ldots, n)$ be τ-measurable functions satisfying the following conditions: *a)* $a_{ij} = a_{ji}$; *b)* there exists a real number $\alpha \geq 1$ such that

$$\frac{1}{\alpha} \sum_{i=1}^{n} \xi_i^2 \leq \sum_{i,j} a_{ij}(x)\, \xi_i\, \xi_j \leq \alpha \sum_{i=1}^{n} \xi_i^2$$

for every $x \in X$ and every $(\xi_i)_{1 \leq i \leq n} \in \mathbb{R}^n$. For any open set U of X, we denote by $\mathscr{H}(U)$, the set of real, continuous functions h on U such that the derivatives of first order of h in the sense of distribution theory belong locally to $L^2(\tau)$ and such that for any real function f of class $\mathscr{C}^\infty$ on U with compact carrier, we have

$$\int_U \sum_{i,j=1}^{n} a_{ij} \frac{\partial h}{\partial x_i} \frac{\partial f}{\partial x_j} d\tau = 0.$$

Then X endowed with the harmonic sheaf $\mathscr{H}$ is a Brelot space (R.-M. Hervé, 1964 [5], 1965 [6]). A generalisation of this result may be the found in R.-M. Hervé-M. Hervé 1969 [1], F.-Y. Maeda 1970 [1] and T.J. Kori 1971 [1].

3.2.11. For any open set U of $\mathbb{R}^2$, let $\mathscr{U}(U)$ be the set of lower semi-continuous, lower finite, numerical functions u on U such that:

a) the function $x \mapsto u(x, 0)$ defined on $\{x \in \mathbb{R} \mid (x, 0) \in U\}$ is locally decreasing (Theorem 2.1.2);

b) the restriction of u to $\{(x, y) \in U \mid y \neq 0\}$ is hyperharmonic for the Laplace equation (Theorem 3.2.1).

X endowed with $\mathscr{U}$ is a $\mathfrak{P}$-harmonic space such that $\mathscr{H}$ possesses the Brelot convergence property and such that the set of regular sets is not a base of X. (In fact the intersection of any regular (or even semi-regular, in the sense of J. Köhn 1968 [1]) set with $\mathbb{R}\times\{0\}$ is empty.)

3.2.12. Let X be equal to $\mathbb{R}^2$. For any open set U of X, we denote by $\mathscr{H}(U)$, the set of real, continuous functions h on U such that: *a)* h is of class $\mathscr{C}^2$ and $\frac{\partial^2 h}{\partial x^2}+\frac{\partial^2 h}{\partial y^2}=0$ on $\{(x,y)\in U \mid x^2+y^2\neq 1\}$; *b)* the function $\theta\mapsto h(\cos\theta,\sin\theta)$ is of class $\mathscr{C}^2$ and $\frac{\partial^2 h}{\partial\theta^2}=0$ on $\{\theta\in R \mid (\cos\theta,\sin\theta)\in U\}$. Then:

a) X endowed with the harmonic sheaf $\mathscr{H}$ is a Bauer space;

b) the open sets U of X containing $\{(x,y)\in\mathbb{R}^2 \mid x^2+y^2=1\}$ and for which $\{(x,y)\in\mathbb{R}^2 \mid x^2+y^2>1\}\cap\partial U$ is a Jordan curve are regular if and only if $\{(x,y)\notin U \mid x^2+y^2<1\}$ contains exactly one point. Hence X contains regular sets which are not MP-sets.

3.2.13. Let X be the topological space

$$\{(x,y,z)\in\mathbb{R}^3 \mid x^2+y^2<1,\ z=0\}\cup\{(x,y,z)\in\mathbb{R}^3 \mid x=y=0,\ 0\le z<1\}.$$

For any open set U of X, we denote by $\mathscr{H}(U)$, the set of continuous, real functions h on U such that $(x,y)\mapsto h(x,y,0)$, $z\mapsto h(0,0,z)$ are solutions of Laplace's equations on the corresponding open sets. Then X endowed with $\mathscr{H}$ is a $\mathfrak{P}$-Bauer space.

3.2.14. Let

$$U:=\{(x,y)\in\mathbb{R}^2 \mid 0<x<1,\ 0<y<1\}\setminus\bigcup_{\substack{n\in\mathbb{N}\\ n\neq 0}}\left\{(x,y)\in\mathbb{R}^2 \,\middle|\, x=\frac{1}{n},\ 0<y\le\frac{1}{2}\right\}.$$

For any $n\in\mathbb{N}$, $n\neq 0$, we set

$$U_n:=U\setminus\left\{(x,y)\in U \,\middle|\, \frac{1}{n+1}<x<\frac{1}{n},\ 0<y\le\frac{1}{2}\right\}$$

and denote by f_n (resp. g_n) the function on ∂U (resp. ∂U_n) which is equal to 1 on

$$\left\{(x,y)\in\partial U \,\middle|\, \frac{1}{n+1}<x<\frac{1}{n},\ y=0\right\}$$

$$\left(\text{resp. on } \left\{(x,y)\in\partial U_n \,\middle|\, \frac{1}{n+1}<x<\frac{1}{n},\ y=\frac{1}{2}\right\}\right)$$

and 0 elsewhere. For any $n\in\mathbb{N}$ we set

$$u_n:=H^U_{f_n},\qquad v_n:=H^{U_n}_{g_n},\qquad \eta_n:=\sup_{x\in\left]\frac{1}{n+1},\frac{1}{n}\right[} u_n(x,\tfrac{1}{2}),$$

where H is considered with respect to the Laplace equation in $\mathbb{R}^2$. Let $a \in U \cap \{(x, y) \in \mathbb{R}^2 \mid y > \frac{1}{2}\}$. Then

$$u_n(a) \leq \eta_n v_n(a), \qquad \lim_{n \to \infty} v_n(a) = 0.$$

Let $(n_k)_{k \in \mathbb{N}}$ be a strictly increasing sequence in $\mathbb{N}$ such that

$$\sum_{k \in \mathbb{N}} v_{n_k}(a) < \infty.$$

We set

$$f := \sum_{k \in \mathbb{N}} \frac{1}{\eta_{n_k}} f_{n_k}.$$

Then

$$H_f^U = \sum_{k \in \mathbb{N}} \frac{1}{\eta_{n_k}} u_{n_k},$$

U is regular, f is resolutive and equal to 0 on a neighbourhood of the point $(\frac{1}{2}, 0)$ but

$$\limsup_{(x, y) \to (\frac{1}{2}, 0)} H_f^U(x, y) = \infty.$$

§ 3.3. The Heat Equation

In this section it will be shown that the *heat equation* (i.e. the equation

$$\triangle h := \sum_{i=1}^{n} \frac{\partial^2 h}{(\partial x_i)^2} - \frac{\partial h}{\partial t} = 0,$$

where t denotes the $n+1-st$ coordinate) gives rise to a potential theory similar to that associated with the Laplace equation. This theory, which is also an important model for the theory of harmonic spaces, was first noticed by J.L. Doob 1956 [1].

Theorem 3.3.1. *Let n be a natural number different from 0. For any open set U of the topological space $\mathbb{R}^n \times \mathbb{R}$, we set*

$$\mathscr{H}(U) := \{h \in \mathscr{C}(U) \mid h \text{ is of class } \mathscr{C}^2 \text{ on } U, \triangle h = 0\}.$$

Then $\mathscr{H}$ is a harmonic sheaf on $\mathbb{R}^n \times \mathbb{R}$ possessing the Doob convergence property and $\mathbb{R}^n \times \mathbb{R}$ endowed with $\mathscr{H}$ is a $\mathfrak{P}$-Bauer space.

For any point x of $\mathbb{R}^n$ we shall denote by x_i, the i-th coordinate, and we set

$$|x| := \left(\sum_{i=1}^{n} (x_i)^2\right)^{\frac{1}{2}}.$$

For any $(a, \alpha), (b, \beta) \in \mathbb{R}^n \times \mathbb{R}$, we set

$$U(a, \alpha, b, \beta) := \{(x, t) \in \mathbb{R}^n \times \mathbb{R} \mid a_i < x_i < b_i,\ \alpha < t < \beta\},$$

$$\partial_{\triangle} U(a, \alpha, b, \beta) := \{(x, t) \in \partial U(a, \alpha, b, \beta) \mid t < \beta\}.$$

Lemma 1. *Let (a, α), (b, β) be points of $\mathbb{R}^n \times \mathbb{R}$ and let f be a real function of class $\mathscr{C}^2$ on $U(a, \alpha, b, \beta)$ whose lower limit at $\partial_{\triangle} U(a, \alpha, b, \beta)$ is positive and such that $\triangle f \le 0$. Then f is positive and if it vanishes at a point (x, t), then it vanishes at every point (x', t') such that $t' \le t$.*

Suppose first $\triangle f < 0$ and assume $\inf_{U(a, \alpha, b, \beta)} f < 0$. There exists then a strictly positive real number ε such that $\inf_{U(a, \alpha, b, \beta - \varepsilon)} f < 0$. Let (x, t) be a point of $U(a, \alpha, b, \beta) \cap \overline{U(a, \alpha, b, \beta - \varepsilon)}$ such that

$$f(x, t) = \inf_{U(a, \alpha, b, \beta - \varepsilon)} f.$$

For any i $(1 \le i \le n)$ the function

$$s \mapsto f(x_1, \ldots, x_{i-1}, s, x_{i+1}, \ldots, x_n, t)$$

defined on a neighbourhood of $x_i \in \mathbb{R}$ is of class $\mathscr{C}^2$ and assumes its infimum at x_i. It follows that

$$\frac{\partial^2 f}{(\partial x_i)^2}(x, t) \ge 0.$$

The function

$$s \mapsto f(x, s)$$

defined on a neighbourhood of $t \in \mathbb{R}$ is of class $\mathscr{C}^2$ and takes its infimum on the interval $]\alpha, t]$ at the point t. It follows

$$\frac{\partial f}{\partial t}(x, t) \le 0$$

and this contradicts the relation $\triangle f < 0$.

Suppose now $\triangle f \le 0$ and assume again $\inf_{U(a, \alpha, b, \beta)} f < 0$. This leads to a contradiction by applying the above argument to the function

$$(x, t) \mapsto f(x, t) + \varepsilon (t - \alpha) \prod_{i=1}^{n} (x_i - a_i)(b_i - x_i),$$

for a sufficiently small strictly positive real number ε. This proves that f is positive.

The last assertion will follow from the next two:

a) If f is strictly positive at (x', t') then for any $t'' > t'$, $t'' < \beta$, f is strictly positive at (x', t'').

b) If f is strictly positive at (x', t') then for any j $(j = 1, \ldots, n)$ and for any $x'' \in \mathbb{R}^n$ such that $x''_i = x'_i$ for every $i \ne j$ and such that $(x'', t') \in U(a, \alpha, b, \beta)$, f is strictly positive at (x'', t').

a) Let a', b' be points of $\mathbb{R}^n$ such that $a_i < a'_i < x'_i < b'_i < b_i$ and such that f is strictly positive on $\{(x,t) \in X \mid a'_i \leq x_i \leq b'_i, t = t'\}$. If ε is a strictly positive real number and g is the function on $U(a', t', b', \beta)$

$$(x,t) \mapsto f(x,t) - \varepsilon \prod_{i=1}^{n} \sin \frac{\pi}{b'_i - a'_i}(x_i - a'_i) \exp\left(-\frac{\pi^2 t}{(b'_i - a'_i)^2}\right),$$

then g is of class $\mathscr{C}^2$ and $\triangle g \leq 0$. For a sufficiently small ε, the lower limit of g at any point of $\partial_{\triangle} U(a', t', b', \beta)$ is positive. By the above considerations, g is positive and we obtain

$$0 < \varepsilon \prod_{i=1}^{n} \sin \frac{\pi}{b'_i - a'_i}(x'_i - a'_i) \exp\left(-\frac{\pi^2 t''}{(b' - a'_i)^2}\right) \leq f(x', t'').$$

b) Let (a', α'), (b', β') be points of $\mathbb{R}^n \times \mathbb{R}$ such that: $a_i < a'_i < x'_i < b'_i < b_i$ for $i \neq j$; $a_j < a'_j = x'_j < b'_j = b_j$, (resp. $a_j = a'_j < x'_j = b'_j < b_j$); $\alpha < \alpha' < t' < \beta' < \beta$; f is strictly positive on

$$\{(x,t) \in \mathbb{R}^n \times \mathbb{R} \mid x_j = x'_j,\ i \neq j \Rightarrow a'_i \leq x_i \leq b'_i,\ \alpha' \leq t \leq \beta'\}.$$

Let ε be a strictly positive real number and let g be the function on $U(a', \alpha', b', \beta')$

$$(x,t) \mapsto f(x,t) - \varepsilon\, h(x,t) \prod_{\substack{i=1 \\ i \neq j}}^{n} \sin \frac{\pi}{b'_i - a'_i}(x_i - a'_i) \exp\left(-\frac{\pi^2 t}{(b'_i - a'_i)^2}\right)$$

where h is the function on $U(a', \alpha', b', \beta')$

$$(x,t) \mapsto \frac{1}{\sqrt{t - \alpha'}}\left(\exp\left(-\frac{(x_j - a_j)^2}{4(t - \alpha')}\right) - \exp\left(-\frac{(x_j - 2b_j + a_j)^2}{4(t - \alpha')}\right)\right)$$

$$\text{resp. } (x,t) \mapsto \frac{1}{\sqrt{t - \alpha'}}\left(\exp\left(-\frac{(x_j - b_j)^2}{4(t - \alpha')}\right) - \exp\left(-\frac{(x_j - 2a_j + b_j)^2}{4(t - \alpha')}\right)\right).$$

Then g is of class $\mathscr{C}^2$ and $\triangle g \leq 0$. For a sufficiently small ε, the lower limit of g at any point of $\partial_{\triangle} U(a', \alpha', b', \beta')$ is positive. By the above considerations g is positive and we get

$$0 < \varepsilon\, h(x'', t') \prod_{\substack{i=1 \\ i \neq j}}^{n} \sin \frac{\pi}{b'_i - a'_i}(x''_i - a'_i) \exp\left(-\frac{\pi^2 t'}{(b'_i - a'_i)^2}\right) \leq f(x'', t'). \quad \square$$

We denoted by I the set $\{1, 2, \ldots, n\}$ and by Γ, the function on $\mathbb{R}^I \times \mathbb{R}$ which is equal to

$$(x,t) \mapsto \frac{1}{(4\pi t)^{n/2}} \exp\left(-\frac{|x|^2}{4t}\right)$$

on $\{(x,t)\in\mathbb{R}^I\times\mathbb{R}\,|\,t>0\}$ and 0 elsewhere. For any $J\subset I$ and any $y\in\mathbb{R}^I$ we denote by $|J|$ the number of elements of J and by y^J the point of $\mathbb{R}^I$ such that $y_i^J=y_i$ for $i\in I\setminus J$ and $y_i^J=-y_i$ for $i\in J$. We denote further for any $J\subset I$ and any $k\in\mathbb{Z}^I$ by Γ_k^J the function on $\mathbb{R}^I\times\mathbb{R}\times\mathbb{R}^I\times\mathbb{R}$

$$(x,t,y,s)\mapsto\Gamma(x-y^J+2k,t-s)$$

and set

$$Q:=\sum_{k\in\mathbb{Z}^I}\sum_{J\subset I}(-1)^{|J|}\Gamma_k^J.$$

It is easily seen that the series defining Q is convergent and that Q is of class $\mathscr{C}^\infty$ outside the set

$$\left\{(x,t,y,s)\,|\,t=s \text{ and } (\exists J)\left(J\subset I,\frac{x-y^J}{2}\in\mathbb{Z}^I\right)\right\}.$$

We denote for any $L,L'\subset I$, $L\cap L'=\phi$, and any $y\in\mathbb{R}^L$, $y'\in\mathbb{R}^{L'}$, by (y,y') the point of $\mathbb{R}^{L\cup L'}$ such that $(y,y')_i=y_i$ for $i\in L$ and $(y,y')_i=y'_i$ for $i\in L'$. We denote further for any $L\subset I$ and any $l\in\{0,1\}$ by $Q^{L,l}$ the function on $\mathbb{R}^I\times\mathbb{R}\times\mathbb{R}^{I\setminus L}\times\mathbb{R}$

$$(x,t,y,s)\mapsto(-1)^{\sum_{i\in I}l_i}\frac{\partial^{|L|}Q}{\prod_{i\in L}\partial y_i}(x,t,(y,l),s).$$

It is obvious that $Q^{L,l}$ is of class $\mathscr{C}^\infty$ outside the set

$$\left\{(x,t,y,s)\,|\,t=s \text{ and } (\exists J)\left(J\subset I,\frac{x-(y,l)^J}{2}\in\mathbb{Z}^I\right)\right\}.$$

For any $(y,s)\in\mathbb{R}^I\times\mathbb{R}$ (resp. $L\subset I$, $l\in\{0,1\}^L$ and $(y,s)\in\mathbb{R}^{I\setminus L}\times\mathbb{R}$) we denote by $Q_{(y,s)}$ (resp. $Q_s^{L,l}$) the function on $\mathbb{R}^I\times\mathbb{R}$

$$(x,t)\mapsto Q(x,t,y,s)\qquad(\text{resp. }(x,t)\mapsto Q^{L,l}(x,t,y,s)).$$

Lemma 2. *Let $L\subset I$ and $l\in\{0,1\}^L$.*

a) $Q(x,t,y,s)=0$ and $Q^{L,l}(x,t,y,s)=0$ if either there exists $j\in I$ such that $x_j\in\mathbb{Z}$ or $t\leq s$;

b) For any $(y,s)\in\mathbb{R}^I\times\mathbb{R}$ (resp. $(y,s)\in\mathbb{R}^{I\setminus L}\times\mathbb{R}$) we have

$$\triangle Q_{(y,s)}=0\qquad(\textit{resp. }\triangle Q_{(y,s)}^{L,l}=0)$$

outside the set

$$\left\{(x,t)\in\mathbb{R}^I\times\mathbb{R}\,|\,t=s \textit{ and } (\exists J)\left(J\subset I,\frac{x-y^J}{2}\in\mathbb{Z}^I\right)\right\}$$

$$\left(\textit{resp. }\left\{(x,t)\in\mathbb{R}^I\times\mathbb{R}\,|\,t=s \textit{ and } (\exists J)\left(J\subset I,\frac{x-(y,l)^J}{2}\in\mathbb{Z}^I\right)\right\}\right).$$

c) For any $(y, s) \in]0, 1[^I \times \mathbb{R}$ *(resp. for any* $(y, s) \in \mathbb{R}^{I \setminus L} \times \mathbb{R}$*) the function* $Q_{(y,s)}$ *(resp.* $Q^{L,l}_{(y,s)}$*) is strictly positive on* $]0, 1[^I \times]s, \infty[$.

d) For any compact sets K', K'' *of* $]0, 1[^I \times]0, \infty[$ *such that*

$$\sup_{(x,t) \in K'} t < \inf_{(x,t) \in K''} t$$

there exists a positive real number δ *such that for any* $y \in [0, 1]^I$ *(resp.* $(y, s) \in [0, 1]^{I \setminus L} \times [0, \infty[$ *we have*

$$\sup_{K'} Q_{(y,0)} \leq \delta \inf_{K''} Q_{(y,0)} \qquad (\textit{resp. } \sup_{K'} Q^{L,l}_{(y,s)} \leq \delta \inf_{K''} Q^{L,l}_{(y,s)}).$$

e) Let $\alpha, \beta \in]0, 1[^I$ *and* $y \in \prod_{i \in I}]\alpha_i, \beta_i[$ *(resp.* $j \in I, L = \{j\}, (\alpha, \gamma), (\beta, \delta) \in]0, 1[^{I \setminus \{j\}} \times]0, \infty[$ *and* $(y, s) \in (\prod_{i \in I}]\alpha_i, \beta_i[) \times]\gamma, \delta[$*). Then*

$$\lim_{\substack{(x,t) \to (y,0) \\ t > 0}} \int_{\prod_{i \in I}]\alpha_i, \beta_i[} Q(x, t, y', 0)\, d\sigma(y') = 1$$

$$\left(\textit{resp.} \lim_{\substack{(x,t) \to ((y,l),s) \\ x \in]0,1[^I}} \int_{(\prod_{\substack{i \in I \\ i \neq j}}]\alpha_i, \beta_i[) \times]\gamma, \delta[} Q^{\{j\},l}_{(y,s)}(x, t, y', s')\, d\sigma(y', s') = 1\right)$$

where σ *denotes the Lebesgue measure.*

a) We have

$$\begin{aligned} Q(x,t,y,s) &= \sum_{J \subset I \setminus \{j\}} (-1)^{|J|} \sum_{k \in \mathbb{Z}^I} \Gamma(x - y^J + 2k, t - s) \\ &\quad - \sum_{J \subset I \setminus \{j\}} (-1)^{|J|} \sum_{k \in \mathbb{Z}^I} \Gamma(x - y^{J \cup \{j\}} + 2k, t - s) \\ &= \sum_{J \subset I \setminus \{j\}} (-1)^{|J|} \Big(\sum_{k \in \mathbb{Z}^I} \Gamma(x - y^J + 2k, t - s) \\ &\qquad - \sum_{k \in \mathbb{Z}^I} \Gamma(x^{\{j\}} - y^J + 2k^{\{j\}}, t - s) \Big) \\ &= \sum_{J \subset I \setminus \{j\}} (-1)^{|J|} \Big(\sum_{k \in \mathbb{Z}^I} \Gamma(x - y^J + 2k, t - s) \\ &\qquad - \sum_{k \in \mathbb{Z}^I} \Gamma(x - y^J + 2k, t - s) \Big) = 0, \end{aligned}$$

$$Q^{L,l}(x,t,y,s) = (-1)^{\sum_{i \in I} l_i} \frac{\partial^{|L|} Q}{\prod_{i \in L} \partial y_i}(x, t, (y, l), s) = 0.$$

b) Follows by mere calculation and interchanging the differential operator with $\sum$.

c) We denote by U_s, the set $]0, 1[^I \times]s, \infty[$. Then the lower limit of the restriction of $Q_{(y,s)}$ (resp. $Q^{L,l}_{(y,s)}$) to U_s at any boundary point of U_s

different from (y, s) (resp. $((y, l), s)$) is 0. Since

$$\lim_{\substack{(x,t)\to(y,s)\\(x,t)\in U_s}} \sum_{\substack{k\in\mathbb{Z}^I,\, J\subset I\\(k,J)\neq(0,\phi)}} (-1)^J \Gamma(x-y^J+2k, t-s)=0$$

we get

$$\liminf_{\substack{(x,t)\to(y,s)\\(x,t)\in U_s}} Q(x,t,y,s)=\liminf_{\substack{(x,t)\to(y,s)\\(x,t)\in U_s}} \Gamma(x-y,t-s)=0,$$

$$\limsup_{\substack{(x,t)\to(y,s)\\(x,t)\in U_s}} Q(x,t,y,s)=\limsup_{\substack{(x,t)\to(y,s)\\(x,t)\in U_s}} \Gamma(x-y,t-s)=\infty.$$

Since

$$J\subset I \quad \text{and} \quad J\not\subset L \Rightarrow \lim_{(x,t)\to((y,l),s)} \sum_{k\in\mathbb{Z}^I} \frac{\partial^{|L|}\Gamma_k^J}{\prod_{i\in L}\partial y_i}(x,t,(y,l),s)=0,$$

$$J\subset L \Rightarrow \lim_{(x,t)\to((y,l),s)} \sum_{\substack{k\in\mathbb{Z}^I\\ k\neq\frac{1}{2}((0,l)^J-(0,l))}} \frac{\partial^{|L|}\Gamma_k^J}{\prod_{i\in L}\partial y_i}(x,t,(y,l),s)=0,$$

$$(-1)^{\sum_{i\in L} l_i}\sum_{J\subset L}(-1)^{|J|}\frac{\partial^{|L|}\Gamma^J_{\frac{1}{2}((0,l)^J-(0,l))}}{\prod_{i\in L}\partial y_i}(x,t,(y,l),s)$$

$$=(-1)^{\sum_{i\in L} l_i}\sum_{J\subset L}(-1)^{|J|}\prod_{i\in L\setminus J}\frac{x_i-l_i}{4(t-s)}\prod_{i\in J}\frac{l_i-x_i}{4(t-s)}\Gamma^J_{\frac{1}{2}((0,l)^J-(0,l))}(x,t,(y,l),s)$$

$$=\sum_{J\subset L}\prod_{i\in L}\frac{|x_i-l_i|}{4(t-s)}\Gamma^J_{\frac{1}{2}((0,l)^J-(0,l))}(x,t,(y,l),s),$$

we get

$$\liminf_{\substack{(x,t)\to((y,l),s)\\(x,t)\in U_s}} Q^{L,l}_{(y,s)}(x,t)$$

$$=\liminf_{\substack{(x,t)\to((y,l),s)\\(x,t)\in U_s}} \sum_{J\subset L}\prod_{i\in L}\frac{|x_i-l_i|}{4(t-s)}\Gamma(x-(y,l)^J+(0,l)^J-(0,l),t-s)=0,$$

$$\limsup_{\substack{(x,t)\to((y,l),s)\\(x,t)\in U_s}} Q^{L,l}_{(y,s)}(x,t)$$

$$=\limsup_{\substack{(x,t)\to((y,l),s)\\(x,t)\in U_s}} \sum_{J\subset L}\prod_{i\in L}\frac{|x_i-l_i|}{4(t-s)}\Gamma(x-(y,l)^J+(0,l)^J-(0,l),t-s)=\infty.$$

The assertion follows now from Lemma 1.

d) We denote by f the function on $K'\times K''\times[0,1]^I$ (resp. on $K'\times K''\times[0,1]^{I\setminus L}\times[0, \sup_{(x,t)\in K'} t])$ such that for any $L'\subset I$ (resp. $L'\subset I$ and $L'\supset L$) and any $(x',t')\in K'$, $(x'',t'')\in K''$, $y\in]0,1[^{I\setminus L'}$, $l'\in\{0,1\}^{L''}$ (resp.

$l' \in \{0,1\}^{L'-L}$)

$$f(x',t',x'',t'',(y,l')) = \frac{Q^{L',l'}_{(y,0)}(x',t')}{Q^{L',l'}_{(y,0)}(x'',t'')}$$

$$\left(\text{resp. } f(x',t',x'',t''(y,l),s) = \frac{Q^{L',(l,l')}_{(y,s)}(x',t')}{Q^{L',(l,l')}_{(y,s)}(x'',t'')}\right).$$

By *c)*, this function is finite and a simple calculation shows that it is continuous. The assertion follows now from the fact that the domain of f is compact.

e) We have

$$\lim_{(x,t)\to(y,0)} \int_{\prod_{i\in I}]\alpha_i,\beta_i[} \sum_{\substack{k\in\mathbb{Z}^I,\, J\subset I\\ (k,J)\neq(0,\phi)}} (-1)^J \Gamma(x-y'^J+2k,t)\,d\sigma(y')=0,$$

$$\lim_{\substack{(x,t)\to(y,0)\\ t>0}} \int_{\prod_{i\in I}]\alpha_i,\beta_i[} \Gamma(x-y',t)\,d\sigma(y')$$

$$=\lim_{\substack{(x,t)\to(y,0)\\ t>0}} \prod_{i\in I}\int_{\alpha_i}^{\beta_i} \frac{1}{2\sqrt{\pi t}}\exp\left(-\frac{(x_i-y_i')^2}{4t}\right)dy_i'$$

$$=\lim_{\substack{(x,t)\to(y,0)\\ t>0}} \prod_{i\in I}\frac{1}{\sqrt{\pi}}\int_{\frac{\alpha_i-x_i}{2\sqrt{t}}}^{\frac{\beta_i-x_i}{2\sqrt{t}}} e^{-u^2}\,du=1$$

$$\left(\text{resp.} \lim_{(x,t)\to((y,l),s)} \int_{(\prod_{i\in I, i\neq j}]\alpha_i,\beta_i[)\times]\gamma,\delta[} \sum_{\substack{k\in\mathbb{Z}^I,\, J\subset I\\ (k,J)\neq(0,\phi)\\ (k,J)\neq(\frac{1}{2}((0,l)^J-(0,l)),\{j\})}} (-1)^J \frac{\partial\Gamma_k^J}{\partial y_j}(x,t,y',s')\,d\sigma(y',s')=0\right)$$

$$\int_{(\prod_{i\in I, i\neq j}]\alpha_i,\beta_i[)\times]\gamma,\delta[} \left(\frac{x_j-l}{2(t-s')}\Gamma(x-(y',l),t-s')\right.$$

$$\left.+\frac{x_j-l}{2(t-s')}\Gamma(x-(y',l),t-s')\right)d\sigma(y',s')$$

$$=\int_{(\prod_{i\in I, i\neq j}]\alpha_i,\beta_i[)\times]\gamma,t[} \frac{x_j-l}{t-s'}\prod_{\substack{i\in I\\ i\neq j}}\frac{1}{2\sqrt{\pi(t-s')}}\exp\left(-\frac{(x_i-y_i')^2}{4(t-s')}\right)$$

$$\cdot\frac{1}{2\sqrt{\pi(t-s')}}\exp\left(-\frac{(x_j-l)^2}{4(t-s')}\right)d\sigma(y',s')$$

$$=\int_\gamma^t\left(\prod_{\substack{i\in I\\ i\neq j}}\int_{\alpha_i}^{\beta_i}\frac{1}{2\sqrt{\pi(t-s')}}\exp\left(-\frac{(x_i-y_i')^2}{4(t-s')}\right)dy_i'\right)$$

$$\cdot\frac{x_j-l}{2\sqrt{\pi}(t-s')^{\frac{3}{2}}}\exp\left(-\frac{(x_j-l)^2}{4(t-s')}\right)ds'$$

$$=\int_{\gamma}^{t}\left(\prod_{\substack{i\in I\\ i\neq j}}\frac{1}{\sqrt{\pi}}\int_{\frac{\alpha_i-x_i}{2\sqrt{t-s'}}}^{\frac{\beta_i-x_i}{2\sqrt{t-s'}}}e^{-u^2}\,du\right)\frac{x_j-l}{2\sqrt{\pi}(t-s')^{\frac{3}{2}}}\exp\left(-\frac{(x_j-l)^2}{4(t-s')}\right)ds'$$

$$=\frac{2}{\sqrt{\pi}}\int_{\frac{|x_i-l|}{2\sqrt{t-\gamma}}}^{\infty}\left(\prod_{\substack{i\in I\\ i\neq j}}\frac{1}{\sqrt{\pi}}\int_{\frac{\alpha_i-x_i}{|x_i-l|}v}^{\frac{\beta_i-x_i}{|x_i-l|}v}e^{-u^2}\,du\right)\frac{x_j-l}{|x_j-l|}e^{-v^2}\,dv,$$

$$\lim_{\substack{(x,t)\to((y,l),s)\\ x\in]0,1[^I}}\int_{(\prod_{\substack{i\in I\\ i\neq j}}]\alpha_i,\beta_i[)\times]\gamma,\delta[}(-1)^l\left[\frac{x_j-l}{2(t-s')}\Gamma(x-(y',l),t-s')\right.$$

$$\left.+\frac{x_j-l}{2(t-s')}\Gamma(x-(y',l),t-s')\right)d\sigma(y',s')=1\bigg].\quad\square$$

Lemma 3. *Let (a,α), (b,β) be points of $\mathbb{R}^n\times\mathbb{R}$ such that $b_i-a_i=b_j-a_j$ for any $i,j\in I$, and let σ be n-dimensional Lebesgue measure $\partial_{\sqcap}U(a,\alpha,b,\beta)$. For any $(x,t)\in U(a,\alpha,b,\beta)$, we denote by $P_{(x,t)}$ the function $\partial_{\sqcap}U(a,\alpha,b,\beta)$ which is equal to*

$$(y,s)\mapsto\frac{1}{(b_i-a_i)^n}Q\left(\frac{x-a}{b_i-a_i},\frac{t-\alpha}{(b_i-a_i)^2},\frac{y-a}{b_i-a_i},0\right)$$

on $\{(y,s)\in\partial_{\sqcap}U(a,\alpha,b,\beta)\,|\,s=\alpha\}$ and for any $l=0,1$ and any $j=1,2\ldots n$ equal to

$$(y,s)\mapsto\frac{1}{(b_j-a_j)^{n+1}}Q^{\{j\},l}\left(\frac{x-a}{b_j-a_j},\frac{t-\alpha}{(b_j-a_j)^2},\frac{y-a}{b_j-a_j},\frac{s-\alpha}{(b_j-a_j)^2}\right)$$

on $\{(y,s)\in\partial_{\sqcap}U(a,\alpha,b,\beta)\,|\,y_j=a_j+l(b_j-a_j),\ s>\alpha\}$. For any σ-integrable function f on $\partial_{\sqcap}U(a,\alpha,b,\beta)$ we denote by H_f, the function on $U(a,\alpha,b,\beta)$

$$(x,t)\mapsto\int P_{(x,t)}f\,d\sigma.$$

Then

a) H_f is an $\mathscr{H}$-function;

b) Let (y,s) be a point of $\partial_{\sqcap}U(a,\alpha,b,\beta)$ such that $s=\alpha$ and $a_i<y_i<b_i$ for any i (resp. such that there exists $j=1,2,\ldots,n$ and $l=0,1$ for which $y_j=a_j+l(b_j-a_j)$, and $\alpha<s<\beta$ and $a_i<y_i<b_i$ for any $i\neq j$). If f is finite and continuous at (y,s) then

$$\lim_{(x,t)\to(y,s)}H_f(x,t)=f(y,s).$$

c) for any compact sets K',K'' of $\mathbb{R}^n\times\mathbb{R}$, for which there exists $\gamma\in R$, such that $K'\subset U(a,\alpha,b,\gamma)$, $K''\subset U(a,\gamma,b,\beta)$, there exists a positive real number δ such that if f is positive

$$\sup_{K'}H_f\leq\delta\inf_{K''}H_f.$$

a) follows from the fact that the operator $\frown$ commutes with integration and from the fact that the function

$$(x,t)\mapsto P_{(x,t)}(y,s)$$

is an $\mathscr{H}$-function for any (y,s) (Lemma 2*b)*).

b) For any strictly positive real number η such that

$$a_i<y_i-\eta<y_i+\eta<b_i \qquad \text{for any } i$$

$$(\text{resp. } \alpha<s-\eta<s+\eta<\beta \text{ and } a_i<y_i-\eta<y_i+\eta<b_i \text{ for } i\neq j)$$

we set

$$V_\eta:=\{(y',s')\in\mathbb{R}^n\times\mathbb{R}\mid y_i-\eta<y'_i<y_i+\eta \text{ for any } i,\ s'=\alpha\}$$

$$\big(\text{resp. } V_\eta:=\{(y',s')\in\mathbb{R}^n\times\mathbb{R}\mid y'_j=y_j,\ s-\eta<s'<s+\eta,\ y_i-\eta<y'_i<y_i+\eta \text{ for any } i\neq j\}.\big)$$

Assume first that f vanishes identically on V_η. Then the assertion follows from the fact that f is σ-integrable and the restriction of $P_{(x,t)}$ to $\partial_\frown U(a,\alpha,b,\beta)\setminus V_\eta$ converges uniformly to 0 when (x,t) converges to (y,s) (Lemma 2*a)*).

Assume now that f is equal to 1 on V_η and 0 elsewhere. Then

$$H_f(x,t)=\frac{1}{(b_i-a_i)^n}\int_{V_\eta} Q\left(\frac{x-a}{b_i-a_i},\frac{t-\alpha}{(b_i-a_i)^2},\frac{y'-a}{b_i-a_i},0\right)d\sigma(y')$$

$$=\int_{\prod_{i\in I}\left]\frac{y_i-\eta-a_i}{b_i-a_i},\frac{y_i+\eta-a_i}{b_i-a_i}\right[} Q\left(\frac{x-a}{b_i-a_i},\frac{t-\alpha}{(b_i-a_i)^2},y',0\right)d\sigma(y')$$

$\Bigg($resp. $H_f(x,t)$

$$=\frac{1}{(b_j-a_j)^{n+1}}\int_{V_\eta} Q^{\{j\},l}\left(\frac{x-a}{b_j-a_j},\frac{t-\alpha}{(b_j-a_j)^2},\frac{y'-a}{b_j-a_j},\frac{s'-\alpha}{(b_j-a_j)^2}\right)d\sigma(y',s')$$

$$=\int_{\left(\prod_{i\in I,\, i\neq j}\left]\frac{y_i-\eta-a_i}{b_i-a_i},\frac{y_i+\eta-a_i}{b_i-a_i}\right[\right)\times\left]\frac{s-\eta-\alpha}{(b_i-a_i)^2},\frac{s+\eta-\alpha}{(b_i-a_i)^2}\right[} Q^{\{j\},l}\left(\frac{x-a}{b_i-a_i},\frac{t-\alpha}{(b_i-a_i)^2},y',s'\right)$$

$$\cdot d\sigma(y',s')\Bigg).$$

The assertion follows now from Lemma 2*e)*.

For the general case, let ε be a strictly positive real number. We choose η so small that the oscillation of f in V_η is smaller than ε. We denote by f' (resp. f'') the function on $\partial_\frown U(a,\alpha,b,\beta)$ which is equal to 0 (resp. 1) on V_η and equal to f (resp. 0) elsewhere. Then

$$|H_f-f(y,s)|\le|H_{f'}|+|f(y,s)(H_{f''}-1)|+\varepsilon H_{f''}.$$

By the above considerations, we get

$$\limsup_{(x,t)\to(y,s)} |H_f(x,t)-f(y,s)| \le \varepsilon$$

and the assertion follows from the fact that ε is arbitrary.

c) Follows immediately from Lemma 2*d)*. □

Part *b)* was proved by L. Schläfli 1870 [1].

For any (a,α), $(b,\beta) = \mathbb{R}^n \times \mathbb{R}$ such that $b_i - a_i = b_j - a_j$ for any $i, j \in I$, we denote by $\omega^{a\alpha b\beta}$ the sweeping on $U(a,\alpha,b,\beta)$, $(x,t) \mapsto P_{(x,t)} \cdot \sigma$, where $P_{(x,t)}$, and σ are defined in Lemma 3 with respect to (a,α), (b,β). We denote also by Ω, the sweeping system $(\omega^{a\alpha b\beta})_{((a,\alpha),(b,\beta))\in(\mathbb{R}^n\times\mathbb{R})^2}$ on $\mathbb{R}^n \times \mathbb{R}$ and by $\mathscr{U}$ the hyperharmonic sheaf on $\mathbb{R}^n \times \mathbb{R}$ generated by Ω.

Lemma 4. *For any (a,α), $(b,\beta) \in \mathbb{R}^n \times \mathbb{R}$ such that $b_i - a_i = b_j - a_j$ for any $i, j \in I$ the sweeping $\omega^{a\alpha b\beta}$ is quasi-regular with respect to $\mathscr{U}$ and $\mathscr{H} = \mathscr{H}_{\mathscr{U}}$.*

For any $l \in \{0,1\}^I$, we set

$$c^l := a + (b_i - a_i)\, l.$$

We denote by $h^{a\alpha b\beta}$, the function on $U(a,\alpha,b,\beta)$

$$(x,t) \mapsto \sum_{l\in\{0,1\}^n} \int_0^1 \frac{1}{\sqrt{\tau}\,(t-\alpha+\tau)^{n/2}} e^{-\frac{|x-c^l|^2}{4(t-\alpha+\tau)}} d\tau$$

$$+ \int_{\mathbb{R}^n} \frac{1}{(t-\alpha)^{n/2}} e^{\frac{|\xi|^2}{4(\beta-\alpha)} - \frac{|x-\xi|^2}{4(t-\alpha)}} d\xi$$

$$+ \sum_{\substack{i,j=1\\ i\neq j}}^{n} \log \frac{((b_i-a_i)^2}{((x_i-a_i)^2+(x_j-a_j)^2)((x_i-b_i)^2+(x_j-a_j)^2)}$$

$$\frac{+(b_j-a_j)^2)^4}{\cdot\, ((x_i-a_i)^2+(x_j-b_j)^2)((x_i-b_i)^2+(x_j-b_j)^2)}.$$

It is easy to see that $h^{a\alpha b\beta}$ is positive, of class $\mathscr{C}^\infty$, and $\triangle h^{a\alpha b\beta} = 0$. Let $\partial_0 U(a,\alpha,b,\beta)$ be the set of $(y,s) \in \partial U(a,\alpha,b,\beta)$ such that one of the following conditions is fulfilled: *a)* $s = \alpha$ and there exists i such that either $y_i = a_i$ or $y_i = b_i$; *b)* $s = \beta$; *c)* there exists i, j, $i \neq j$, such that

$$(y_i, y_j) \in \{(a_i,a_j), (b_i,a_j), (a_i,b_j)(b_i,b_j)\}.$$

A simple calculation shows that

$$\lim_{(x,t)\to(y,s)} h^{a\alpha b\beta}(x,t) = \infty$$

for any $(y,s) \in \partial_0 U(a,\alpha,b,\beta)$.

Let U be an open set of X. From Lemma 3 *a)* we have that

$$\mathscr{U}(U)\cap(-\mathscr{U}(U))\subset\mathscr{H}(U).$$

Let $h\in\mathscr{H}(U)$, choose $(a,\alpha),(b,\beta)\in\mathbb{R}^n\times\mathbb{R}$ such that $b_i-a_i=b_j-a_j$ for any $i,j\in I$ and such that $\overline{U(a,\alpha,b,\beta)}\subset U$, and let ε be a strictly positive real number. From the above considerations and from Lemma 3 *a)* and *b)*

$$\square(h-\omega^{a\alpha b\beta}h+\varepsilon h^{a\alpha b\beta})=0,$$

$$\liminf_{(x,t)\to(y,s)}\big(h(x,t)-\omega^{a\alpha b\beta}h(x,t)+\varepsilon h^{a\alpha b\beta}(x,t)\big)\geq 0$$

for any $(y,s)\in\partial U(a,\alpha,b,\beta)$. Hence (Lemma 1)

$$h-\omega^{a\alpha b\beta}h+\varepsilon h^{a\alpha b\beta}\geq 0.$$

Since ε and (a,α), (b,β) are arbitrary, we see that $h\in\mathscr{U}(U)$ and therefore

$$\mathscr{H}(U)\subset\mathscr{U}(U)\cap(-\mathscr{U}(U)).$$

By Lemma 3 *b)* any filter on $U(a,\alpha,b,\beta)$ converging to a point of $\partial U(a,\alpha,b,\beta)\setminus\partial_0 U(a,\alpha,b,\beta)$ is regular for the sweeping $\omega^{a\alpha b\beta}$. The quasi-regularity of this sweeping follows from this fact by using the function $h^{a\alpha b\beta}$. □

We prove now the theorem. Let us show first that $\mathbb{R}^n\times\mathbb{R}$ endowed with $\mathscr{U}$ is a harmonic space. Since for any $a\in\mathbb{R}^n$ the function on $\mathbb{R}^n\times\mathbb{R}$ $(x,t)\mapsto e^{\sum_{i=1}^n a_i x_i+t\sum_{i=1}^n (a_i)^2}$ is an $\mathscr{H}$-function, the axiom of positivity is satisfied and $\mathscr{U}(\mathbb{R}^n\times\mathbb{R})$ separates the points of $\mathbb{R}^n\times\mathbb{R}$. By the Brelot-Bauer theorem, every open set of $\mathbb{R}^n\times\mathbb{R}$ in an MP-set with respect to $\mathscr{U}$. By Proposition 2.1.1, the axioms of resolutivity and of completeness are also valid.

The above considerations show that the set of strictly positive harmonic functions on $\mathbb{R}^n\times\mathbb{R}$ separates the points of $\mathbb{R}^n\times\mathbb{R}$. Hence (Corollary 3.1.3), there exists a base of regular sets and therefore $\mathbb{R}^n\times\mathbb{R}$ endowed with $\mathscr{H}$ is a Bauer space (Corollary 3.1.2), and even a $\mathfrak{P}$-Bauer space (Proposition 2.3.2).

We show now that $\mathscr{H}$ has the Doob convergence property. Let U be an open set of $\mathbb{R}^n\times\mathbb{R}$ and $(h_m)_{m\in\mathbb{N}}$ be an increasing sequence of harmonic functions on U such that its supremum, h, is finite on a dense subset A of U. We may assume $h_0\geq 0$. Choose $((a,\alpha),(b,\beta))\in U^2$ with $\overline{U(a,\alpha,b,\beta)}\subset U$. Let $(a',\alpha')\in U(a,\alpha,b,\beta)$, $(b',\beta')\in U(a,\alpha,b,\beta)$ and let (x_0,t_0) be a point of $A\cap U(a,\alpha,b,\beta)$ with $t_0>\beta'$. By Lemma 4, $h_m=\omega^{a\alpha b\beta}h_m$ on $U(a,\alpha,b,\beta)$ for any $m\in\mathbb{N}$. By Lemma 3 *c)* there exists a positive real number δ such that

$$\sup_{\overline{U(a',\alpha',b',\beta')}} h = \limsup_{\substack{m\to\infty \\ \overline{U(a',\alpha',b',\beta')}}} h_m \le \lim_{m\to\infty} \delta h_m(x_0,t_0) = \delta h(x_0,t_0) < \infty .$$

Hence h is bounded on $\overline{U(a',\alpha',b',\beta')}$ and we have that

$$h = \lim_{m\to\infty} h_m = \lim_{m\to\infty} \omega^{a'\alpha' b'\beta'} h_m = \omega^{a'\alpha' b'\beta'} h$$

on $U(a',\alpha',b',\beta')$. By Lemma 3 *a)* h is harmonic on $U(a',\alpha',b',\beta')$ and therefore on U. $\square$

Exercises

Throughout these exercises we shall denote by n a natural number different from 0.

3.3.1. Any bounded solution of the heat equation on $\mathbb{R}^n \times \mathbb{R}$ is constant.

3.3.2. Let U be an open set of $\mathbb{R}^n \times \mathbb{R}$ and let f be a real function of class $\mathscr{C}^2$ on U. The following assertions are equivalent:

a) f is hyperharmonic for the heat equation;

b) for any $((a,\alpha),(b,\beta)) \in (\mathbb{R}^n \times \mathbb{R})^2$ such that $\overline{U(a,\alpha,b,\beta)} \subset U$, we have $\omega^{a\alpha b\beta} f \le f$ on $U(a,\alpha,b,\beta)$;

c) $\frown f \le 0$;

d) for any positive, real, continuous function g on U, of class $\mathscr{C}^\infty$ and with compact carrier, we have

$$\int_U f\left(\sum_{i=1}^n \frac{\partial^2 g}{(\partial x^i)^2} + \frac{\partial g}{\partial t}\right) d\tau \le 0,$$

where τ is Lebesgue measure on $\mathbb{R}^{n+1}$.

($a \Rightarrow b$ follows from Theorem 1.2.2; $b \Leftrightarrow c$ follows from Lemma 1 and Lemma 3 *a)* and Theorem 3.3.1 with the aid of the function $h^{a\alpha b\beta}$; $b \,\&\, c \Rightarrow a$ follows from the Brelot-Bauer theorem; $c \Leftrightarrow d$ follows from Stokes' formula).

3.3.3. Let U be an open set of $\mathbb{R}^n \times \mathbb{R}$ and (y,s) be a boundary point of U. Any filter on U converging to (y,s) has a barrier and is therefore regular if one of the following conditions is fulfilled:

a) there exists a neighbourhood V of (y,s) such that

$$U \cap V \subset \{(x,t) \in \mathbb{R}^n \times \mathbb{R} \mid t > s\};$$

b) there exists a neighbourhood V of (y, s) and a point $(a, \alpha) \in \mathbb{R}^n \times \mathbb{R}$ such that $a \neq 0$ and

$$U \cap V \subset \left\{ (x, t) \in \mathbb{R}^n \times \mathbb{R} \,\middle|\, \sum_{i=1}^{n} a_i(x_i - y_i) + \alpha(t - s) > 0 \right\};$$

c) there exists a real function f, of class $\mathscr{C}^2$, defined on a neighbourhood of (y, s) and a neighbourhood V of (y, s) such that

$$U \cap V \subset \{(x, t) \in \mathbb{R}^n \times \mathbb{R} \mid \varphi(x, t) > 0\},$$

$$\varphi(y, s) = 0, \qquad \sum_{i=1}^{n} \left(\frac{\partial \varphi}{\partial x_i} \right)^2 (y, s) > 0.$$

(Use Proposition 2.4.7 and Exercise 3.3.2.) Deduce that any ball is regular.

3.3.4. This exercise showes that the theory developed for the heat equation leads to simpler proofs of the properties of the Laplace equation.

a) Let U be a relatively compact open set of $\mathbb{R}^n$. For any $\alpha \in \mathbb{R}$ let f_α be the function on $\partial(U \times]\alpha, \infty[)$ which is equal to 1 on $U \times \{\alpha\}$ and 0 elsewhere. Then, for any $(x, t) \in U \times \mathbb{R}$

$$\lim_{\alpha \to -\infty} H_{f_\alpha}^{U \times]\alpha, \infty[}(x, t) = 0,$$

where H was constructed with respect to the heat equation. The function h_α on $U \times \mathbb{R}$ which is equal to 1 on $U \times]-\infty, \alpha]$ and $H_{f_\alpha}^{U \times]\alpha, \infty[}$ on $U \times]\alpha, \infty[$ is hyperharmonic for the heat equation.

b) Let U be a relatively compact, open set of $\mathbb{R}^n$. Any solution of the heat equation on $U \times \mathbb{R}$ is positive if it is lower bounded and its limit at any boundary point of $U \times \mathbb{R}$ is positive. (Use *a)* and the function on $U \times \mathbb{R}$, $(x, t) \mapsto e^t$.)

c) Any convex, relatively compact, open set of $\mathbb{R}^n$ is regular for the Laplace equation (Let U be such a set and $f \in \mathscr{C}(\partial U)$. Let f' be the function on $\partial(U \times \mathbb{R})$, $(x, t) \mapsto f(x)$. Use the preceding exercise and the proposition of Bouligand in order to show that $H_{f'}^{U \times \mathbb{R}}$ converges to f' at any boundary point. By *b)* the function on $\mathbb{R}$, $t \mapsto H_{f'}^{U \times \mathbb{R}}(x, t)$, is constant for any $x \in U$. Hence $x \mapsto H_{f'}^{U \times \mathbb{R}}(x, 0)$ is a solution of the Laplace equation).

d) For any open, connected set $U \subset \mathbb{R}^n$ and any compact subset K of U, there exists positive real number α such that for every positive solution of the Laplace equation, h, on U and for all $x, y \in K$, we have $h(x) \leq \alpha h(y)$. (Use Lemma 3 *c)* of Theorem 3.3.1.)

3.3.5. Let X be an open set of $\mathbb{R}^n \times \mathbb{R}$ and let L be the differential operator on X,

$$Lh := \sum_{i,j=1}^{n} a_{ij} \frac{\partial^2 h}{\partial x_i \partial x_j} + \sum_{i=1}^{n} b_i \frac{\partial h}{\partial x_i} + c h - \frac{\partial h}{\partial t}$$

such that (a_{ij}) is a symmetric matrix whose associated quadratic form is strictly positive definite at every point of X. For any open set U of X, let $\mathscr{H}(U)$ be the set of real functions h, on U, which are of class $\mathscr{C}^2$ with respect to x, of class $\mathscr{C}^1$ with respect to t, and for which $Lh = 0$.

a) If, for every compact set K of X, there exist two continuous, increasing real functions, δ and η, on $\mathbb{R}_+$ which are 0 at 0, satisfy

$$\int_0 \frac{\sqrt{\delta(t)}}{t} dt < \infty$$

and such that

$$|a_{ij}(x,t) - a_{ij}(x',t')| < \delta(|x-x'|^2) + \eta(|t-t'|)$$
$$|b_i(x,t) - b_i(x',t')| < \delta(|x-x'|^2) + \eta(|t-t'|)$$
$$|c(x,t) - c(x',t')| < \delta(|x-x'|^2) + \eta(|t-t'|)$$

for any $(x,t), (x',t') \in K$, then X endowed with the sheaf $\mathscr{H}$ is a Bauer space (P. Mustaţă 1970 [unpublished]).

b) If b_i, c $(i=1,\ldots,n)$ are Hölder continuous and if $\frac{\partial a_{ij}}{\partial x_k}$ exist and are Hölder continuous for any $i,j,k \in \{1,\ldots,n\}$, then X endowed with $\mathscr{H}$ is a Bauer space and the sheaf $\mathscr{H}$ has the Doob convergence property (S. Guber 1967 [1]).

3.3.6. Let X be an open set of $\mathbb{R}^n$ and $\mathscr{H}$ be a harmonic sheaf on X such that: *a)* any $\mathscr{H}$-function is of class $\mathscr{C}^2$; *b)* $\mathscr{H}$ is non-degenerate at every point of X: *c)* the set of $\mathscr{H}$-regular sets is a base of X. Then there exists a system of real functions a_{ij}, b_i, c $(i,j=1,\ldots,n)$ on X such that: *a)* $a_{ij} = a_{ji}$; *b)* (a_{ij}) is a non-zero positive definite matrix at any point of X; *c)* for any $\mathscr{H}$-function h, we have

$$\sum_{i,j=1}^{n} a_{ij} \frac{\partial^2 h}{\partial x_i \partial x_j} + \sum_{i=1}^{n} b_i \frac{\partial h}{\partial x_i} + c h = 0;$$

d) there exists an open, dense set U, of X, such that a_{ij}, b_i, c are continuous on U, and such that any solution of the above equation on any open subset of U, is an $\mathscr{H}$-function; *e)* if there exists a non-empty open set V of U and a coordinate system $(\eta_i)_{1\le i\le n}$ on V such that the above operator has the form

$$\sum_{i,j=1}^{n-1} a'_{ij}\frac{\partial^2}{\partial\eta_i\,\partial\eta_j}+\sum_{i=1}^{n} b'_i\frac{\partial}{\partial\eta_i}+c$$

then $\mathscr{H}$ does not have the Brelot convergence property and

$$\overline{\{x\in V\,|\,b'_n(x)\neq 0\}}\supset V;$$

f) if: *α)* $n=2$, *β)* any $\mathscr{H}$-function is of class $\mathscr{C}^3$, *γ)* the functions a_{ij}, b_i, c are of class $\mathscr{C}^1$, *δ)* $\mathscr{H}$ has the Brelot convergence property, then the matrix (a_{ij}) is strictly positive definite (J.-M. Bony 1967 [1], 1970 [6]).

3.3.7. Let X be an open set of $\mathbb{R}^n$, let a be a continuous real function on X, and let $(A_i)_{1\le i\le j}$, B be first order differential operators of class $\mathscr{C}^\infty$ on X. For any open set U of X, let $\mathscr{H}(U)$ be the set of real functions h, of class $\mathscr{C}^2$ on U and satisfying

$$\sum_{i=1}^{j}(A_i)^2\,h+B\,h+a\,h=0.$$

If the Lie algebra generated by the operators A_i, (resp. A_i and B) for the operation

$$\left(\sum_{i=1}^{n} a_i\frac{\partial}{\partial x_i}\right)\cdot\left(\sum_{i=1}^{n} b_i\frac{\partial}{\partial x_i}\right)=\sum_{i=1}^{n}\left(\sum_{j=1}^{n} a_j\frac{\partial b_i}{\partial x_j}-b_j\frac{\partial a_i}{\partial x_j}\right)\frac{\partial}{\partial x_i}$$

is of rank n at any point of X, then X endowed with $\mathscr{H}$ is a Brelot space (resp. Bauer space for which $\mathscr{H}$ possesses the Doob convergence property provided the matrix of the differential operator does not vanish at any point of X) (L. Hörmander 1967 [1], J.-M. Bony 1968 [3], 1969 [4], [5], 1970 [6]).

3.3.8. Let (r,θ) be the polar coordinates of $\mathbb{R}^2$, let 0 be the origin, and for any open set U of $\mathbb{R}^2$, let $\mathscr{H}(U)$ be the set of real functions, h, on U, of class $\mathscr{C}^2$ such that: *a)* $\dfrac{\partial^2 h}{\partial\theta^2}-\dfrac{\partial h}{\partial r}=0$ on $U\setminus\{0\}$; *b)* if $0\in U$, then h is constant on any open circle centered at 0 and contained in U. Then $\mathscr{H}$ is a harmonic sheaf on $\mathbb{R}^2$ which satisfies the conditions of Exercise 3.1.1 and $F=\{0\}$ (hence F is non-empty). (Let $r_1, r_2\in\mathbb{R}$,

$0 < r_1 < r_2$; set $U := \{(r, \theta) \mid r_1 < r < r_2\}$; prove, as in Lemma 1 of Theorem 3.3.1, that for any $h \in \mathscr{H}(U)$, we have

$$\sup_U h = \limsup_{r \to r_1} h(r, \theta).$$

Use this result to show that $\mathscr{H}$ is a harmonic sheaf. *a), c), d)* are trivial by Theorem 3.3.1 as well as *b)* on $\mathbb{R}^2 \setminus \{0\}$. Any set of the form

$$\{(r, \theta) \mid r < \alpha + \beta\theta, \quad \theta \in [0, 2\pi[\} \quad (\alpha, \beta > 0)$$

is regular with respect to $\mathscr{H}$ by Exercise 3.3.3.)

PART TWO

Chapter 4

Convex Cones of Continuous Functions on Baire Topological Spaces

The principal results concerning the natural and specific order on the convex cone of hyperharmonic functions on a harmonic space as well as the balayage of positive, hyperharmonic functions may be derived from some of the elementary properties of this cone. This fact enables us to treat these problems in an axiomatic setting which makes the theory more general and more elegant.

Throughout this chapter we shall denote by X, a Baire topological space and by $\mathscr{W}$, a convex cone of continuous, lower finite, numerical functions on X. In addition, we shall always assume that $\mathscr{W}$ contains the identically ∞ function. In order to develop the theory, three axioms will be needed: two of them will be assumed from the start but the third will be introduced later (p. 104). From $0 \cdot \infty = 0$, it follows that the identically 0 function always belongs to $\mathscr{W}$. In the applications, X will be a harmonic space endowed with its fine topology (p. 116) and $\mathscr{W}$ will be the convex cone of hyperharmonic functions on X.

§ 4.1. Natural Order and Specific Order

We consider two order relations on $\mathscr{W}$. The first one, called the *natural order*, denoted by $\leq$, is the order relation

$$u \leq v \quad \text{if} \quad u(x) \leq v(x) \quad \text{for all } x \in X .$$

The second one called the *specific order*, denoted by $\preccurlyeq$, is the order relation

$$u \preccurlyeq v \text{ if there exists } w \in \mathscr{W}, \quad w \geq 0, \text{ such that } u + w = v .$$

The specific order on harmonic spaces was introduced by M. Brelot 1958 [11]. The natural order is the order relation induced on $\mathscr{W}$ by the usual order relation on the set of all numerical functions on X. Of course

$$u \preccurlyeq v \Rightarrow u \leq v .$$

We shall use the notation

$$\wedge\mathscr{V},\ \bigwedge_{\iota\in I} u_\iota,\ u\wedge v,$$

$$\vee\mathscr{V},\ \bigvee_{\iota\in I} u_\iota,\ u\vee v,$$

$$\text{(resp. } \curlywedge\mathscr{V},\ \curlywedge_{\iota\in I} u_\iota,\ u\curlywedge v,$$

$$\curlyvee\mathscr{V},\ \curlyvee_{\iota\in I} u_\iota,\ u\curlyvee v)$$

for the natural (resp. specific) infimum and supremum of a set $\mathscr{V}\subset\mathscr{W}$, of a family $(u_\iota)_{\iota\in I}$ in $\mathscr{W}$ and of $u, v\in\mathscr{W}$ if it exists.

We shall assume that the following axioms are satisfied:

Axiom of Lower Semi-Continuous Regularization. *$\mathscr{W}$ is lower directed with respect to the natural order and for any non-empty subset $\mathscr{V}$ of $\mathscr{W}$ possessing a natural minorant, we have*

$$\widehat{\inf_{u\in\mathscr{V}} u}\in\mathscr{W}. \text{ [1]}$$

Axiom of Upper Directed Sets. *The supremum of any non empty upper directed set (with respect to the natural order) of $\mathscr{W}$ belongs to $\mathscr{W}$.*

It is obvious that $\mathscr{W}$ is upper complete with respect to its natural order. For any finite subset $\mathscr{V}$ of $\mathscr{W}$, $\wedge\mathscr{V}$ exists and is equal to $\inf_{u\in\mathscr{V}} u$. For any $\mathscr{V}\subset\mathscr{W}$ for which $\wedge\mathscr{V}$ exists,

$$\wedge\mathscr{V}=\widehat{\inf_{u\in\mathscr{V}} u}$$

and, if $\mathscr{V}$ is upper directed,

$$\vee\mathscr{V}=\sup_{u\in\mathscr{V}} u.$$

Proposition 4.1.1. *If $u\in\mathscr{W}_+$ then the sets $\{x\in X\,|\,u(x)=0\}$, $\{x\in X\,|\,u(x)<\infty\}$ are open.*

The set $\{n\,u\,|\,n\in\mathbb{N}\}$ is upper directed and its supremum is continuous, equal to 0 on $\{x\in X\,|\,u(x)=0\}$ and equal to ∞ elsewhere. We have

$$\overline{\{x\in X\,|\,u(x)<\infty\}}=\left\{x\in X\,\middle|\,\left(\bigwedge_{n\in\mathbb{N}}\frac{1}{n}u\right)(x)=0\right\}. \quad \square$$

Proposition 4.1.2. *For any set $\mathscr{V}\subset\mathscr{W}$ for which $\wedge\mathscr{V}$ exists, the set $\{x\in X\,|\,(\wedge\mathscr{V})(x)<\inf_{u\in\mathscr{V}} u(x)\}$ is meagre.*

---

[1] We remember that for any numerical function f on X, $\hat{f}$ denotes the greatest lower semi-continuous minorant of f.

Since $\wedge\mathscr{V}$ is continuous and $\inf_{u\in\mathscr{V}} u$ is upper semi-continuous, the set

$$\left\{x\in X \,\middle|\, \frac{(\wedge\mathscr{V})(x)}{1+(\wedge\mathscr{V})(x)}+\varepsilon\leq\frac{\inf_{u\in\mathscr{V}} u(x)}{1+\inf_{u\in\mathscr{V}} u(x)}\right\},$$

with the convention $\frac{\infty}{1+\infty}=1$, is closed and nowhere dense for every strictly positive real number ε. The assertion now follows from the relation

$$\{x\in X\,|\,(\wedge\mathscr{V})(x)<\inf_{u\in\mathscr{V}} u(x)\}$$

$$\subset\bigcup_{n\in\mathbb{N}}\left\{x\in X \,\middle|\, \frac{(\wedge\mathscr{V})(x)}{1+(\wedge\mathscr{V})(x)}+\frac{1}{n}\leq\frac{\inf_{u\in\mathscr{V}} u(x)}{1+\inf_{u\in\mathscr{V}} u(x)}\right\}. \quad \square$$

Corollary 4.1.1. *Let $\mathscr{V}',\mathscr{V}''$ be subsets of $\mathscr{W}$ which have natural minorants in $\mathscr{W}$. If we set*

$$\mathscr{V}'+\mathscr{V}'':=\{u'+u''\,|\,u'\in\mathscr{V}',\ u''\in\mathscr{V}''\}$$

then

$$\wedge(\mathscr{V}'+\mathscr{V}'')=\wedge\mathscr{V}'+\wedge\mathscr{V}''.$$

Obviously

$$\inf_{u\in\mathscr{V}'+\mathscr{V}''} u=\inf_{u'\in\mathscr{V}'} u'+\inf_{u''\in\mathscr{V}''} u''.$$

Hence, by the proposition, the continuous functions $\wedge(\mathscr{V}'+\mathscr{V}'')$, $\wedge\mathscr{V}'+\wedge\mathscr{V}''$ are equal outside a meager set. Since X is a Baire space, these functions coincide. $\square$

Corollary 4.1.2. *Let $(\mathscr{V}_n)_{n\in\mathbb{N}}$ be a sequence of subsets of $\mathscr{W}$ such that $\wedge\mathscr{V}_n$ exists for any $n\in\mathbb{N}$, and let $\mathscr{V}\subset\mathscr{W}$. We set*

$$f_n:=\inf_{u\in\mathscr{V}_n} u,\qquad f:=\inf_{u\in\mathscr{V}} u.$$

If $(f_n)_{n\in\mathbb{N}}$ is an increasing sequence converging to f, then

$$\lim_{n\to\infty}\wedge\mathscr{V}_n=\wedge\mathscr{V}.$$

By the proposition, the set

$$\{x\in X\,|\,(\wedge\mathscr{V})(x)<f(x)\}\cup\Big(\bigcup_{n\in\mathbb{N}}\{x\in X\,|\,(\wedge\mathscr{V}_n)(x)<f_n(x)\}\Big)$$

is meagre. Outside this set, the continuous functions $\lim_{n\to\infty}\wedge\mathscr{V}_n$ and $\wedge\mathscr{V}$ are equal and (X being a Baire space) this implies that they coincide. $\square$

Proposition 4.1.3. *Let $(\mathscr{V}_n)_{n\in\mathbb{N}}$ be a sequence of subsets of $\mathscr{W}_+$ and let $\mathscr{V}$ be the set of functions of $\mathscr{W}_+$ of the form*

$$\sum_{n\in\mathbb{N}} u_n,$$

when $u_n\in\mathscr{V}_n$ for all $n\in\mathbb{N}$. Then

$$\wedge\mathscr{V}=\sum_{n\in\mathbb{N}}\wedge\mathscr{V}_n.$$

Since for any $u\in\mathscr{V}$, we have

$$\sum_{n\in\mathbb{N}}\wedge\mathscr{V}_n\leq u$$

we get

$$\sum_{n\in\mathbb{N}}\wedge\mathscr{V}_n\leq\wedge\mathscr{V}.$$

We set for any $n\in\mathbb{N}$,

$$A_n:=\{x\in X\,|\,(\wedge\mathscr{V}_n)(x)<\inf_{v\in\mathscr{V}_n} v(x)\}.$$

Let $x\in X\setminus\bigcup_{n\in\mathbb{N}} A_n$. If

$$\sum_{n\in\mathbb{N}}(\wedge\mathscr{V}_n)(x)=\infty$$

then

$$\sum_{n\in\mathbb{N}}(\wedge\mathscr{V}_n)(x)=(\wedge\mathscr{V})(x).$$

Assume

$$\sum_{n\in\mathbb{N}}(\wedge\mathscr{V}_n)(x)<\infty$$

and let ε be a strictly positive real number. For any $n\in\mathbb{N}$, there exists $u_n\in\mathscr{V}_n$ such that

$$u_n(x)<(\wedge\mathscr{V}_n)(x)+\frac{\varepsilon}{2^n}.$$

Hence we obtain

$$(\wedge\mathscr{V})(x)\leq\sum_{n\in\mathbb{N}}u_n(x)\leq\sum_{n\in\mathbb{N}}(\wedge\mathscr{V}_n)(x)+2\varepsilon,$$

and since ε is arbitrary, it follows that

$$\sum_{n\in\mathbb{N}}(\wedge\mathscr{V}_n)(x)=(\wedge\mathscr{V})(x).$$

Therefore

$$\sum_{n\in\mathbb{N}}\wedge\mathscr{V}_n=\wedge\mathscr{V}$$

on $X \setminus \bigcup_{n\in\mathbb{N}} A_n$. Since $\bigcup_{n\in\mathbb{N}} A_n$ is meagre and X is a Baire topological space, the equality holds everywhere. □

Let f be a numerical function on X. We set

$$Rf := \inf_{\substack{u\in\mathscr{W}\\ u\geq f}} u.$$

If f dominates a function of $\mathscr{W}$ then $\widehat{Rf}\in\mathscr{W}$. If moreover, f is lower semi-continuous, then $Rf=\widehat{Rf}$. We have

$$f\leq g \Rightarrow Rf\leq Rg, \qquad R(f+g)\leq Rf+Rg, \qquad R(\alpha f)=\alpha Rf$$

for any positive, real number α.

Proposition 4.1.4. *Let $\mathscr{V}$ be a specifically upper directed (resp. specifically lower directed and naturally lower bounded) non-empty subset of $\mathscr{W}$. Then $\curlyvee\mathscr{V}$ (resp. $\curlywedge\mathscr{V}$) exists,*

$$\curlyvee\mathscr{V}=\vee\mathscr{V} \qquad (\textit{resp. } \curlywedge\mathscr{V}=\wedge\mathscr{V}),$$

and for any $x\in X$ (resp. $x\in X$ such that $\inf_{u\in\mathscr{V}} u(x)<\infty$), we have

$$(\curlyvee\mathscr{V})(x)=\sup_{u\in\mathscr{V}} u(x) \qquad \left(\textit{resp. } (\curlywedge\mathscr{V})(x)=\inf_{u\in\mathscr{V}} u(x)\right).$$

Let $v\in\mathscr{V}$. For any $u\in\mathscr{V}$ such that $u\succcurlyeq v$ (resp. $u\preccurlyeq v$), we denote by f_u, the function on X which is equal to $u-v$ where v is finite (resp. $v-u$ where u is finite) and 0 elsewhere. Then f_u is positive and lower semi-continuous. Hence $Rf_u\in\mathscr{W}$ and we have

$$u=v+Rf_u \qquad (\text{resp. } v=u+Rf_u).$$

The set $\{Rf_u \mid u\in\mathscr{V},\ u\succcurlyeq v \ (\text{resp. } u\preccurlyeq v)\}$ is upper directed with respect to the natural order and thus

$$\vee\mathscr{V}=v+\bigvee_{\substack{u\in\mathscr{V}\\ u\succcurlyeq v}} Rf_u \qquad \Big(\text{resp. } v=\inf_{u\in\mathscr{V}} u+\sup_{\substack{u\in\mathscr{V}\\ u\preccurlyeq v}} Rf_u,\ v=\wedge\mathscr{V}+\bigvee_{\substack{u\in\mathscr{V}\\ u\preccurlyeq v}} Rf_u\Big)$$

(Corollary 4.1.1). Hence $\vee\mathscr{V}$ (resp. $\wedge\mathscr{V}$) is a specific majorant (resp. minorant) of $\mathscr{V}$. The proof that it is the smallest (resp. greatest) one is similar.

Let $x\in X$. If $\mathscr{V}$ specifically upper directed, then the relation

$$(\curlyvee\mathscr{V})(x)=\sup_{u\in\mathscr{V}} u(x)$$

is obvious. Assume that $\mathscr{V}$ is specifically lower directed and $\inf_{u\in\mathscr{V}} u(x)<\infty$. Choose $v\in\mathscr{V}$ such that $v(x)<\infty$. From

$$v(x)=\inf_{u\in\mathscr{V}} u(x)+\sup_{\substack{u\in\mathscr{V}\\ u\preccurlyeq v}} Rf_u(x)=\wedge\mathscr{V}(x)+\bigvee_{\substack{u\in\mathscr{V}\\ u\preccurlyeq v}} Rf_u(x)$$

we deduce

$$(\curlywedge \mathscr{V})(x) = \inf_{u \in \mathscr{V}} u(x). \quad \square$$

Proposition 4.1.5. *The following assertions are equivalent:*

a) if u, v', v'' are functions of $\mathscr{W}_+$ such that $u \leq v' + v''$ then there exist $u', u'' \in \mathscr{W}_+$ such that

$$u = u' + u'', \qquad u' \leq v', \qquad u'' \leq v'';$$

b) let u, v be functions of $\mathscr{W}_+$ and let f be the function on X which is equal to $\sup(u-v, 0)$ *where v is finite and* 0 *elsewhere; then $Rf \in \mathscr{W}_+$ and $Rf \preccurlyeq u$.*

(G. Mokobodzki-D. Sibony 1968 [4].)

$a \Rightarrow b$. Since f is positive and lower semi-continuous, $Rf \in \mathscr{W}_+$. Since

$$u \leq v + Rf$$

there exist $u', u'' \in \mathscr{W}_+$ such that

$$u = u' + u'', \qquad u' \leq v, \qquad u'' \leq Rf.$$

Hence

$$u \leq v + u'', \qquad f \leq u'', \qquad Rf \leq u'', \qquad Rf = u'' \preccurlyeq u.$$

$b \Rightarrow a$. Let f be the function on X which is equal to $\sup(u - v', 0)$ where v' is finite and 0 elsewhere. Then $Rf \leq v''$ and there exists $w \in \mathscr{W}_+$, such that

$$u = w + Rf.$$

Since

$$u \leq v' + Rf$$

we have

$$u = \inf(v' + Rf, w + Rf) = \inf(v', w) + Rf. \quad \square$$

We shall assume from now on that the following axiom is also satisfied:

Axiom of Natural Decomposition. *If u, v', v'' are functions of $\mathscr{W}_+$ such that $u \leq v' + v''$, there exist $u', u'' \in \mathscr{W}_+$ such that*

$$u = u' + u'', \qquad u' \leq v', \qquad u'' \leq v''.$$

Proposition 4.1.6. *Let $u, v \in \mathscr{W}$. The function on X which is equal to v on $F := \{x \in X \mid u(x) < \infty\}$ and ∞ elsewhere belongs to $\mathscr{W}$.*

The set $\{\alpha u \mid \alpha \in \mathbb{R}, 0 < \alpha < 1\}$ dominates the function $\inf(u, 0) \in \mathscr{W}$. Hence $w := \bigwedge_{0<\alpha<1} \alpha u$ exists. Obviously w is negative on F and ∞ elsewhere. The set $\{v + \alpha w \mid \alpha \in \mathbb{R}, 0 < \alpha < 1\}$ is upper directed and its supremum, which is obviously in $\mathscr{W}$, is identical with the function introduced in the proposition. $\square$

Proposition 4.1.7. *Let* $u \in \mathscr{W}$. *We set* $X' := \{x \in X \mid u(x) < \infty\}$, $\mathscr{W}' = \mathscr{W}|_{X'}$. *Then* X' *is a Baire space and* $\mathscr{W}'$ *is a convex cone of continuous, lower finite, numerical functions on* X', *contains the identically* ∞ *function, and satisfies the axiom of lower regularisation, the axiom of upper directed sets and the axiom of natural decomposition.*

X' is a Baire space since it is an open set in a Baire space.

Let $v' \in \mathscr{W}'$. By the preceding proposition, there exists $v \in \mathscr{W}$ such that $v|_{X'} = v'$ and such that $v = \infty$ on $X \setminus \overline{X'}$. All assertions of the proposition follow easily from this remark. □

The following theorem was proved for Brelot spaces by R.-M. Hervé 1960 [2].

Theorem 4.1.1 (R.-M. Hervé). $\mathscr{W}$ *is upper complete with respect to the specific order.*

Let $(u_\iota)_{\iota \in I}$ be a family in $\mathscr{W}$ and $\mathscr{V}$ be the set of its specific majorants. We set $v := \curlywedge \mathscr{V}$. For any $\iota \in I$ and any $w \in \mathscr{V}$, choose w_ι in $\mathscr{W}_+$ such that

$$w = u_\iota + w_\iota .$$

By Corollary 4.1.1, we have

$$v = \curlywedge \mathscr{V} = u_\iota + \curlywedge_{w \in \mathscr{V}} w_\iota \succcurlyeq u_\iota .$$

Let $w \in \mathscr{W}$. We have to show that $v \preccurlyeq w$. Suppose first that w is finite everywhere. From

$$w_\iota - v_\iota = w - v$$

we deduce that there exists a function v'_ι in $\mathscr{W}_+$ such that

$$w_\iota = v'_\iota + R(w - v)$$

(Proposition 4.1.5). We have

$$w = u_\iota + v'_\iota + R(w - v).$$

Hence $w - R(w - v) \in \mathscr{V}$ and therefore

$$v \leq w - R(w - v).$$

Since

$$w - v \leq R(w - v)$$

we get

$$w - v = R(w - v), \qquad v \preccurlyeq w .$$

Let us drop now the assumption that w is finite everywhere. We set

$$X' := \{x \in X \mid w(x) < \infty\}.$$

If we restrict the whole theory to X' and apply Propositions 4.1.6 and 4.1.7, we see that $v|_{X'}$ is the natural infimum of the set of specific majorants (in $\mathscr{W}|_{X'}$) of the family $(u_\iota|_{X'})_{\iota \in I}$. Because of this, there must exist $u \in \mathscr{W}$ such that

$$w|_{X'} = v|_{X'} + u|_{X'}, \qquad u|_{X'} \geq 0.$$

By Proposition 4.1.6, we may assume $u = \infty$ on $X \setminus \overline{X}'$ Then

$$w = v + u, \qquad u \geq 0. \quad \square$$

Proposition 4.1.8. *Let $\mathscr{V} \subset \mathscr{W}_+$ and $u \in \mathscr{W}_+$ be a specific majorant of $\mathscr{V}$. Then $\bigvee \mathscr{V} \preccurlyeq u$* (N. Boboc-A. Cornea 1970 [6]).

Assume first that u is finite everywhere. We set

$$f := \sup(u - \bigvee \mathscr{V}, 0).$$

By Proposition 4.1.5, there exists $w \in \mathscr{W}_+$ such that

$$u = Rf + w.$$

From

$$u = f + \bigvee \mathscr{V} \leq Rf + \bigvee \mathscr{V}$$

we get immediately

$$w \leq \bigvee \mathscr{V}.$$

For any $v \in \mathscr{V}$, let v' be the function in $\mathscr{W}_+$ such that

$$u = v + v'.$$

Obviously $f \leq v'$ and therefore $Rf \leq v'$. We obtain for any $v \in \mathscr{V}$,

$$v + Rf \leq v + v' = u = Rf + w,$$

$$v \leq w, \qquad \bigvee \mathscr{V} \leq w, \qquad \bigvee \mathscr{V} = w \preccurlyeq u.$$

We drop now the assumption that u is finite everywhere and set

$$X' := \{x \in X \mid u(x) < \infty\}.$$

By the above considerations and Proposition 4.1.7, there exists $w \in \mathscr{W}$ such that

$$u|_{X'} = \bigvee (\mathscr{V}|_{X'}) + w|_{X'}, \qquad w|_{X'} \geq 0.$$

From Proposition 4.1.6, it follows immediately that

$$\bigvee (\mathscr{V}|_{x'}) = (\bigvee \mathscr{V})|_{X'}$$

and that there exists $w' \in \mathscr{W}_+$ such that

$$w'|_{X'} = w|_{X'}.$$

Hence

$$\vee \mathscr{V} \preccurlyeq u. \quad \square$$

Corollary 4.1.3. *For any subset $\mathscr{V}$ of $\mathscr{W}_+$, we have*

$$\vee \mathscr{V} \preccurlyeq \curlyvee \mathscr{V}. \quad \square$$

Exercises

4.1.1. Corollary 4.1.1 is true without the hypothesis that X is a Baire space.

4.1.2. Let I be an upper directed ordered set and $(u_\iota)_{\iota \in I}$, $(v_\iota)_{\iota \in I}$ be families in $\mathscr{W}$ such that for any $\iota \in I$ we have $u_\iota \preccurlyeq v_\iota$, and for any $\iota, \kappa \in I$, $\iota \leq \kappa$, we have $u_\iota \leq u_\kappa$, $v_\iota \leq v_\kappa$ (resp. $u_\iota \geq u_\kappa$, $v_\iota \geq v_\kappa$). Then

$$\bigvee_{\iota \in I} u_\iota \preccurlyeq \bigvee_{\iota \in I} v_\iota \qquad (\text{resp. } \bigwedge_{\iota \in I} u_\iota \preccurlyeq \bigwedge_{\iota \in I} v_\iota).$$

(For any $\iota \in I$, let w_ι be an element of $\mathscr{W}_+$ such that $u_\iota + w_\iota = v_\iota$. We set, for any $\iota \in I$,

$$w'_\iota := \bigwedge_{\kappa \geq \iota} w_\kappa .$$

For any $\iota, \kappa \in I$, $\iota \leq \kappa$, we have

$$\bigvee_{\iota \in I} u_\iota + w_\kappa \geq v_\iota \qquad (\text{resp. } \bigwedge_{\iota \in I} u_\iota + w'_\kappa \leq v_\iota, \; v_\kappa \leq u_\iota + w_\kappa).$$

Hence

$$\bigvee_{\iota \in I} u_\iota + w'_\iota \geq v_\iota \geq u_\iota + w'_\iota, \qquad (\text{resp. } \bigwedge_{\iota \in I} u_\iota + \bigvee_{\iota \in I} w'_\iota \leq v_\iota, \; \bigwedge_{\iota \in I} v_\iota \leq u_\iota + \bigvee_{\iota \in I} w'_\iota)$$

$$\bigvee_{\iota \in I} u_\iota + \bigvee_{\iota \in I} w'_\iota \geq \bigvee_{\iota \in I} v_\iota \geq \bigvee_{\iota \in I} u_\iota + \bigvee_{\iota \in I} w'_\iota,$$

$$(\text{resp. } \bigwedge_{\iota \in I} u_\iota + \bigvee_{\iota \in I} w'_\iota \leq \bigwedge_{\iota \in I} v_\iota \leq \bigwedge_{\iota \in I} u_\iota + \bigvee_{\iota \in I} w'_\iota).)$$

4.1.3. Let $u, v \in \mathscr{W}$. Then $u + v \in -\mathscr{W}$ if and only if $u, v \in -\mathscr{W}$.

4.1.4. Let $\mathscr{V}$ be the set of functions p, of $\mathscr{W}_+$, which are finite on a dense subset of X and such that $u \in \mathscr{W}_+ \cap (-\mathscr{W})$ and $u \leq p \Rightarrow u = 0$; then: *a)* $\mathscr{V}$ is a convex cone; *b)* $u \in \mathscr{W}_+$, $p \in \mathscr{V}$, $u \leq p \Rightarrow u \in \mathscr{V}$; *c)* if the specific supremum of a subset of $\mathscr{V}$ is finite on a dense set, then it belongs to $\mathscr{V}$.

4.1.5. Let $\mathscr{W}_s$ be the set of functions u, of $\mathscr{W}_+$, which are finite on a dense subset of X and such that $v \in \mathscr{W}_+$ and $u \leq v \Rightarrow u \preccurlyeq v$. Then: *a)* $\mathscr{W}_s$ is a convex cone; *b)* $u \in \mathscr{W}_s$, $v \in \mathscr{W}_+$, $v \preccurlyeq u \Rightarrow v \in \mathscr{W}_s$; *c)* if the natural supremum of a subset of $\mathscr{W}_s$ is finite on a dense set, then it belongs to $\mathscr{W}_s$.

§ 4.2. Balayage

Throughout this section we shall assume that the axiom of lower semi-continuous regularization, the axiom of upper directed sets and the axiom of natural decomposition are satisfied by $\mathscr{W}$.

The bibliography quoted in this section refers to equivalent results proved on harmonic spaces.

Let u be a function of $\mathscr{W}_+$ and A be a subset of X. If f is the function on X which is equal to u on A and 0 on $X \setminus A$, we set

$$R_u^A := R f.$$

R_u^A is called *the reduit of u on A. The function $\hat{R}_u^A$ is called the balayage of u on A.* These functions were introduced and intensively studied in classical potential theory by M. Brelot 1945 [8]. Obviously $\hat{R}_u^A \in \mathscr{W}_+$ and

$$R_u^A = u \quad \text{on } A,$$

$$A \subset B, \quad u \leq v \Rightarrow R_u^A \leq R_v^B,$$

$$R_{u+v}^A \leq R_u^A + R_v^A, \quad R_u^{A \cup B} \leq R_u^A + R_u^B,$$

$$\hat{R}_u^A = R_u^A \quad \text{if } A \text{ is open}.$$

Proposition 4.2.1. *Let $A \subset X$ and $u \in \mathscr{W}_+$. If u is finite on A, then*

$$R_u^A = \inf_U R_u^U,$$

where U runs through the set of all open sets which contain A and on which u is finite.

Let $v \in \mathscr{W}_+$, $v \geq u$ on A, and let α be a real number, $\alpha > 1$. We set

$$U := \{x \in X \mid u(x) < \alpha v(x)\} \cup \{x \in X \mid u(x) = 0\}.$$

According to Proposition 4.1.1, U is an open set containing A and u is finite on U. Hence

$$\inf_U R_u^U \leq \alpha v$$

and since v and α are arbitrary, we have

$$\inf_U R_u^U \leq R_u^A.$$

The converse inequality is trivial. □

Theorem 4.2.1. *For any subset A of X and any $u, v \in \mathscr{W}_+$, we have*

$$R_{u+v}^A = R_u^A + R_v^A, \quad \hat{R}_{u+v}^A = \hat{R}_u^A + \hat{R}_v^A$$

(C. Constantinescu-A. Cornea 1963 [2], N. Boboc-C. Constantinescu-A. Cornea 1965 [4]; these relations where also proved by R.-M. Hervé 1962 [4] for A closed or open).

The second assertions follows immediately from the first one by Corollary 4.1.1.

Assume that A is open and that u, v are finite on A. Since

$$R_{u+v}^A \leq R_u^A + R_v^A,$$

there exist $u', v' \in \mathscr{W}_+$ such that

$$R_{u+v}^A = u' + v', \qquad u' \leq R_u^A, \qquad v' \leq R_v^A.$$

Since

$$R_{u+v}^A = u + v$$

on A, we get

$$u' = u, \qquad v' = v$$

on A. Hence

$$R_u^A \leq u', \qquad R_v^A \leq v', \qquad R_{u+v}^A = R_u^A + R_v^A.$$

The assumption that A is open may now be dropped by applying the preceding proposition.

Let us consider now the general case. We denote by B the subset of A where $u+v$ is finite. Let x be a point where R_{u+v}^A is finite and let w be a function in $\mathscr{W}_+$ which is finite at x and such that

$$w \geq u + v$$

on A. For any strictly positive real number ε and any $w' \in \mathscr{W}_+$ such that

$$w' \geq u$$

on B, we have

$$w' + \varepsilon w \geq u$$

on A. Hence

$$w' + \varepsilon w \geq R_u^A.$$

But ε and w' are arbitrary and thus

$$R_u^B(x) \geq R_u^A(x).$$

Similarly,

$$R_v^B(x) \geq R_v^A(x).$$

We have, by the above considerations,

$$R_{u+v}^A(x) \geq R_{u+v}^B(x) = R_u^B(x) + R_v^B(x) \geq R_u^A(x) + R_v^A(x).$$

Hence

$$R_{u+v}^A \geq R_u^A + R_v^A$$

and the proof is complete. □

Theorem 4.2.2. *For any $u \in \mathscr{W}_+$ and for any two subsets A, B of X, we have*

$$R_u^{A \cup B} + R_u^{A \cap B} \leq R_u^A + R_u^B, \qquad \hat{R}_u^{A \cup B} + \hat{R}_u^{A \cap B} \leq \hat{R}_u^A + \hat{R}_u^B$$

(N. Boboc-C. Constantinescu-A. Cornea 1965 [4]).

The second inequality follows from the first one by Corollary 4.1.1. Suppose that A and B are open. Let

$$u' := R_u^{A \cup B}, \qquad u'' := R_u^{A \cap B}.$$

Obviously

$$u' = R_{u'}^{A \cup B}, \qquad u'' = R_{u''}^{A \cap B},$$

$$u' + u'' \leq R_u^A + R_u^B$$

on $A \cup B$. Hence, by the preceding theorem,

$$R_u^{A \cup B} + R_u^{A \cap B} = u' + u'' = R_{u'}^{A \cup B} + R_{u''}^{A \cup B} = R_{u'+u''}^{A \cup B} \leq R_u^A + R_u^B.$$

Suppose now that A and B are arbitrary and let us denote by A' (resp. B'), the subset of A (resp. B) where u is finite. If x is a point where $R_u^A + R_u^B$ is finite, then $R_u^{A \cup B}$, $R_u^{A \cap B}$ are also finite at x. Let $v \in \mathscr{W}_+$,

$$v \geq u$$

on $A \cup B$ and finite at x. For any strictly positive real number ε and any $w \in \mathscr{W}_+$ such that

$$w \geq u$$

on $A' \cup B'$, we have

$$w + \varepsilon v \geq u$$

on $A \cup B$. Hence

$$w(x) + \varepsilon v(x) \geq R_u^{A \cup B}(x).$$

Since ε and w are arbitrary, we have

$$R_u^{A' \cup B'}(x) \geq R_u^{A \cup B}(x).$$

Similarly,

$$R_u^{A' \cap B'}(x) \geq R_u^{A \cap B}(x).$$

We have further

$$\begin{aligned} R_u^A(x) + R_u^B(x) &\geq R_u^{A'}(x) + R_u^{B'}(x) = \inf_{A''} R_u^{A''}(x) + \inf_{B''} R_u^{B''}(x) \\ &= \inf_{A'', B''} (R_u^{A''}(x) + R_u^{B''}(x)) \geq \inf_{A'', B''} (R_u^{A'' \cup B''}(x) + R_u^{A'' \cap B''}(x)) \\ &\geq R_u^{A' \cup B'}(x) + R_u^{A' \cap B'}(x) \geq R_u^{A \cup B}(x) + R_u^{A \cap B}(x), \end{aligned}$$

where A'' (resp. B'') runs through the set of open sets which contain A' (resp. B') and on which u is finite. □

Proposition 4.2.2. *Let $(A_n)_{n\in\mathbb{N}}$ be an increasing sequence of subsets of X, $A := \bigcup_{n\in\mathbb{N}} A_n$ and $u \in \mathscr{W}_+$, finite on A. Then*

$$\lim_{n\to\infty} R_u^{A_n} = R_u^A.$$

Let $x \in X$. Obviously

$$\lim_{n\to\infty} R_u^{A_n}(x) \le R_u^A(x).$$

In order to prove the converse inequality, it is sufficient to assume

$$\lim_{n\to\infty} R_u^{A_n}(x) < \infty.$$

Let ε be a strictly positive real number. We shall define inductively, an increasing sequence of open sets $(U_n)_{n\in\mathbb{N}}$ such that u is finite on U_n and

$$A_n \subset U_n, \qquad R_u^{U_n}(x) \le R_u^{A_n}(x) + \sum_{i=0}^{n} \frac{\varepsilon}{2^i}$$

for any $n \in \mathbb{N}$. Suppose U_n has been constructed. By Proposition 4.2.1, there exists an open set U' such that u is finite on U' and

$$A_{n+1} \subset U', \qquad R_u^{U'}(x) \le R_u^{A_{n+1}}(x) + \frac{\varepsilon}{2^{n+1}}.$$

Setting

$$U_{n+1} := U' \cup U_n,$$

we get by the preceding theorem,

$$R_u^{U_{n+1}}(x) + R_u^{U' \cap U_n}(x) \le R_u{}^{U'}(x) + R_u^{U_n}(x).$$

Hence

$$\begin{aligned} R_u^{U_{n+1}}(x) &\le R_u^{U'}(x) + R_u^{U_n}(x) - R_u^{U' \cap U_n}(x) \\ &\le R_u^{A_{n+1}}(x) + \frac{\varepsilon}{2^{n+1}} + R_u^{A_n}(x) + \sum_{i=0}^{n} \frac{\varepsilon}{2^i} - R_u^{A_n}(x) \\ &\le R_u^{A_{n+1}}(x) + \sum_{i=0}^{n+1} \frac{\varepsilon}{2^i}. \end{aligned}$$

Let us denote

$$U := \bigcup_{n\in\mathbb{N}} U_n.$$

We have

$$R_u^A(x) \le R_u^U(x) = \lim_{n\to\infty} R_u^{U_n}(x) \le \lim_{n\to\infty} R_u^{A_n}(x) + \varepsilon,$$

and ε being arbitrary implies that

$$R_u^A(x) \le \lim_{n\to\infty} R_u^{A_n}(x). \quad \square$$

Theorem 4.2.3. *Let $u \in \mathscr{W}_+$, A be a subset of X and $(f_n)_{n\in\mathbb{N}}$ be an increasing sequence of positive, numerical functions on X which are equal* *to* 0 *on* $X \setminus A$ *and such that for any* $x \in A$

$$u(x) = \lim_{n \to \infty} f_n(x).$$

Then

$$\lim_{n \to \infty} R f_n = R_u^A, \qquad \lim_{n \to \infty} \widehat{R f_n} = \hat{R}_u^A$$

(N. Boboc-C. Constantinescu-A. Cornea 1965 [4]).

The second relation follows from the first one by Corollary 4.1.2. Let $x \in X$. Since the inequality

$$\lim_{n \to \infty} R f_n(x) \leq R_u^A(x)$$

is obvious, it is sufficient to prove only the converse one.

Suppose that u is infinite on A. If, for any $n \in \mathbb{N}$,

$$R f_n(x) = 0,$$

then

$$R_u^A(x) = 0.$$

Indeed, for any strictly positive real number ε we may choose a sequence $(u_n)_{n \in \mathbb{N}}$ in $\mathscr{W}_+$ such that $u_n \geq f_n$ on A

$$\sum_{n \in \mathbb{N}} u_n(x) < \varepsilon.$$

The function

$$\sum_{n \in \mathbb{N}} u_n$$

belongs to $\mathscr{W}_+$ and is infinite on A. Hence

$$R_u^A(x) \leq \sum_{n \in \mathbb{N}} u_n(x) < \varepsilon,$$

and since ε is arbitrary, we get

$$R_u^A(x) = 0 \leq \lim_{n \to \infty} R f_n(x).$$

We may assume therefore,

$$0 < R f_k(x) < +\infty$$

for a $k \in \mathbb{N}$. Choose $v \in \mathscr{W}_+$ such that

$$v \geq f_k \quad \text{on } A, \qquad v(x) < \infty.$$

Let α be a strictly positive number. We set

$$B_n := \{y \in A \mid f_n(y) > \alpha v(y)\},$$

$$B := \bigcup_{n \in \mathbb{N}} B_n.$$

Obviously v is finite on B and infinite on $A \setminus B$. Since v is finite at x, $R_v^{A \setminus B}$ vanishes at x and we obtain

$$R f_k(x) \le R_v^A(x) \le R_v^B(x) + R_v^{A \setminus B}(x) = R_v^B(x).$$

By the preceding proposition, we have

$$\lim_{n \to \infty} R f_n(x) \ge \alpha \lim_{n \to \infty} R_v^{B_n}(x) = \alpha R_v^B(x) \ge \alpha R f_k(x).$$

Since α is arbitrary,

$$\lim_{n \to \infty} R f_n(x) = \infty \ge R_u^A(x).$$

Suppose now that u is arbitrary. Let α be a real number with $0 < \alpha < 1$, and let us denote by

$$C_n := \{y \in A \mid \alpha u(y) < f_n(y)\} \cup \{y \in A \mid u(y) = 0\},$$
$$C := \bigcup_{n \in \mathbb{N}} C_n.$$

Obviously

$$C = \{y \in A \mid u(y) < \infty\}.$$

We have, by the preceding proposition,

$$\lim_{n \to \infty} R f_n(x) \ge \alpha \lim_{n \to \infty} R_u^{C_n}(x) = \alpha R_u^C(x).$$

Since α is arbitrary

$$\lim_{n \to \infty} R f_n(x) \ge R_u^C(x).$$

Since u is infinite on $A \setminus C$, we have either

$$R_u^{A \setminus C}(x) = 0$$

or

$$R_u^{A \setminus C}(x) = \infty.$$

In the first case, we have

$$R_u^A(x) \le R_u^C(x) + R_u^{A \setminus C}(x) = R_u^C(x) \le \lim_{n \to \infty} R f_n(x),$$

whereas in the second case, the first part of the proof implies

$$\lim_{n \to \infty} R f_n(x) \ge R_u^{A \setminus C}(x) = \infty \ge R_u^A(x). \quad \square$$

Corollary 4.2.1. *Let* $(u_n)_{n \in \mathbb{N}}$ *be an increasing sequence in* $\mathscr{W}_+$ *and let* $u := \lim_{n \to \infty} u_n$. *Then, for any subset* A *of* X, *we have*

$$\lim_{n \to \infty} \hat{R}_{u_n}^A = \hat{R}_u^A. \quad \square$$

Corollary 4.2.2. *Let $(A_n)_{n\in\mathbb{N}}$ be an increasing sequence of subsets of X and $A:=\bigcup_{n\in\mathbb{N}} A_n$. Then, for any $u\in\mathscr{W}_+$, we have*

$$\lim_{n\to\infty} \hat{R}_u^{A_n}=\hat{R}_u^A. \quad \square$$

Exercises

4.2.1. Let $(u_\iota)_{\iota\in I}$ be an upper directed family in $\mathscr{W}_+$ and $(U_\lambda)_{\lambda\in\Lambda}$ an upper directed family of open sets. Then

$$R_u^U = \bigvee_{\substack{\iota\in I\\ \lambda\in\Lambda}} R_{u_\iota}^{U_\lambda}$$

where

$$U:=\bigcup_{\lambda\in\Lambda} U_\lambda, \quad u:=\bigvee_{\iota\in I} u_\iota.$$

4.2.2. Let $\mathscr{V}$ be a specifically upper directed subset of $\mathscr{W}_+$ such that $\curlyvee\mathscr{V}$ is finite on a dense set. Then, for any $A\subset X$, $(\hat{R}_V^A)_{V\in\mathscr{V}}$ is specifically upper directed and

$$\hat{R}^A_{\curlyvee\mathscr{V}} = \curlyvee_{v\in\mathscr{V}} \hat{R}^A_v.$$

The assertion is not true in general, if $\curlyvee\mathscr{V}$ is infinite on an open set (Exercise 5.3.5 *e*)).

4.2.3. Let $u, v\in\mathscr{W}_+$ and $A:=\{x\in X|v(x)<\infty\}$. Then

$$\hat{R}_u^A=\hat{R}_u^{\bar{A}}\preccurlyeq u.$$

(Choose $u_1, u_2\in\mathscr{W}_+$ such that

$$u=u_1+u_2, \quad u_1\le\hat{R}_u^A, \quad u_2\le\hat{R}_u^{X\setminus A}.$$

Since $\hat{R}_u^{X\setminus A}=0$ on A, we have $u_1=u$ on A. It follows that $u_1=\hat{R}_u^A$).

4.2.4. Let A be a subset of X and $u\in\mathscr{W}_+$. If we set

$$B:=\{x\in A|u(x)<\infty\},$$

then

$$R_u^A=R_u^B+R_u^{A\setminus B}.$$

4.2.5. Let A be a subset of X, $u\in\mathscr{W}_+$, and let $\mathfrak{A}$ be the set of subsets B, of X, containing A and such that the set $\{x\in A|u(x)<\infty\}$ is contained in the interior of B. Then

$$R_u^A=\inf_{B\in\mathfrak{A}} R_u^B.$$

4.2.6. Let $u, v \in \mathscr{W}_+$ and A be a subset of X. Then

$$\hat{R}_u^A \curlyvee \hat{R}_v^A \preccurlyeq (u + \hat{R}_v^A) \wedge (v + \hat{R}_u^A).$$

(It is sufficient to prove the relation $\hat{R}_u^A \preccurlyeq (u + \hat{R}_v^A) \wedge (v + \hat{R}_u^A)$. Assume first that A is open and v is finite on A. Choose $u', v' \in \mathscr{W}_+$ such that

$$(u + \hat{R}_v^A) \wedge (v + \hat{R}_u^A) = u' + v', \qquad u' \leq \hat{R}_u^A, \qquad v' \leq v.$$

Then $u' = u$ on A and therefore

$$\hat{R}_u^A = u' \preccurlyeq (u + \hat{R}_v^A) \wedge (v + \hat{R}_u^A).$$

Now let A be arbitrary and assume that $\hat{R}_v^A$ is finite on a dense set of X. Set

$$B := \{x \in A \mid v(x) < \infty\}.$$

Then

$$\hat{R}_u^A = \hat{R}_u^B, \qquad \hat{R}_v^A = \hat{R}_v^B.$$

The above relation may be proved in this case by using Proposition 4.2.1, Corollary 4.1.1 and Exercise 4.1.2. The general case follows now with the aid of Proposition 4.1.7.)

4.2.7. Let $u, v \in \mathscr{W}_+$ and $A \subset X$. Then the set

$$\{x \in X \mid \hat{R}_u^A(x) < u(x),\ \hat{R}_v^A(x) = v(x) < \infty\}.$$

is open. (Set $B := \{x \in X \mid \hat{R}_v^A(x) < \infty\}$. For any $n \in \mathbb{N}$, let w_n be a function of $\mathscr{W}_+$ such that

$$(u + n\,\hat{R}_v^A) \wedge (\hat{R}_u^A + n\,v) = n\,\hat{R}_v^A + w_n$$

and such that $w_n = \infty$ on $X \setminus \bar{B}$ (Exercise 4.2.6). Then $(w_n)_{n \in \mathbb{N}}$ is increasing; let w be its limit. We have

$$\{x \in B \mid \hat{R}_u^A(x) < u(x),\ \hat{R}_v^A(x) = v(x)\} = \{x \in B \mid \hat{R}_u^A(x) < u(x),\ \hat{R}_u^A(x) = w(x)\},$$
$$\{x \in B \mid \hat{R}_u^A(x) < u(x),\ \hat{R}_v^A(x) < v(x)\} = \{x \in B \mid \hat{R}_u^A(x) < u(x),\ u(x) = w(x)\}$$

and therefore

$$\{x \in X \mid \hat{R}_u^A(x) < u(x),\ \hat{R}_v^A(x) = v(x) < \infty\}$$
$$= \{x \in X \mid \hat{R}_u^A(x) < u(x),\ w(x) < u(x),\ v(x) < \infty\}.)$$

4.2.8. For all subsets, A, B of X and for any $u \in \mathscr{W}_+$, we have

$$\hat{R}_u^{A \cup B} \preccurlyeq \hat{R}_u^A + \hat{R}_u^B.$$

(See the proof of Proposition 5.3.4.)

Chapter 5

The Convex Cone of Hyperharmonic Functions

Throughout this chapter X will denote a harmonic space.

§ 5.1. The Fine Topology

The aim of this section is to show that the convex cone of hyperharmonic functions on X satisfies the axioms of the preceding chapter if X is endowed with a suitable topology, called fine topology.

The fine topology on X is the coarsest topology on X which is finer than the initial topology of X and in which any hyperharmonic function on any open set of X is continuous. This topology was introduced in classical potential theory by M. Brelot 1944 [6], following a suggestion of H. Cartan. The fine topology is the coarsest topology on X in which the sets of the form

$$(U, u, \alpha) := \{x \in U \mid u(x) < \alpha\}$$

are open, where U denotes an open set of X, u a hyperharmonic function on U and α a real number. We shall write: *fine neighbourhood, fine open set, fine continuous function,* etc., instead of neighbourhood, open set, continuous function, etc. when we are referring to the fine topology.

Proposition 5.1.1. *Let A be a subset of X and $x \in X$. A is a fine neighbourhood of x if and only if either A is a neighbourhood of x (in the initial topology) or there exists a hyperharmonic function u, defined on an open neighbourhood of x, such that*

$$u(x) < \liminf_{A \not\ni y \to x} u(y).$$

If A is a neighbourhood of x in the initial topology, it is obviously a fine neighbourhood of x. Suppose now that there exists a hyperharmonic function u defined on an open neighbourhood of x and possessing the above property. Let α be a real number such that

$$u(x)<\alpha<\liminf_{A\not\ni y\to x} u(y)$$

and U be an open neighbourhood of y on which u is defined and such that $u\geq\alpha$ on $U\setminus A$. Then

$$x\in(U, u, \alpha)\subset A$$

and A is a fine neighbourhood of x.

Suppose now that A is a fine neighbourhood of x. Then there exists a finite system (U_i, u_i, α_i), $i=1, 2, \ldots, n$, such that

$$A\supset\bigcap_{i=1}^{n}(U_i, u_i, \alpha_i)\ni x.$$

Let u be the hyperharmonic function defined on $U:=\bigcap_{i=1}^{n} U_i$,

$$u:=\sum_{i=1}^{n} u_i$$

and $\mathfrak{U}$ be an ultrafilter on $U\setminus A$, converging to x, such that

$$\lim_{\mathfrak{U}} u=\liminf_{U\setminus A\ni y\to x} u(y).$$

Then there exists an j such that

$$U\setminus(U_j, u_j, \alpha_j)\in\mathfrak{U}.$$

Hence

$$\lim_{\mathfrak{U}} u=\sum_{i=1}^{n}\lim_{\mathfrak{U}} u_i\geq\sum_{i=1}^{n} u_i(x)+\alpha_j-u_j(x)>u(x). \quad \square$$

Proposition 5.1.2. *Any point of X possesses a fundamental system of fine neighbourhoods which are compact in the initial topology* (C. Constantinescu-A. Cornea 1963 [2]).

Let x be a point of X and A be a fine neighbourhood of x. We may suppose $x\in\overline{X\setminus A}$. There exists, then, a hyperharmonic function u defined on a neighbourhood of x and a real number α such that

$$\liminf_{A\ni y\to x} u(y)>\alpha>u(x).$$

Let L be a compact neighbourhood of x such that u is defined on L and $u>\alpha$ on $L\setminus A$. The set

$$K:=\{y\in L\,|\,u(y)\leq\alpha\}$$

fulfills the required conditions. $\square$

Corollary 5.1.1. *X endowed with the fine topology is a Baire space* (A. Cornea 1966 [unpublished]).

Let $(G_n)_{n\in\mathbb{N}}$ be a sequence of fine open, fine dense sets of X and let G be a fine open set of X. We may construct inductively a sequence, $(K_n)_{n\in\mathbb{N}}$, of compact sets of X such that K_{n+1} is contained in the fine interior of K_n and $K_n \subset G \cap G_n$ for any $n \in \mathbb{N}$. The assertion then follows from

$$\emptyset \neq \bigcap_{n\in\mathbb{N}} K_n \subset G \cap \Big(\bigcap_{n\in\mathbb{N}} G_n\Big). \quad \square$$

Proposition 5.1.3. *Let $x \in X$, K be a fine neighbourhood of x, compact in the initial topology, and $\mathfrak{F}_x$ be the section filter of the set of resolutive neighbourhoods of x, ordered by the converse inclusion relation. Then*

$$\lim_{V,\mathfrak{F}_x} \mu_x^V(K) = 1, \qquad \lim_{V,\mathfrak{F}_x} \mu_x^V(X \setminus K) = 0$$

(R.-M. Hervé 1962 [4]).

Let h be a positive harmonic function, defined on an open, relatively compact neighbourhood U of x, with $h(x)=1$. We want to show that

$$\lim_{\substack{V,\mathfrak{F}_x \\ V\subset U}} \int_{X\setminus K} h\, d\mu_x^V = 0.$$

If K is a neighbourhood of x (in the initial topology) the assertion is trivial. Assume the contrary and let u be a hyperharmonic function, defined on an open neighbourhood of x, for which

$$u(x) < \liminf_{K \not\ni y \to x} u(y)$$

(Proposition 5.1.1). Let β be a real number such that

$$u(x) < \beta < \liminf_{K \not\ni y \to x} u(y).$$

By choosing U sufficiently small, we may assume that u is defined on U and greater than βh on $U \setminus K$. For any real number $\alpha < u(x)$, we have

$$\begin{aligned} u(x) - \alpha &\geq \limsup_{\substack{V,\mathfrak{F}_x \\ V\subset U}} \int (u - \alpha h)\, d\mu_x^V \\ &\geq \limsup_{\substack{V,\mathfrak{F}_x \\ V\subset U}} \Big(\int_{X\setminus K} (\beta - \alpha)\, h\, d\mu_x^V + \inf_V (u - \alpha h)\, \mu_x^V(K) \Big) \\ &\geq (\beta - \alpha) \limsup_{\substack{V,\mathfrak{F}_x \\ V\subset U}} \int_{X\setminus K} h\, d\mu_x^V. \end{aligned}$$

Since α is arbitrary, we obtain

$$\lim_{\substack{V,\mathfrak{F}_x \\ V\subset U}} \int_{X\setminus K} h\, d\mu_x^V = 0.$$

For any resolutive set V, $x \in V$, $\overline{V} \subset U$, we have

$$(\inf_V h)\, \mu_x^V(K) \leq \int h\, d\mu_x^V = 1 \leq \sup_V h \cdot \mu_x^V(K) + \int_{X \setminus K} h\, d\mu_x^V,$$

$$(\inf_V h)\, \mu_x^V(X \setminus K) \leq \int_{X \setminus K} h\, d\mu_x^V.$$

Hence

$$\limsup_{V, \mathfrak{F}_x} \mu_x^V(K) \leq 1 \leq \liminf_{V, \mathfrak{F}_x} \mu_x^V(K),$$

$$0 \leq \limsup_{V, \mathfrak{F}_x} \mu_x^V(X \setminus K) \leq 0. \quad \square$$

Corollary 5.1.2. *There exist no isolated points in the fine topology.* □

Let $\mathfrak{B}$ be a base of X of relatively compact, resolutive sets. A locally lower bounded, numerical function f on X is called $\mathfrak{B}$-*nearly hyperharmonic* if, for any $V \in \mathfrak{B}$, we have $\mu^V f \leq f$.

Proposition 5.1.4. *Let $\mathfrak{B}$ be a base of X of relatively compact resolutive sets and f be a $\mathfrak{B}$-nearly hyperharmonic function on X. Then the lower semi-continuous regularisation of f, with respect to the initial topology, is hyperharmonic and coincides with the lower semi-continuous regularisation of f with respect to the fine topology* (M. Brelot 1959 [14]).

Let $V \in \mathfrak{B}$. By Proposition 1.1.4 the function $\mu^V f$ is lower semi-continuous. Hence $\hat{f} \geq \mu^V f \geq \mu^V \hat{f}$. This proves that $\hat{f}$ is hyperharmonic (Corollary 2.3.4).

Let x be a point of X, and $\mathfrak{F}_x$ be the section filter of the set $\{V \in \mathfrak{B} \mid x \in V\}$ ordered by the converse inclusion relation. Let α be any real number strictly smaller than the fine lower limit of f at x and A be a fine neighbourhood of x such that $f > \alpha$ on A. We may assume that A is compact for the initial topology (Proposition 5.1.2). We have (Proposition 5.1.3)

$$\hat{f}(x) \geq \lim_{V, \mathfrak{F}_x} \int^* f\, d\mu_x^V \geq \alpha \lim_{V, \mathfrak{F}_x} \mu_x^V(A) + \lim_{V, \mathfrak{F}_x} (\inf_V f)\, \mu_x^V(X \setminus A) \geq \alpha.$$

Since α is arbitrary, we deduce that the fine lower limit of f at x is smaller than $\hat{f}$. The converse inequality is trivial. □

Corollary 5.1.3. *Let f be a fine lower semi-continuous, locally lower bounded, numerical function on X. Then Rf is a hyperharmonic function.*

Obviously $Rf \geq f$ and therefore $Rf = \widehat{Rf}$. □

Theorem 5.1.1. *If we endow a harmonic space with the fine topology, then the convex cone of hyperharmonic functions on this space satisfies* *the axiom of lower semi-continuous regularisation, the axiom of upper directed sets and the axiom of natural decomposition.*

A harmonic space endowed with the fine topology is a Baire topological space (Corollary 5.1.1). The axiom of lower semi-continuous regularisation follows from Proposition 5.1.4 and the axiom of upper directed sets is trivial. We now prove the axiom of natural decomposition.

Let u, v be positive hyperharmonic functions on X and let f be the function on X which is equal to $\sup(u-v, 0)$, where v is finite, and 0 elsewhere. Since f is fine lower semi-continuous, Rf is a hyperharmonic function (Corollary 5.1.3). Let V be a quasi-regular MP-set and $g \in \mathscr{C}(\partial V)$, $g \leq u$ on ∂V. Choose $x \in V$ and let u_0 be a positive hyperharmonic function on V, finite at x and converging to ∞ along any non-regular ultrafilter on V. Let ε be a strictly positive real number and let u^* be the function on X which is equal to Rf on $X \setminus V$ and $\inf(Rf, \mu^V(Rf)+u-\mu^V g+\varepsilon u_0)$ on V. By Proposition 2.1.3, u^* is a hyperharmonic function. We have on V,

$$\mu^V g \leq \mu^V u \leq \mu^V v + \mu^V(Rf) \leq v + \mu^V(Rf),$$

$$u = \mu^V g + (u - \mu^V g) \leq v + \mu^V(Rf) + u - \mu^V g.$$

Hence $u \leq v + u^*$ on X and thus $f \leq u^*$, $Rf \leq u^*$. We deduce that

$$Rf(x) \leq \mu^V(Rf)(x) + u(x) - \mu^V g(x) + \varepsilon u_0(x)$$

on V. Since ε, x and g are arbitrary it follows that

$$Rf + \mu^V u \leq \mu^V(Rf) + u$$

on V.

Let f' be the function on X which is equal to $u - Rf$, where Rf is finite, and ∞ elsewhere. From the above inequality, we see that for any quasi-regular set V we have $\mu^V f' \leq f'$ on V. Since the quasi-regular sets form a base of X (Theorem 2.4.3) and $f' \geq 0$ the fine lower semi-continuous regularisation $\hat{f}$, of f, is hyperharmonic (Proposition 5.1.4). Since

$$u = Rf + f'$$

and since u and Rf are fine continuous, we see that

$$u = Rf + \hat{f}', \qquad Rf \preccurlyeq u.$$

The axiom of natural decomposition follows now from Proposition 4.1.5. □

Remark 1. This theorem enables us to apply all results of the preceding chapter to the convex cone of hyperharmonic functions on a harmonic space. *All of the notions and notation introduced there with* *respect to the convex cone $\mathscr{W}$ will also be applied to harmonic spaces.* We remark that if $(u_\iota)_{\iota\in I}$ is locally lower bounded family of hyperharmonic functions on X, then $\bigwedge_{\iota\in I} u_\iota$ exists and by Proposition 5.1.4 it is equal to the lower semi-continuous regularization of the function $\inf_{\iota\in I} u_\iota$.

Remark 2. Interesting properties of the fine topology, especially those concerned with connectivity, were recently obtained, under supplementary conditions, by B. Fuglede 1970 [2] (see Exercises 9.1.6, 9.2.3, 9.2.4).

Exercises

5.1.1. The fine topologies on X with respect to $\mathscr{U}$ and $\mathscr{U}_f$ from (Exercise 2.1.1) coincide.

5.1.2. The fine and the initial topologies coincide in the examples from Exercise 3.1.4.

5.1.3. In the example from Theorem 2.1.2, the fine topology is neither locally compact nor locally connected. The same is true for the Bauer space of the heat equation.

5.1.4. Any point of a harmonic space possesses a fundamental system of fine neighbourhoods which are compact and connected in the initial topology (C. Constantinescu-A. Cornea 1963 [2]).

5.1.5. Any fine open set is uncountable. (Use Proposition 5.1.3.)

5.1.6. Let A be a fine open set and u be a positive hyperharmonic function. Then

$$R_u^A = \bigvee_K \hat{R}_u^K,$$

where K runs through the set of compact subsets of A. (Use Proposition 5.1.2.)

5.1.7. Let u, v be hyperharmonic functions on X, $v \geq 0$, and let f be a locally lower bounded, numerical function such that $u = v + f$. Then $\widehat{Rf}$ is a specific minorant of u.

5.1.8. Let $\mathscr{V}$ be a set of hyperharmonic functions on X. For any open set U, we set $v_U := \curlyvee \mathscr{V}|_U$. If U, V are open sets of X such that $U \subset V$, then v_U is a specific minorant of $v_V|_U$ and if this latter function is superharmonic, then there exists a positive harmonic function h on U such that

$$v_V|_U = v_U + h.$$

We have

$$\curlyvee \mathscr{V}(x) = \sup_U v_U(x),$$

where U runs through the set of all relatively compact open sets containing x.

5.1.9. Let u, v be superharmonic functions on X. Then $u \curlyvee v$ is superharmonic if and only if $u \curlywedge v$ exists. In this case we have:

a) $u+v=u\curlyvee v+u\curlywedge v$;

b) for any superharmonic function w on X

$$u\curlyvee v+w=(u+w)\curlyvee(v+w),$$

$$u\curlywedge v+w=(u+w)\curlywedge(v+w).$$

5.1.10. Let $\mathscr{V}$ be a lower directed set of superharmonic functions on X such that $\curlywedge \mathscr{V}=0$. If X has a countable base, then there exists a sequence $(u_n)_{n\in\mathbb{N}}$ in $\mathscr{V}$ such that $\sum\limits_{n\in\mathbb{N}} u_n$ is superharmonic.

5.1.11. Let u be a superharmonic function on X and v be a hyperharmonic function with the property that there exists a base $\mathfrak{B}$ of X, of relatively compact, resolutive sets for which

$$u+\mu^V v \leq v+\mu^V u$$

on V for any $V\in\mathfrak{B}$. Let f be the function on X which is equal to $v-u$, where u is finite, and ∞ elsewhere. If f is locally lower bounded then $\hat{f}$ is hyperharmonic and we have

$$v=u+\hat{f}.$$

5.1.12. Let $\mathfrak{B}$ be a base of X of relatively compact, resolutive sets. *a)* If the infimum of a set of $\mathfrak{B}$-nearly hyperharmonic functions on X is locally lower bounded, then it is $\mathfrak{B}$-nearly hyperharmonic. *b)* If f, g are $\mathfrak{B}$-nearly hyperharmonic functions then $f+g$ is a $\mathfrak{B}$-nearly hyperharmonic function and

$$\widehat{f+g}=\hat{f}+\hat{g}.$$

c) If $(f_n)_{n\in\mathbb{N}}$ is an increasing sequence of $\mathfrak{B}$-nearly hyperharmonic functions on X then its limit is $\mathfrak{B}$-nearly hyperharmonic and

$$\widehat{\lim_{n\to\infty} f_n}=\lim_{n\to\infty}\hat{f}_n.$$

5.1.13. Let A be a subset of X and B, C be subsets of X such that $B\cup C=X\setminus A$ and $\bar{B}\cap\bar{C}\subset A$. Then, for any positive hyperharmonic function u, on X, we have $\hat{R}_u^A=\hat{R}_u^{A\cup C}$ on B.

§ 5.2. Capacity

This section is devoted to the proof of a generalised version of the Choquet theorem on capacitibility (G. Choquet 1952 [1]). Throughout this section, we shall let $\mathscr{F}$ denote a set of positive, lower semi-continuous, numerical functions on a given locally compact space Z such that:

a) the least upper bound of any upper directed subset of $\mathscr{F}$ belongs to $\mathscr{F}$;

b) the lower semi-continuous regularisation of the limit of any decreasing sequence in $\mathscr{F}$ belongs to $\mathscr{F}$.

Let Y be a locally compact space. We call *a capacity on Y (with values in $\mathscr{F}$)*, a map γ defined on the set of subsets of Y with values in $\mathscr{F}$ satisfying the following conditions:

c_1) $A \subset B \subset Y \Rightarrow \gamma(A) \leq \gamma(B)$;

c_2) for any increasing sequence, $(A_n)_{n \in \mathbb{N}}$, of sets of Y we have

$$\lim_{n \to \infty} \gamma(A_n) = \gamma\Big(\bigcup_{n \in \mathbb{N}} A_n\Big);$$

c_3) for any decreasing sequence, $(K_n)_{n \in \mathbb{N}}$, of compact subsets of Y we have

$$\widehat{\lim_{n \to \infty} \gamma(K_n)} = \gamma\Big(\bigcap_{n \in \mathbb{N}} K_n\Big).$$

The classical (real valued) capacity is obtained taking as Z a one point space and identifying the set $[0, \infty]$ with the set of positive functions on Z. A subset A of Y is called *γ-capacitable* if

$$\gamma(A) = \sup_K \gamma(K),$$

where K runs through the set of compact subsets of A.

Proposition 5.2.1. *Let Y' be a locally compact space and let φ be a continuous map of Y' into Y. Then, for any capacity γ on Y, the map γ'*

$$A \mapsto \gamma(\varphi(A))$$

defined on the set of subsets of Y' is a capacity on Y'. If a subset A of Y' is γ'-capacitable, then $\varphi(A)$ is γ-capacitable.

The properties *c_1)* and *c_2)* follow from

$$A \subset B \subset Y' \Rightarrow \varphi(A) \subset \varphi(B),$$

$$\varphi\Big(\bigcup_{n \in \mathbb{N}} A_n\Big) = \bigcup_{n \in \mathbb{N}} \varphi(A_n),$$

and $c_3)$ follows from

$$\varphi\Big(\bigcap_{n\in\mathbb{N}} K_n\Big)=\bigcap_{n\in\mathbb{N}}\varphi(K_n)$$

which is true for any decreasing sequence of compact sets of Y' since φ is continuous.

The last assertion follows from

$$\gamma\big(\varphi(A)\big)=\gamma'(A)=\sup_K \gamma'(K)=\sup_K \gamma\big(\varphi(K)\big)\leq \sup_L \gamma(L)\leq \gamma\big(\varphi(A)\big),$$

where K (resp. L) runs through the set of compact subsets of A (resp. $\varphi(A)$). □

Proposition 5.2.2. *Every $K_{\sigma\delta}$-set of a locally compact space is capacitable with respect to any capacity.*

Let A be a $K_{\sigma\delta}$-set of a locally compact space Y and let γ be a capacity on Y. There exists then, a family $(K_{m,n})_{m\in\mathbb{N},\, n\in\mathbb{N}}$ of compact sets of Y such that for any $m\in\mathbb{N}$, the sequence $(K_{m,n})_{n\in\mathbb{N}}$ is increasing and such that

$$A=\bigcap_{m\in\mathbb{N}}\bigcup_{n\in\mathbb{N}} K_{m,n}.$$

Let $x\in Z$. It is sufficient to show that

$$\big(\gamma(A)\big)(x)=\sup_K\big(\gamma(K)\big)(x),$$

where K runs through the set of compact subsets of A. This relation is trivial if

$$\big(\gamma(A)\big)(x)=0.$$

Assume, therefore, that

$$\big(\gamma(A)\big)(x)>0.$$

and let $\alpha\in\big]0,\big(\gamma(A)\big)(x)\big[$. Since $\gamma(A)$ is lower semi-continuous there exists a compact neighbourhood L of x such that

$$\gamma(A)>\alpha$$

on L.

We construct inductively a sequence $(n_m)_{m\in\mathbb{N}}$ of natural numbers such that for any $i\in\mathbb{N}$

$$\gamma\Big(\Big(\bigcap_{m\leq i} K_{m,n_m}\Big)\cap A\Big)>\alpha$$

on L. Assume that the construction was performed for all $m<i$. Now $\Big(K_{i,j}\cap\Big(\bigcap_{m<i}K_{m,n_m}\Big)\cap A\Big)_{j\in\mathbb{N}}$ is an increasing sequence of subsets of Y whose union is $\Big(\bigcap_{m<i}K_{m,n_m}\Big)\cap A$, L is compact, and the functions of $\mathscr{F}$ are

lower semi-continuous. Hence there exists by $c_2)$, a natural number n_i such that K_{i,n_i} satisfies the required condition.

By $c_1)$ and $c_3)$, we get

$$\gamma\Big(\bigcap_{m\in\mathbb{N}} K_{m,n_m}\Big)=\widehat{\lim_{i\to\infty}\gamma\Big(\bigcap_{m\le i} K_{m,n_m}\Big)}\ge\widehat{\lim_{i\to\infty}\gamma\Big(\Big(\bigcap_{m\le i} K_{m,n_m}\Big)\cap A\Big)}\ge\alpha$$

on the interior of L. In particular

$$\Big(\gamma\Big(\bigcap_{m\in\mathbb{N}} K_{m,n_m}\Big)\Big)(x)\ge\alpha.$$

The assertion follows since $\bigcap_{m\in\mathbb{N}} K_{m,n_m}$ is a compact subset of A and since α is arbitrary. □

A subset A of a locally compact space Y is called *K-analytic* if there exists a locally compact space Y', a $K_{\sigma\delta}$-set A' of Y' and a continuous map φ from Y' into Y such that

$$A=\varphi(A').$$

Corollary 5.2.1. *Every K-analytic set is capacitable for any capacity.* □

Proposition 5.2.3. *Let A be a relatively compact, K-analytic set of a locally compact space Y. Then, there exists a compact space Y', a $K_{\sigma\delta}$-set A' of Y' and a continuous map φ from Y' into Y such that*

$$A=\varphi(A').$$

We prove first the following lemma.

Lemma. *Let X be a locally compact space and $\mathcal{Q}$ be a set of continuous maps on X with every $\varphi\in\mathcal{Q}$ having a compact range, X_φ. Then there exists a compact space $X^*_{\mathcal{Q}}$, containing X as dense open subset, such that every $\varphi\in\mathcal{Q}$ can be extended to a continuous map φ^* from $X^*_{\mathcal{Q}}$ into X_φ and such that for any $x, y\in X^*_{\mathcal{Q}}\setminus X$, $x\neq y$, there exists $\varphi\in\mathcal{Q}$ for which $\varphi^*(x)\neq\varphi^*(y)$.*

We may assume that $\mathcal{Q}$ contains the canonical embedding φ_0 of X into its Alexandroff compactification. We denote by ψ the map of X into the compact space $\prod_{\varphi\in\mathcal{Q}} X_\varphi$ defined by

$$(\psi(x))(\varphi)=\varphi(x)$$

for any $\varphi\in\mathcal{Q}$. Since $\mathcal{Q}$ contains φ_0, ψ defines a homeomorphism of X onto $\psi(X)$ and therefore we may identify X with $\psi(X)$.

Let $X^*_{\mathcal{Q}}$ be the closure of $\psi(X)$ in $\prod_{\varphi\in\mathcal{Q}} X_\varphi$. It is obvious that for any $\varphi\in\mathcal{Q}$, the restriction to $X^*_{\mathcal{Q}}$ of the φ-projection is a continuous extention of φ to $X^*_{\mathscr{Q}}$. It is also obvious that $\psi(X)$ is dense in $X^*_{\mathscr{Q}}$. But being dense in $X^*_{\mathscr{Q}}$ and locally compact implies that $\psi(X)$ is open. The last assertion is obvious. □

We prove now the proposition. By hypothesis, there exists a locally compact space Y'', a $K_{\sigma\delta}$-set A' of Y'' and a continuous map ψ of Y'' into Y such that

$$A=\psi(A').$$

The topological space $\psi^{-1}(\bar{A})$ is locally compact. By the lemma, there exists a compact space Y', containing $\psi^{-1}(\bar{A})$ as a dense subspace, and a continuous map φ from Y' to $\bar{A}$ which is equal to ψ on $\psi^{-1}(\bar{A})$. Since $\psi^{-1}(\bar{A})$ is closed in Y'', A' is a $K_{\sigma\delta}$-set of the topological space $\psi^{-1}(\bar{A})$ and therefore a $K_{\sigma\delta}$-set of Y' which proves the proposition. □

Proposition 5.2.4. *A countable union and a countable intersection of K-analytic sets are K-analytic.*

Let $(A_n)_{n\in\mathbb{N}}$ be a sequence of K-analytic sets of a locally compact space Y. There exists for any $n\in\mathbb{N}$, a locally compact space Y_n, a continuous map φ_n from Y_n into Y, and a $K_{\sigma\delta}$-set B_n on Y_n such that

$$A_n=\varphi_n(B_n).$$

Let Y' be the topological sum of the topological spaces $(Y_n)_{n\in\mathbb{N}}$, B be the union of $(B_n)_{n\in\mathbb{N}}$ and φ be the map of Y' into Y which coincides with φ_n on Y_n for any $n\in\mathbb{N}$. Then B is a $K_{\sigma\delta}$-set of Y', φ is continuous and

$$\bigcup_{n\in\mathbb{N}} A_n=\varphi(B).$$

Hence $\bigcup_{n\in\mathbb{N}} A_n$ is a K-analytic set.

In order to prove that $\bigcap_{n\in\mathbb{N}} A_n$ is K-analytic, we assume first that there exists $n_0\in\mathbb{N}$ such that A_n is relatively compact for $n\geq n_0$. By the preceding proposition, we may assume that for any $n\in\mathbb{N}$ with $n\geq n_0$, Y_n is compact. Let ψ be the map of the locally compact space $\prod_{n\in\mathbb{N}} Y_n$ into the topological space $Y^{\mathbb{N}}$

$$(y_n)_{n\in\mathbb{N}}\mapsto(\varphi_n(y_n))_{n\in\mathbb{N}}.$$

For any $n\in\mathbb{N}$, we denote by π_n, the n-th projection of $Y^{\mathbb{N}}$ into Y, i.e. the map

$$(y_m)_{m\in\mathbb{N}}\mapsto y_n.$$

We denote by

$$Y':=\{z\in\prod_{n\in\mathbb{N}} Y_n \,|\, n, m\in\mathbb{N}\Rightarrow\pi_n\circ\psi(z)=\pi_m\circ\psi(z)\}.$$

Since ψ and π_n are continuous and since Y is a Hausdorff space, Y' is a closed set and therefore a locally compact subspace of $\prod_{n\in\mathbb{N}} Y_n$. On the other hand, $\prod_{n\in\mathbb{N}} B_n$ is a $K_{\sigma\delta}$-set of $\prod_{n\in\mathbb{N}} Y_n$ and therefore $Y'\cap(\prod_{n\in\mathbb{N}} B_n)$ is a $K_{\sigma\delta}$-set of Y'. Since

$$\pi_1\circ\psi\Big(Y'\cap(\prod_{n\in\mathbb{N}} B_n)\Big)=\bigcap_{n\in\mathbb{N}} A_n,$$

it follows that $\bigcap_{n\in\mathbb{N}} A_n$ is a K-analytic set.

Let us prove now the general case. From the definition of a K-analytic set, there exists a sequence $(K_j)_{j\in\mathbb{N}}$ of compact subsets of Y such that

$$\bigcup_{n\in\mathbb{N}} A_n\subset\bigcup_{j\in\mathbb{N}} K_j.$$

By the above proof we note that for any $n\in\mathbb{N}$ and $j\in\mathbb{N}$, the set $A_n\cap K_j$ is K-analytic. Hence, again by the above proof, it follows that

$$K_j\cap(\bigcap_{n\in\mathbb{N}} A_n)=\bigcap_{n\in\mathbb{N}}(A_n\cap K_j)$$

is a K-analytic set. The assertion follows now from the first part of the proof and from the relation

$$\bigcap_{n\in\mathbb{N}} A_n=\bigcup_{j\in\mathbb{N}}\Big(K_j\cap(\bigcap_{n\in\mathbb{N}} A_n)\Big). \quad \square$$

Corollary 5.2.2. *In any locally compact space with countable base every Borel set is K-analytic and therefore capacitable for any capacity.*

Let Y be a locally compact space with a countable base and let $\mathfrak{A}$ be the set of subsets A of Y such that A and $Y\setminus A$ are K-analytic. By the proposition, the union of a countable family of sets from $\mathfrak{A}$ belongs to $\mathfrak{A}$ and every closed set belongs to $\mathfrak{A}$. Since the complement of any set of $\mathfrak{A}$ belongs to $\mathfrak{A}$, it follows that any Borel set belongs to $\mathfrak{A}$. The last assertion follows from Corollary 5.2.1. $\square$

§ 5.3. Supplementary Results on the Balayage of Positive Superharmonic Functions

Proposition 5.3.1. *Let u be a positive superharmonic function on X and $A\subset X$. Then R_u^A is harmonic on $X\setminus\bar{A}$. R_u^A and $\hat{R}_u^A$ concide on $X\setminus\bar{A}$ and on the fine interior of A.*

The first assertion follows from Perron's proposition and Corollary 2.1.2. The second one follows from Proposition 5.1.4. $\square$

Corollary 5.3.1. *If p is a potential on X, then*

$$\bigwedge_K R_p^{X \setminus K} = 0,$$

where K runs through the set of compact sets of X. If, moreover, p is continuous at every point where it vanishes, then

$$\bigwedge_K R_p^{X \setminus K} = 0,$$

where K runs through the set of compact sets of $\{x \in X \mid p(x) > 0\}$.

The first assertion follows immediately from the proposition. The second one follows from the first one using Corollary 2.2.2. □

Proposition 5.3.2. *Let A be a subset of X, u be a positive, superharmonic function on X which is finite on A, and K be a compact G_δ-subset of $X \setminus A$. Then for any strictly positive real number ε, there exists a fine open set G, containing A, such that*

$$R_u^G \leq \hat{R}_u^A + \varepsilon$$

on K.

Assume first that $\bar{A} \cap K = \varnothing$. Let U be an open set containing A such that $\bar{U} \cap K = \varnothing$. By Proposition 4.2.1

$$R_u^A = \inf_G R_u^G,$$

where G runs through the set of fine open sets containing A. Obviously

$$R_u^A = \inf_G R_u^{G \cap U}.$$

By Proposition 5.3.1, R_u^A and $R_u^{G \cap U}$ are finite and continuous on K for any G. By Dini's theorem, there exists a fine open set G, containing A, such that

$$R_u^G \leq R_u^A + \varepsilon$$

on K. Again by Proposition 5.3.1, we have $R_u^A = \hat{R}_u^A$ on K. Hence

$$R_u^G \leq \hat{R}_u^A + \varepsilon$$

on K.

Let us prove now the general case. Since K is of type G_δ, there exists an increasing sequence $(A_n)_{n \in \mathbb{N}}$ of subsets of A, $A_0 = \varnothing$, whose union is equal to A and such that $\bar{A}_n \cap K = \varnothing$ for any $n \in \mathbb{N}$. We shall construct inductively, an increasing sequence $(G_n)_{n \in \mathbb{N}}$ of fine open sets such that $A_n \subset G_n$ and

$$R_u^{G_n} \leq \hat{R}_u^{A_n} + \sum_{i \leq n} \frac{\varepsilon}{2^{i+1}}$$

for any $n \in \mathbb{N}$. We take $G_0 = \varnothing$ and assume that G_n was constructed. By the first part of the proof, there exists a fine open set G' containing

A_{n+1} and such that

$$R_u^{G'} \leq \hat{R}_u^{A_{n+1}} + \frac{\varepsilon}{2^{n+2}}.$$

We set

$$G_{n+1} := G_n \cup G'.$$

Using Theorem 4.2.2 we get

$$\begin{aligned}\hat{R}_u^{A_n} + R_u^{G_{n+1}} &\leq R_u^{G_n \cap G'} + R_u^{G_n \cup G'} \leq R_u^{G_n} + R_u^{G'} \\ &\leq \hat{R}_u^{A_n} + \sum_{i \leq n} \frac{\varepsilon}{2^{i+1}} + \hat{R}_u^{A_{n+1}} + \frac{\varepsilon}{2^{n+2}}\end{aligned}$$

on K. Since $\hat{R}_u^{A_n}$ is finite on K, we deduce that

$$R_u^{G_{n+1}} \leq \hat{R}_u^{A_{n+1}} + \sum_{i \leq n+1} \frac{\varepsilon}{2^{i+1}}$$

on K. If we set

$$G := \bigcup_{n \in \mathbb{N}} G_n$$

we obtain

$$R_u^G = \lim_{n \to \infty} R_u^{G_n} \leq \lim_{n \to \infty} \hat{R}_u^{A_n} + \sum_{n \in \mathbb{N}} \frac{\varepsilon}{2^{n+1}} \leq \hat{R}_u^A + \varepsilon$$

on K. □

Corollary 5.3.2. *Let u be a positive superharmonic function on X and A be a subset of X. For any compact set $K \subset X \setminus A$ of type G_δ and for any strictly positive real number ε there exists a positive superharmonic function v on X which is equal to u on A and smaller than $\hat{R}_u^A + \varepsilon$ on K. In particular $R_u^A = \hat{R}_u^A$ on K* (N. Boboc-C. Constantinescu-A. Cornea 1965 [4]).

Since K is of type G_δ, there exists a decreasing sequence $(U_n)_{n \in \mathbb{N}}$ of open sets of X such that

$$K = \bigcap_{n \in \mathbb{N}} \overline{U}_n.$$

By Proposition 5.3.1 $\hat{R}_u^{X \setminus U_n}$ is finite and continuous on K. There exists therefore a sequence $(\alpha_n)_{n \in \mathbb{N}}$ of strictly positive real numbers such that

$$\alpha_n \hat{R}_u^{X \setminus U_n} < \frac{\varepsilon}{2^{n+2}}$$

on K for any $n \in \mathbb{N}$. By the proposition, there exists a fine open set G containing the set $\{x \in A \mid u(x) < \infty\}$ such that

$$R_u^G \leq \hat{R}_u^A + \frac{\varepsilon}{2}$$

on K. Hence the function

$$v := \inf\left(u, R_u^G + \sum_{n\in\mathbb{N}} \alpha_n \hat{R}_u^{X\setminus U_n}\right)$$

possesses the required properties. □

Remark. In this corollary, it is not possible to remove the hypothesis that K is of type G_δ (Exercise 5.3.5 *e)*).

Proposition 5.3.3. *If U is an MP-set, then for any positive hyperharmonic function u on X, we have $R_u^{X\setminus U} = \overline{H}_u^U$ on U.*

Let $v \in \overline{\mathscr{U}}_u^U$. The function on X which is equal to u on $X\setminus U$ and $\inf(u, v)$ on U is a positive hyperharmonic function on X dominating u on $X\setminus U$ (Proposition 2.1.2). Hence $R_u^{X\setminus U} \leq v$ on U. Since v is arbitrary, we get $R_u^{X\setminus U} \leq \overline{H}_u^U$. The converse inequality is immediate. □

Proposition 5.3.4. *Let A, B be two subsets of X and u be a positive superharmonic function on X. We set*

$$R_u^{A,B} := \inf_{\substack{A'\in\mathfrak{A}\\ B'\in\mathfrak{B}}} R_{(A',B';u)}^{A'\cup B'}$$

where $\mathfrak{A}$ (resp. $\mathfrak{B}$) denotes the set of fine open sets containing the set $\{x\in A \mid u(x)<\infty\}$ (resp. $\{x\in B \mid u(x)<\infty\}$) and $(A', B'; u) := R_u^{A'} \wedge R_u^{B'}$. Then:

a) $\hat{R}_u^{A\cup B} + \hat{R}_u^{A,B} = \hat{R}_u^A + \hat{R}_u^B$.

b) For all subsets C, D of X and any positive superharmonic function v on X such that $A\subset C$, $B\subset D$, $u\leq v$ we have

$$\hat{R}_u^{A,B} \leq \hat{R}_v^{C,D}.$$

(C. Constantinescu 1967 [6].)

a) We set

$$A_0 := \{x\in A \mid u(x)<\infty\}, \quad B_0 := \{x\in A \mid u(x)<\infty\}.$$

Since u is finite on a dense set we get

$$\hat{R}_u^A = \hat{R}_u^{A_0} = \widehat{\inf_{A'\in\mathfrak{A}} R_u^{A'}}, \qquad \hat{R}_u^B = \hat{R}_u^{B_0} = \widehat{\inf_{B'\in\mathfrak{B}} R_u^{B'}},$$

$$\hat{R}_u^{A\cup B} = \hat{R}_u^{A_0\cup B_0} = \widehat{\inf_{\substack{A'\in\mathfrak{A}\\ B'\in\mathfrak{B}}} R_u^{A'\cup B'}}$$

(Proposition 4.2.1). For any $A'\in\mathfrak{A}$, $B'\in\mathfrak{B}$, we have

$$u + (A', B'; u) = R_u^{A'} + R_u^{B'}$$

on $A' \cup B'$. Hence

$$R_u^{A' \cup B'} + R_{(A',\, B';\, u)}^{A' \cup B'} = R_u^{A'} + R_u^{B'}$$

(Theorem 4.2.1). The required relation follows from the above equality using Corollary 4.1.1.

b) Since the relation

$$R_v^{A,\, B} \leq R_v^{C,\, D}$$

is trivial, it is sufficient to prove the inequality

$$R_u^{A,\, B} \leq R_v^{A,\, B}.$$

Let A' (resp. B') be a fine open set containing the set $\{x \in A \,|\, v(x) < \infty\}$ (resp. $\{x \in B \,|\, v(x) < \infty\}$) and let ε be a strictly positive real number. We set

$$G := \{x \in X \,|\, u(x) < \varepsilon\, v(x)\},$$

$$A'' := A' \cup G, \qquad B'' := B' \cup G.$$

Obviously A'' (resp. B'') is a fine open set containing the set $\{x \in A \,|\, u(x) < \infty\}$ (resp. $\{x \in B \,|\, v(x) < \infty\}$). We have

$$(A'', B''; u) \leq (A', B'; u) + \varepsilon\, v,$$

$$\hat{R}_u^{A,\, B} \leq R_{(A'',\, B'';\, u)}^{A'' \cup B''} \leq R_{(A',\, B';\, u)}^{A' \cup B' \cup G'} + \varepsilon\, v \leq R_{(A',\, B';\, u)}^{A' \cup B'} + 2\varepsilon\, v \leq R_{(A',\, B';\, v)}^{A',\, B'} + 2\varepsilon\, v.$$

Since A', B' are arbitrary, we obtain

$$\hat{R}_u^{A,\, B} \leq \hat{R}_v^{A,\, B} + 2\varepsilon\, v$$

and the required relation holds since ε is arbitrary. □

Theorem 5.3.1. *Let $(p_\iota)_{\iota \in I}$ be a family of finite, continuous potentials on X such that*

$$p := \sum_{\iota \in I} p_\iota$$

is finite on a dense set.

a) For any subset A of X we have

$$\hat{R}_p^A = \bigwedge_U \hat{R}_p^U,$$

where U runs through the set of neighbourhoods of A.

b) The map $A \mapsto \hat{R}_p^A$ defined on the set of subsets of X is a capacity on X with values in the set of positive hyperharmonic functions on X. (M. Brelot 1967 [21].)

a) Let $\iota \in I$ and let u be a positive hyperharmonic function on X which is greater than p_ι on A. For any compact set $L \subset \{x \in X \,|\, p_\iota(x) > 0\}$ and for any $\alpha \in \mathbb{R}$, $\alpha > 1$, the set

$$U_0 := \{x \in X \,|\, p_\iota(x) < \alpha\, u(x)\} \cup (X \setminus L)$$

is open and contains A. Hence

$$\bigwedge_U \hat{R}^U_{p_\iota} \leq R^{U_0}_{p_\iota} \leq \alpha u + R^{X \setminus L}_{p_\iota}.$$

Since α, u, and L are arbitrary, we deduce that

$$\bigwedge_U \hat{R}^U_{p_\iota} \leq \hat{R}^A_{p_\iota}$$

(Corollary 5.3.1).

Let J be a finite subset of I. By the above considerations, we have

$$\hat{R}^A_p \leq \bigwedge_U \hat{R}^U_p \leq \sum_{\iota \in J} \bigwedge_U \hat{R}^U_{p_\iota} + \sum_{\iota \in I \setminus J} p_\iota \leq \sum_{\iota \in J} \hat{R}^A_{p_\iota} + \sum_{\iota \in I \setminus J} p_\iota \leq \hat{R}^A_p + \sum_{\iota \in I \setminus J} p_\iota.$$

But J is arbitrary and p is finite on a dense set. Thus

$$\hat{R}^A_p = \bigwedge_U \hat{R}^U_p.$$

b) By Theorem 5.1.1 and the remark following it, the set of positive hyperharmonic functions on X satisfies conditions *a)* *b)* of the set of values of a capacity introduced in § 5.2.

The relation

$$A \subset B \Rightarrow \hat{R}^A_p \leq \hat{R}^B_p$$

is trivial. If $(A_n)_{n \in \mathbb{N}}$ is an increasing sequence of subsets of X, then from the preceding theorem and from Corollary 4.2.2, we get

$$\hat{R}^A_p = \lim_{n \to \infty} \hat{R}^{A_n}_p$$

where

$$A := \bigcup_{n \in \mathbb{N}} A_n.$$

The assertion follows now easily from *a)*. □

Corollary 5.3.3. *Assume that X is a $\mathfrak{P}$-harmonic space and has a countable base and let A be a K-analytic set of X. Then, for any positive hyperharmonic function u on X, we have*

$$\hat{R}^A_u = \sup_K \hat{R}^K_u,$$

where K runs through the set of compact subsets of A (N. Boboc-A. Cornea 1968 [4]).

Let $(p_n)_{n \in \mathbb{N}}$ be an increasing sequence of finite, continuous potentials on X whose limit is equal to u (Corollary 2.3.1). By the theorem, we have

$$\hat{R}^A_{p_n} = \sup_K \hat{R}^K_{p_n} \leq \sup_K \hat{R}^K_u,$$

where K runs through the set of compact subsets of A. From Corollary 4.2.1 it follows that

$$\hat{R}^A_u = \lim_{n \to \infty} \hat{R}^A_{p_n} \leq \sup_K \hat{R}^K_u,$$

and the corollary is proved, since the converse inequality is trivial. □

Proposition 5.3.5. *If X is a $\mathfrak{P}$-harmonic space, then the balayage of any positive superharmonic function on X, on any compact set of X, is a potential.*

Let u be a positive, superharmonic function on X and let K be a compact set of X. Let L be a compact neighbourhood of K. Since $\hat{R}_u^K$ is harmonic on $X \setminus K$ (Proposition 5.3.1), there exists a potential p on X greater than $\hat{R}_u^K$ on ∂L. From

$$\hat{R}_u^K = \overline{H}_u^{X \setminus K}$$

on $X \setminus K$ (Proposition 5.3.3) we get

$$\hat{R}_u^K = \overline{H}_u^{X \setminus K} = \overline{H}_{\hat{R}_u^K}^{X \setminus L} \leq p$$

on $X \setminus L$ (Proposition 2.4.4). Since $\hat{R}_u^K$ is dominated by a potential outside a compact set, $\hat{R}_u^K$ must be a potential. □

Exercises

5.3.1. Let u be a positive, superharmonic function on X, U be an open set of X and A be a subset of X containing $X \setminus U$. We set

$$v := (u - \hat{R}_u^{X \setminus U})|_U.$$

Then

$$\hat{R}_u^A = \hat{R}_u^{X \setminus U} + {}^U\hat{R}_v^{A \cap U},$$

where ${}^U\hat{R}_v^{A \cap U}$ denotes the balayage of the function v on the set $A \cap U$, taken on the harmonic space U. (Use Proposition 2.1.2.)

5.3.2. Let A be a subset of X, u be a positive, superharmonic function on X and h be the greatest harmonic minorant of $\hat{R}_u^A$. Then:

a) $\hat{R}_h^A = h$.

b) $\hat{R}_{u-h}^A$ is a potential.

c) If X is a $\mathfrak{P}$-harmonic space and A is relatively compact, then $h = 0$.

d) If X is a $\mathfrak{P}$-harmonic space then

$$h = \bigwedge_K \hat{R}_u^{A \setminus K},$$

where K runs through the set of compact sets of X.

(First prove the assertion *a)* for A fine open. Use then Proposition 4.2.1 and Exercise 4.2.2.)

5.3.3. A positive superharmonic function u on X is called *extremal* if every positive superharmonic function specifically smaller than u is proportional to u (this notion was introduced for harmonic functions by R.S. Martin 1941 [1]). Any extremal superharmonic function is either a potential or a harmonic function. If u is an extremal, superharmonic function on X, then for any $A \subset X$, $\hat{R}_u^A$ is either a potential or equal to u (use the preceding exercise).

5.3.4. Let U be an open set of X and h be a harmonic function on X which is an extremal superharmonic function on X (Exercise 5.3.3). If $\hat{R}_h^{X \setminus U}$ is a potential, then there exists exactly one component V, of U, such that $\hat{R}_h^{X \setminus V}$ is a potential. (First prove that there exists $x \in U$ such that $\hat{R}_h^{X \setminus U}(x) < h(x)$ and let V be the component of U containing x. Use then Exercise 5.1.13 and Exercise 5.3.3.)

5.3.5. Let X be the set

$$(\{(x, y) \in \mathbb{R}^2 \mid 1 < x^2 + y^2 < 2\} \times [0, \omega_1[) \cup [0, \omega_1],$$

where ω_1 denotes the first uncountable ordinal number. We introduce on X the following topology: a subset U of X is open if and only if the following conditions are satisfied: *a)* the set

$$\{(x, y) \in \mathbb{R}^2 \mid 1 < x^2 + y^2 < 2, \ ((x, y), \xi) \in U\}$$

is open in $\mathbb{R}^2$ for any $\xi \in [0, \omega_1[$; *b)* if $0 \in U$, then there exists a strictly positive real number ε such that

$$\{(x, y) \in \mathbb{R}^2 \mid 1 < x^2 + y^2 < 1 + \varepsilon\} \times \{0\} \subset U;$$

c) for any $\xi \in [0, \omega_1[$ such that $\xi + 1 \in U$, there exists a strictly positive real number ε such that

$$\{(x, y) \in \mathbb{R}^2 \mid 2 - \varepsilon < x^2 + y^2 < 2\} \times \{\xi\} \subset U,$$
$$\{(x, y) \in \mathbb{R}^2 \mid 1 < x^2 + y^2 < 1 + \varepsilon\} \times \{\xi + 1\} \subset U;$$

d) if ξ is a limit ordinal number of $[0, \omega_1[$, then there exists a strictly positive real number ε and an ordinal number $\eta < \xi$ such that

$$\{(x, y) \in \mathbb{R}^2 \mid 1 < x^2 + y^2 < 1 + \varepsilon\} \times \{\xi\} \subset U,$$
$$\{(x, y) \in \mathbb{R}^2 \mid 1 < x^2 + y^2 < 2\} \times]\eta, \xi[\subset U,$$
$$]\eta, \xi] \subset U;$$

e) if $\omega_1 \in U$ then there exists an ordinal number $\eta < \omega_1$ such that

$$\{(x, y) \in \mathbb{R}^2 \mid 1 < x^2 + y^2 < 2\} \times]\eta, \omega_1[\subset U, \quad]\eta, \omega_1] \subset U.$$

We set $X' := X \setminus \{\omega_1\}$, $X'' := X \setminus \{0\}$ and denote, for any open set U of X' (resp. X'') by $\mathscr{H}'(U)$ (resp. $\mathscr{H}''(U)$), the set of real continuous functions h on U such that: *a)* for any $\xi \in [0, \omega_1[$ the function

$$(x, y) \mapsto h((x, y), \xi)$$

on $\{(x, y) \in \mathbb{R}^2 \mid 1 < x^2 + y^2 < 2, ((x, y), \xi) \in U\}$ is a solution of the Laplace equation; *b)* for any $\xi \in U \cap [0, \omega_1[$ (resp. $\xi + 1 \in U \cap]0, \omega_1[$) we have

$$\int_0^{2\pi} (f((r\cos\theta, r\sin\theta), \xi) - f(\xi))\, d\theta = 0$$

$$\left(\text{resp. } \int_0^{2\pi} (f((r\cos\theta, r\sin\theta), \xi) - f(\xi+1))\, d\theta = 0\right)$$

for any real number r such that

$$\{(x, y) \in \mathbb{R}^2 \mid 1 < x^2 + y^2 \le r^2\} \times \{\xi\} \subset U$$

$$\left(\text{resp. } \{(x, y) \in \mathbb{R}^2 \mid r^2 \le x^2 + y^2 < 2\} \times \{\xi\} \subset U\right).$$

a) X' (resp. X'') is a locally compact space and $\mathscr{H}'$ (resp. $\mathscr{H}''$) is a harmonic sheaf on X' (resp. X'') possessing the Doob convergence property.

b) X' (resp. X'') endowed with $\mathscr{H}'$ (resp. $\mathscr{H}''$) is a $\mathfrak{P}$-Bauer space.

c) For any hyperharmonic function defined on the intersection with X' (resp. X'') of an open neighbourhood of ω_1, there exists a neighbourhood of ω_1, such that the hyperharmonic function is constant on the intersection of X' (resp. X'') with it.

d) Every harmonic function on X' (resp. X'') is constant; every potential on X' has a compact carrier.

e) Let A be a set such that for any $\xi \in [0, \omega_1[$, the set

$$A \cap \{((x, y), \xi) \in \mathbb{R}^2 \times [0, \omega_1[\}$$

contains exactly one point; then for any positive hyperharmonic function u on X' (resp. X'') R_u^A is the greatest constant function dominated by u (resp. $R_u^A(\omega_1) = u(\omega_1)$, $\hat{R}_u^A(\omega_1) = 0$).

5.3.6. Prove the Theorem 5.3.1 by assuming that for any $\iota \in I$, p_ι is a positive, finite, continuous, superharmonic function provided that the A of assertion *a)* is relatively compact and X is an $\mathfrak{S}$-harmonic space (the last condition is necessary; see Exercise 5.3.7).

5.3.7. Let X be the topological space

$$\{(x, y, z) \in \mathbb{R}^3 \mid x^2 + y^2 = z^2\}$$

and let f be the function on $\mathbb{R}$ which is equal to $\log\frac{1}{z}$ for $0 < z \le \frac{1}{e}$ and 1 elsewhere. For any open set U, of X, we denote by $\mathscr{H}(U)$, the set of continuous real functions h on U such that the function on $\{(x, y) \in \mathbb{R}^2 \mid \textit{there exists } z \leq 0 \text{ (resp. } z>0) \textit{ such that } (x, y, z) \in U\}$ and defined by

$$(x, y) \mapsto f(z)\, h(x, y, z)$$

is a solution of the Laplace equation in $\mathbb{R}^2$. X endowed with $\mathscr{H}$ is a Bauer space. Every positive superharmonic function on X is identically 0 on $\{(x, y, z) \in X \mid z \leq 0\}$ and proportional to the function $(x, y, z) \mapsto \dfrac{1}{f(z)}$ on $\{(x, y, z) \in X \mid z>0\}$. Hence every positive superharmonic function is harmonic and X is not an $\mathfrak{S}$-harmonic space. If we take $K=\{(0,0,0)\}$, then for any positive superharmonic function u on X,

$$\hat{R}_u^K = 0, \quad R_u^U = u$$

for any neighbourhood U of K.

Chapter 6

Absorbent Sets, Polar Sets and Semi-Polar Sets

Throughout this chapter X will denote a harmonic space.

§ 6.1. Absorbet Sets

A closed set F of a harmonic space X is called an *absorbent set* (*of* X) if the function on X which is equal to 0 on F and ∞ outside F is hyperharmonic. This notion was introduced by H. Bauer 1963 [6]. Obviously any absorbent set is fine open and fine closed.

Proposition 6.1.1. *Let F be a closed set of X. The following assertions are equivalent:*

a) F is absorbent;

b) there exists a hyperharmonic function on X which is equal to 0 on F and strictly positive outside F;

c) for any resolutive set V of X, we have $\mu_x^V(X \setminus F)=0$ for every $x \in F \cap V$;

d) for any $x \in F$ and any neighbourhood U of x, there exists a resolutive neighbourhood V, of x, contained in U such that $\mu_x^V(X \setminus F)=0$.

(H. Bauer 1966 [9].)

$a \Rightarrow b$ is trivial.

$b \Rightarrow c$. Let u be a hyperharmonic function on X which is equal to 0 on F and strictly positive outside F. Let f be a positive function of $\mathscr{K}(\partial V)$ whose carrier is contained in $\partial V \setminus F$. Then there exists a positive real number α such that $\alpha u \geq f$ on ∂V. We deduce that

$$0 \leq \mu^V f(x) \leq \alpha u(x) = 0$$

and since f is arbitrary, we obtain $\mu_x^V(X \setminus F)=0$.

$c \Rightarrow d$ is trivial.

$d \Rightarrow a$. The function on X equal to 0 on F and equal to ∞ outside F is hyperharmonic by Corollary 2.3.4. ▯

Proposition 6.1.2. *Let $(F_\iota)_{\iota\in I}$ be a family of absorbent sets of X. Then $\bigcap_{\iota\in I} F_\iota$, $\overline{\bigcup_{\iota\in I} F_\iota}$ are absorbent sets.*

For any $\iota\in I$, let u_ι be the hyperharmonic function on X which is equal to 0 on F_ι and ∞ on $X\setminus F_\iota$. The assertion follows from

$$\bigcap_{\iota\in I} F_\iota=\{x\in X\,|\,\bigvee_{\iota\in I} u_\iota(x)=0\},$$

$$\overline{\bigcup_{\iota\in I} F_\iota}=\{x\in X\,|\,\bigwedge_{\iota\in I} u_\iota(x)=0\}. \quad \square$$

Proposition 6.1.3. *A harmonic space is elliptic if and only if any absorbent set of any open connected subspace U of X is either empty or equal to U.*

The "if" part follows from Proposition 3.1.4 $a\Rightarrow e$. The "only if" part follows from Proposition 6.1.1 $b\Rightarrow a$ and Proposition 3.1.4 $e\Rightarrow a$. $\square$

Corollary 6.1.1. *Any absorbent set of a connected Brelot space is either empty or equal to the whole space.*

The assertion follows immediately from the proposition with the aid of Theorem 3.1.2. $\square$

Proposition 6.1.4. *Let u be a hyperharmonic function on X and $A:=\{x\in X\,|\,u(x)<\infty\}$. Then for any relatively compact, resolutive set V and any $x\in V\cap\bar{A}$ we have*

$$\mu_x^V(X\setminus A)=0.$$

Hence $\bar{A}$ is an absorbent set (N. Boboc-C. Constantinescu-A. Cornea 1965 [3]).

The function $x\mapsto\mu_x^V(X\setminus A)$ is a continuous function on V (Proposition 1.1.4) which is obviously 0 on $A\cap V$. Hence it vanishes on $\bar{A}\cap V$. $\square$

Proposition 6.1.5. *Let μ be a measure on X, F the smallest absorbent set of X containing the carrier of μ, U an open set of X contained in F and K a compact subset of U. If the sheaf of harmonic functions on X possesses the Doob convergence property, then there exists a real number α such that for any positive, hyperharmonic function u on X which is harmonic on U we have*

$$\sup_K u\leq\alpha\int^{*} u\,d\mu$$

(G. Mokobodzki 1964 [unpublished]).

Assuming the assertion is not true, then for any $n\in\mathbb{N}$ there exists a positive hyperharmonic function u_n on X, harmonic on U, such that

$$\int u_n\,d\mu\leq\frac{1}{n^2}\quad\text{and}\quad \sup_K u_n\geq n.$$

Then $u := \sum_{n \in \mathbb{N}} u_n$ is a hyperharmonic function on X and

$$\int u \, d\mu < \infty .$$

Hence the set $\overline{\{x \in X \mid u(x) < \infty\}}$ contains the carrier of μ and therefore contains U (Proposition 6.1.4). But the Doob converge property implies that u is harmonic on U and this contradicts the relations

$$\sup_K u \geq \sup_K u_n \geq n$$

for any $n \in \mathbb{N}$. ▯

Corollary 6.1.2. *Let x be a point of a connected Brelot space X, U an open set of X and K a compact subset of U. Then there exists a real number α such that for any positive, hyperharmonic function u on X, which is harmonic on U, we have*

$$\sup_K u \leq u(x).$$

The assertion follows immediately from the fact that on a connected Brelot space, any non-empty absorbent set is equal to X. ▯

Exercises

6.1.1. Let $(F_\iota)_{\iota \in I}$ be a family of absorbent sets.

a) For any relatively compact, resolutive set V, the set $X \setminus (\bigcup_{\iota \in I} F_\iota)$ is of μ_x^V-measure zero for any $x \in V \cap \overline{\bigcup_{\iota \in I} F_\iota}$.

b) If F is an absorbent set such that $F \cap F_\iota = \varnothing$ for any $\iota \in I$, then

$$F \cap \overline{\bigcup_{\iota \in I} F_\iota} = \varnothing .$$

c) If $(F_\iota)_{\iota \in I}$ are pairwise disjoint, then the family is locally finite. (C. Constantinescu 1966 [3].)

6.1.2. The smallest absorbent set containing a connected set is connected.

6.1.3. Let F be an absorbent set of an $\mathfrak{S}$-harmonic space, X. If $X \setminus F$ is an MP-set, then all of its boundary points are regular (H. Bauer 1963 [6]). (Use Exercise 2.4.1.)

6.1.4. Let $\mathscr{V}$ be a lower directed set of hyperharmonic functions on X such that for any relatively compact, resolutive set V and for any $v \in \mathscr{V}$, we have $v_V \in \mathscr{V}$. Let f be its infimum.

a) For any $v \in \mathscr{V}$, there exists a unique positive hyperharmonic function v^* on X such that

$$v^*(x) = \infty \qquad \text{if } f(x) = -\infty,$$
$$v^*(x) + f(x) = v(x) \qquad \text{if } f(x) \neq -\infty$$

and such that v^* is infinite on the interior of the set $\{x \in X \mid v(x) = \infty\}$.

(The unicity follows from Proposition 6.1.4. For any relatively compact, resolutive set V and any $w \in \mathscr{V}$, $w \leq v$, there exists a positive, hyperharmonic function w' on V such that $v = \mu^V w + w'$ on V (Proposition 1.1.4 and Exercise 5.1.11). Take v^* equal to $\bigvee_{\substack{w \in \mathscr{V} \\ w \leq v}} w'$ on V.)

b) For any relatively compact, resolutive set V, we have $f = \mu^V f$ on $\{x \in V \mid f(x) \neq \infty\}$.

c) f is harmonic if it is locally bounded.

d) $\hat{f}$ is harmonic if it is superharmonic.

6.1.5. Let U be an MP-set and let $(f_n)_{n \in \mathbb{N}}$ be an increasing sequence of numerical functions on ∂U. Then

$$\lim_{n \to \infty} \overline{H}^U_{f_n} = \overline{H}^U_{\lim\limits_{n \to \infty} f_n} \quad \text{on} \quad \{x \in U \mid \lim_{n \to \infty} \overline{H}^U_{f_n}(x) > -\infty\}.$$

(Use the preceeding exercise.)

6.1.6. Let U be an MP-set of a harmonic space and let f be a numerical function on ∂U. Let U' be an open subset of U and f' be the function on $\partial U'$ which is equal to f on $\partial U \cap \partial U'$ and $\overline{H}^U_f$ on $U \cap \partial U'$. Then $\overline{H}^U_f = \overline{H}^{U'}_{f'}$ on $\{x \in U' \mid \overline{H}^U_f(x) < \infty\}$. (Use Exercise 6.1.4.)

6.1.7. Let U be an open set of X and consider the following assertions:

a) U is an MP-set;

b) every compact, absorbent subset of U is empty.

If there exists a strictly negative, hyperharmonic function on U, then $a \Rightarrow b$ (H. Bauer 1963 [6]). If for every $x \in U$ there exists a strictly positive, hyperharmonic function on X, finite at x, then $b \Rightarrow a$.

6.1.8. Let F be an absorbent set of X, and U any open set of F (with respect to the induced topology on F). Denote by $\mathscr{U}_F(U)$ the set of numerical functions u on U such that the function on (the open set) $U \cup (X \setminus F)$ which is equal to u on U and ∞ on $X \setminus F$ is hyperharmonic.

a) $\mathscr{U}_F$ is a hyperharmonic sheaf.

b) If u is a hyperharmonic function on an open set U of X, then $u|_{U \cap F} \in \mathscr{U}_F(U \cap F)$.

c) If U is an MP-set of X, then $U \cap F$ is an MP-set with respect to $\mathcal{U}_F$.

d) If V is a resolutive set of X, then $V \cap F$ is a resolutive set with respect to $\mathcal{U}_F$, and for any $x \in V \cap F$, the harmonic measure on $V \cap F$ at x with respect to $\mathcal{U}_F$ is equal to μ_x^V.

e) The harmonic sheaf $\mathcal{H}_{\mathcal{U}_F}$ on F possesses the Bauer convergence property.

f) F endowed with $\mathcal{U}_F$ is a harmonic space.

g) If V is a quasi-regular (resp. regular) MP-set of X then $V \cap F$ is a quasi-regular (resp. regular) MP-set of F.

h) If u is a superharmonic function (resp. potential) on X, then $u|_F$ is a superharmonic function (resp. potential) on F.

i) A lower semi-continuous, numerical function on X is hyperharmonic if it is hyperharmonic on $X \setminus F$ and its restriction to F belongs to $\mathcal{U}_F(F)$.

j) The topology induced on F by the fine topology on X is the fine topology of the harmonic space $(F, \mathcal{U}_F)$.

k) If $A \subset F$ and if u is a positive, hyperharmonic function on X, then R_u^A (resp. $\hat{R}_u^A$) is equal on F to the reduit function (resp. to the balayaged function) of $u|_F$ on A on the harmonic space $(F, \mathcal{U}_F)$.

(C. Constantinescu 1966 [3].)

6.1.9. The absorbent sets in the example of Theorem 2.1.2 are exactly the sets $[\alpha, \infty[$ for any $\alpha \in \mathbb{R}$.

6.1.10. The absorbent sets for the heat equation are exactly the sets of the form $\{(x, t) \in \mathbb{R}^n \times \mathbb{R} | t \leq t_0\}$ for any $t_0 \in \mathbb{R}$.

6.1.11. Let u be a hyperharmonic function on X and A be a subset of X. For any relatively compact, resolutive set V such that $V \cap A = \emptyset$, we have $\mu^V R_u^A = R_u^A$ on $\{x \in V | R_u^A(x) < \infty\}$; if A is fine open, this equality holds on the whole of V. R_u^A is harmonic on V if it is locally bounded on V and $\hat{R}_u^A$ is harmonic on V if it is superharmonic on V.

6.1.12. Let u be a positive, hyperharmonic function on X. Then for any absorbent set F, we have $R_u^F \preccurlyeq u$ (C. Constantinescu 1966 [3]).

6.1.13. Let F be a closed, nowhere dense set of X such that the union of the absorbent sets, disjoint from F, is dense in X. Then any positive hyperharmonic function on X which is harmonic on $X \setminus F$, is harmonic on X (C. Constantinescu 1966 [3]).

6.1.14. Let u be an extremal superharmonic function on X and let F be an absorbent set. Then R_u^F is either equal to u or identically 0.

§ 6.2. Polar Sets

The polar sets play in potential theory, the role of sets which may be neglected in many problems. They were introduced into classical potential theory by M. Brelot 1941 [4], where they coincided with the sets of zero exterior capacity (H. Cartan 1942 [1]).

Let U be an open set of a harmonic space, A a subset of U and u a positive, hyperharmonic function on U; we denote by ${}^{U}R_u^A$ (resp. ${}^{U}\hat{R}_u^A$) the reduit function (resp. the balayaged function) of u on A with respect to the harmonic subspace U.

A subset A of X is called *polar* if there exists an open covering $\mathfrak{W}$ of X such that for every $W \in \mathfrak{W}$, we have ${}^{W}\hat{R}_\infty^{A \cap W} = 0$. By Corollary 4.2.1 and Theorem 5.1.1, this definition is equivalent to the following one: there exists an open covering $\mathfrak{W}$ of X such that for every $W \in \mathfrak{W}$, there exists a positive hyperharmonic function u on W which is strictly positive on $A \cap W$ and such that

$$ {}^{W}\hat{R}_u^{A \cap W} = 0. $$

Any subset of a polar set is polar. If A is a polar set of X and if U is an open set of X, then $A \cap U$ is a polar set of the harmonic subspace U.

Proposition 6.2.1. *Let u be a hyperharmonic function on X and finite on a dense subset. Then $\{x \in X \mid u(x) = \infty\}$ is polar.*

For any relatively compact, resolutive set V, any $f \in \mathscr{C}(\partial V)$ with $f \le u$ on ∂V, and any strictly positive real number ε, we have

$$ {}^{V}\hat{R}_\infty^{A \cap V} \le \varepsilon (u - \mu^V f) $$

on V, where $A := \{x \in X \mid u(x) = \infty\}$. Since ε is arbitrary and u is finite on a dense subset, it follows that

$$ {}^{V}\hat{R}_\infty^{A \cap V} = 0. \quad \square $$

Proposition 6.2.2. *A K-analytic set of X is polar if all of its compact subsets are polar* (H. Bauer 1965 [7]).

Let A be a K-analytic set of X whose compact subsets are polar. Let W be an open, σ-compact set on which there exists a finite, strictly positive, continuous potential p. By Proposition 5.2.4 $A \cap W$ is a K-analytic set and by Corollary 5.2.1 $A \cap W$ is capacitable for any capacity. We obtain

$$ \hat{R}_p^{A \cap W} = \sup_K \hat{R}_p^K = 0, $$

where K runs through the set of compact subsets of $A \cap W$ (Theorem 5.3.1 *b*)). $\square$

Proposition 6.2.3. *The fine closure of a polar set is polar and nowhere dense in the fine topology. Every polar set of a harmonic space with countable base is fine closed.*

Let A be a polar set and $\mathfrak{W}$ be the open covering of X mentioned in the definition of polar sets. For any $W \in \mathfrak{W}$, the set

$$\{x \in W \mid {}^W R_\infty^{A \cap W}(x) = \infty\},$$

is a fine closed subset of W containing A. It is nowhere dense since ${}^W\hat{R}_\infty^{A \cap W} = 0$.

The last assertion follows immediately from Corollary 5.3.2. □

Remark. Not every polar set is fine closed (Exercise 6.2.15).

Proposition 6.2.4. *Let A be a polar set of a harmonic space X. Then*

$$\hat{R}_p^A = 0$$

for every potential p on X.

Assume first that $\bar{A}$ is contained in an open set W of X such that

$${}^W\hat{R}_\infty^A = 0.$$

Let u be a positive, hyperharmonic function on X, which is greater than p on A, let v be a positive, hyperharmonic function on X which is greater than u on $X \setminus W$, and let w be a positive, hyperharmonic function on W which is equal to ∞ on A. Then the function on X which is equal to u on $X \setminus W$ and $\inf(u, v+w)$ on W is hyperharmonic (Proposition 2.1.2). Since it is greater than p on A, we obtain

$$\hat{R}_p^A \leq v + w$$

on W. But w is arbitrary and thus

$$\hat{R}_p^A \leq v$$

on X and since v is arbitrary

$$\hat{R}_p^A \leq \hat{R}_u^{X \setminus W} \leq u, \qquad \hat{R}_p^A \leq \bigwedge_u \hat{R}_u^{X \setminus W} \leq \bigwedge_u u = \hat{R}_p^A.$$

Now the set $\{\hat{R}_u^{X \setminus W}|_W \mid u\}$ is a Perron set and therefore it follows that $\hat{R}_p^A$ is harmonic on W. Since it is harmonic on $X \setminus \bar{A}$ (Proposition 5.3.1), it is harmonic on X. Being a minorant of p, it vanishes identically and we get

$$\hat{R}_p^A = 0.$$

Let us now consider the general case. Let K be a compact set of X. Then there exists a finite family $(W_\iota)_{\iota \in I}$ of open sets of X and a finite family $(K_\iota)_{\iota\in I}$ of compact sets such that for any $\iota\in I$

$$K_\iota\subset W_\iota,\qquad K=\bigcup_{\iota\in I}K_\iota,\qquad {}^{W_\iota}\hat{R}_\infty^{A\cap W_\iota}=0.$$

By the above considerations

$$\hat{R}_p^{A\cap K}\leq\sum_{\iota\in I}\hat{R}_p^{A\cap K_\iota}=0.$$

Hence

$$\hat{R}_p^A\leq\hat{R}_p^{A\cap K}+\hat{R}_p^{A\setminus K}=\hat{R}_p^{A\setminus K},\qquad \hat{R}_p^A=\hat{R}_p^{A\setminus K}.$$

It follows that $\hat{R}_p^A$ is harmonic on the interior of K. Since K is arbitrary, $\hat{R}_p^A$ is harmonic on X, and since it is dominated by p, it vanishes identically. □

Corollary 6.2.1. *If A, B are subsets of a harmonic space X such that $(A-B)\cup(B-A)$ is a polar set, then*

$$\hat{R}_p^A=\hat{R}_p^B$$

for any potential p on X. □

Corollary 6.2.2. *Let X be a $\mathfrak{P}$-harmonic space. If A is a polar set of X contained in a σ-compact set, then*

$$\hat{R}_\infty^A=0.$$

Let $(K_n)_{n\in\mathbb{N}}$ be an increasing sequence of compact sets such that

$$A\subset\bigcup_{n\in\mathbb{N}}K_n.$$

Then there exists an increasing sequence $(p_n)_{n\in\mathbb{N}}$ of potentials on X, such that for any $n\in\mathbb{N}$, p_n is strictly positive on K_n. Thus

$$\hat{R}_\infty^A=\lim_{n\to\infty}\hat{R}_{np_n}^A=0$$

(Corollary 4.2.1). □

Remark. The assertion is no longer true without the hypothesis that A is contained in a σ-compact set (Exercise 5.3.5 *e*)). □

Corollary 6.2.3. *The union of a countable family of polar sets is polar.*

Let $(A_n)_{n\in\mathbb{N}}$ be a sequence of polar sets and let $A:=\bigcup_{n\in\mathbb{N}}A_n$. For any σ-compact $\mathfrak{P}$-set W of X, we have

$${}^W\hat{R}_\infty^{A\cap W}=\lim_{m\to\infty}{}^W\hat{R}_\infty^{(\bigcup_{n\leq m}A_n)\cap W}\leq\lim_{m\to\infty}\sum_{n\leq m}{}^W\hat{R}_\infty^{A_n\cap W}=0$$

(Corollary 4.2.2). The corollary follows now from Theorem 2.3.3. □

Corollary 6.2.4. *Let X be a $\mathfrak{P}$-harmonic space and A be a polar set of X. Then, for any relatively compact, open set V of X and any $x\in V$, we have*

$$\mu_x^V(A)=0.$$

Let $\mathscr{V}$ be the set of positive, hyperharmonic functions on X which are equal to ∞ on $A\cap\partial V$. For any $v\in\mathscr{V}$, we have

$$\mu_x^V(A)\leq v(x).$$

By the proposition

$$(\wedge\mathscr{V})(x)=0$$

and therefore

$$\mu_x^V(A)=0$$

(Proposition 1.1.4). □

Proposition 6.2.5. *If A is a polar set of a connected harmonic space X, then $X\setminus A$ is connected* (M. Brelot 1941 [4], 1959 [14], H. Bauer 1965 [7]).

Assume the contrary. Then there exist two open non-empty sets U, V of X such that

$$X\setminus A\subset U\cup V,\qquad (U\setminus A)\cap(V\setminus A)=\emptyset.$$

Obviously $U\cap V=\emptyset$ (Proposition 6.2.3) and therefore $\overline{U}\cap\overline{V}\subset A$.

Now assume that X is a σ-compact $\mathfrak{P}$-harmonic space and let $\mathscr{V}$ be the set of positive hyperharmonic functions on X which are equal to ∞ on A. For any $v\in\mathscr{V}$, we denote by v^* the function on X which is equal to v on U and ∞ elsewhere. Since $\partial U\subset A$, v^* is hyperharmonic (Proposition 2.1.2). We have

$$\bigwedge_{v\in\mathscr{V}} v^*=\hat{R}_\infty^A=0$$

on U (Corollary 6.2.2). Since $\bigwedge_{v\in\mathscr{V}} v^*$ is obviously ∞ on $X\setminus\overline{U}$, we deduce that $\overline{U}$ is absorbent. Similarly $\overline{V}$ is absorbent. Since $\overline{U}\cap\overline{V}$ is fine open and A is nowhere dense (Proposition 6.2.3) we get $\overline{U}\cap\overline{V}=\emptyset$, which is a contradiction, since X is connected.

Now let X be a general harmonic space, and choose $x\in\overline{U}\cap\overline{V}$. Let W be a connected, σ-compact $\mathfrak{P}$-set containing x. By the above considerations, $W\setminus A$ is connected and is therefore contained either in $U\setminus A$ or in $V\setminus A$. If $W\setminus A\subset U\setminus A$ then $W\cap V=\emptyset$ (Proposition 6.2.3) which implies the contradictory relation

$$x\in W\cap\overline{V}=\emptyset. \quad \square$$

Theorem 6.2.1. *Let F be a closed, polar set of X and u be a hyperharmonic function on $X\setminus F$. If the function u^* on X, which is equal to u on $X\setminus F$ and*

$$y\mapsto \liminf_{x\to y} u(x)$$

on F is lower finite, then it is hyperharmonic (M. Brelot 1941 [4], 1959 [14]).

Let V be a relatively compact, resolutive, σ-compact $\mathfrak{P}$-set and $f \in \mathscr{C}(\partial V)$, $f \leq u^*$ on ∂V. Let v be a positive, hyperharmonic function on V which is equal to ∞ on $F \cap V$. The function on V which is equal to $u+v$ on $V \setminus F$ and ∞ on $F \cap V$ is hyperharmonic (Proposition 2.1.2). It obviously belongs to $\overline{\mathscr{U}}_f^V$, and we thus have

$$\mu^V f \leq u+v,$$

on $V \setminus F$. Since v and f are arbitrary, we deduce that

$$\mu^V u^* \leq u$$

on $V \setminus F$ (Corollary 6.2.2). It follows immediately that

$$\mu^V u^* \leq u^*$$

on V. Since V is arbitrary, u^* must be hyperharmonic (Corollary 2.3.4). □

Corollary 6.2.5. *Let F be a closed, polar set of X and h a harmonic function on $X \setminus F$ such that*

$$\limsup_{x \to y} |h(x)| < \infty$$

at any point y of F. Then h is (uniquely) extendable to a harmonic function on X (G. Bouligand 1926 [1]).

Let $\underline{h}$ and $\bar{h}$ be the functions on X which are equal to h on $X \setminus F$ and equal to

$$y \mapsto \liminf_{x \to y} h(x), \qquad y \mapsto \limsup_{x \to y} h(x)$$

respectively on F. Let V be a relatively compact subset of a $\mathfrak{P}$-set. By the proposition

$$\mu^V \underline{h} \leq \underline{h} \leq \bar{h} \leq \mu^V \bar{h}$$

on V. Since $\underline{h} = \bar{h}$ on $\partial V \setminus F$ we have

$$\mu^V \underline{h} = \mu^V \bar{h}$$

on V (Corollary 6.2.4). Hence $\underline{h} = \bar{h}$ on V and they are therefore harmonic (Theorem 2.3.3). □

Remark. H. A. Schwarz 1872 [2] proved this corollary in classical potential theory for the case when F consists of a single point.

Theorem 6.2.2. *Let X be a harmonic space. If every point of X is polar, then X endowed with the sheaf of harmonic functions is a Bauer space.*

Let U be a $\mathfrak{P}$-set of X on which there exists a strictly positive, harmonic function h and let x, y be points of U, $x \neq y$. Since $\{y\}$ is a polar set, there exists a positive, hyperharmonic function u on U, infinite at y and finite at x (Corollary 6.2.2 and Corollary 5.3.2). Hence

$$\frac{u(x)}{h(x)} < \frac{u(y)}{h(y)}.$$

It follows that U endowed with the sheaf of harmonic functions has a base of regular sets (Corollary 3.1.3). The theorem follows now from Corollary 3.1.2. □

Exercises

6.2.1. If A is a polar set of a $\mathfrak{P}$-harmonic space with a countable base, then there exists a potential on the space which is ∞ on A (M. Brelot 1958 [12]). (Use Exercise 5.1.10.) If A is of type G_δ, then there exists a potential on X which is ∞ on A and finite on $X \setminus A$.

6.2.2. Let U be an open, K_σ-set of X and let A be a subset of U. If A is a polar set of the harmonic subspace U, then A is a polar set of X (Use Theorem 2.3.2). The assertion is no longer true if U is not a K_σ-set (Use Exercise 5.3.5 *e*)).

6.2.3. Let $A \mapsto \hat{R}_p^A$ be the capacity introduced in Theorem 5.3.1 *b*). If A is a polar set of X, then it is of capacity 0. If X is a $\mathfrak{P}$-harmonic space and A is of capacity 0 for every such capacity, then A is a polar set.

6.2.4. Let F be a closed, nowhere dense set of a harmonic space and let U be an MP-neighbourhood of F. If $\overline{H}_f^{U \setminus F} = 0$, where f is the characteristic function of F, then F is a polar set (H. Bauer 1962 [5]). The hypothesis that F is nowhere dense may be dropped if the space is elliptic, but it is necessary for a general harmonic space (Use Exercise 3.2.13).

6.2.5. If F is a closed, non-polar set of a connected, elliptic $\mathfrak{S}$-harmonic space X then $X \setminus F$ is a $\mathfrak{P}$-harmonic space (P. J. Myrberg 1933 [1], C. Constantinescu-A. Cornea 1963 [1]). (Let U be a component of $X \setminus F$. Then ∂U is non-polar and there exists a connected quasi-regular MP-set V of X such that $V \cap \partial U$ is non-polar. Let u be a strictly positive, superharmonic function on X and let f (resp. g) be the function on $\partial(U \cap V)$ which is equal to u (resp. 0) on $U \cap \partial V$ and 0 (resp. u) elsewhere. Then

$$H_f^{U \cap V} + H_g^{U \cap} V = H_u^{U \cap V} \leq u.$$

Since X is elliptic, V connected and $V \setminus U$ non-polar, ${}^V R_u^{V \setminus U}$ is strictly positive. Since $H_g^{U \cap V}$ is equal to ${}^V R_u^{V \setminus U}$ on $U \cap V$ we get

$$H_f^{U \cap V} < u$$

on $U \cap V$. Choose $v \in \overline{\mathscr{U}}_f^{U \cap V}$ such that

$$v < u$$

at a point of $U \cap V$. The function on U which is equal to u on $U \setminus V$ and $\inf(u, v)$ on $U \cap V$ is a positive, superharmonic function on U, not proportional to u. Hence U is a $\mathfrak{P}$-set (Exercise 3.1.10).)

6.2.6. Let X be a connected, elliptic $\mathfrak{S}$-harmonic space. Then there exists a sequence $(K_n)_{n \in \mathbb{N}}$ of compact sets of X such that $X \setminus \bigcup_{n \in \mathbb{N}} K_n$ is polar (C. Constantinescu-A. Cornea 1963 [1]).

6.2.7. A connected, elliptic $\mathfrak{S}$-harmonic space which possesses locally a countable base has a countable base (C. Constantinescu-A. Cornea 1963 [1]).

6.2.8. The set of non-relatively compact components of an open set of a connected, elliptic $\mathfrak{S}$-harmonic space is at most countable (C. Constantinescu-A. Cornea 1963 [1]).

6.2.9. A connected $\mathfrak{S}$-Brelot space is σ-compact (A. Cornea 1967 [1]).

6.2.10. If F is an absorbent set of X and A a polar set contained in F, then A is a polar set of the harmonic space $(F, \mathscr{U}_F)$ constructed in Exercise 6.1.8. The converse is not true. (Use Exercise 3.2.13.)

6.2.11. In the example of Theorem 2.1.2, the empty set is the only polar set. A point of $\mathbb{R}^n$ ($n \in \mathbb{N}$, $n \neq 0$) is polar for the Laplace equation if and only if $n > 1$. Any point is polar for the heat equation.

6.2.12. Any point of the example of Exercise 3.2.8 is polar if $n > 1$ and the function (a_{ij}) are Dini continuous at the point (N. Boboc-P. Mustaţă 1967 [2]).

6.2.13. A subset of the harmonic space defined in Exercise 3.2.7 is polar if and only if it is polar for the Laplace equation (R.-M. Hervé 1962 [4], N. Boboc-P. Mustaţă 1967 [2], 1968 [5]).

6.2.14. A family $(A_\iota)_{\iota \in I}$ of subsets of X is called *evanescent* if there exists a covering $\mathfrak{W}$ of X, with open $\mathfrak{P}$-sets, such that for any $W \in \mathfrak{W}$ and any locally bounded potential p on W,

$$\bigwedge_{\iota \in I} {}^W\hat{R}_p^{A_\iota \cap W} = 0.$$

a) If the family $(A_\iota)_{\iota \in I}$ is evanescent, then for any locally bounded potential p on X, we have

$$\bigwedge_{\iota \in I} \hat{R}_p^{A_\iota} = 0.$$

b) Let $((A^n_\iota)_{\iota\in I_n})_{n\in\mathbb{N}}$ be a sequence of evanescent families and let $I=\prod_{n\in\mathbb{N}} I_n$. For any $\iota:=(\iota_n)_{n\in\mathbb{N}}$ of I we set

$$A_\iota := \bigcup_{n\in\mathbb{N}} A^n_{\iota_n}.$$

Then $(A_\iota)_{\iota\in I}$ is a evanescent family (use Proposition 4.1.3 and Theorem 5.1.1).

6.2.15. The set A from Exercise 5.3.5 *e)* is polar but not fine closed for X''.

6.2.16. Let F be a closed, nowhere dense set of a $\mathfrak{P}$-harmonic space. If every positive harmonic function h on $X\setminus F$ which satisfies

$$\limsup_{x\to y} h(x) < \infty$$

for any $y\in F$, is extendable to a harmonic function on X, then F is a polar set. If X is elliptic, the same assertion holds for compact F ($F\neq X$) without the condition that F is nowhere dense (O.D. Kellogg 1929 [2]). (Let p be a finite, continuous potential on X. Then there exists a harmonic function h on X which is equal to $\hat{R}^F_p$ on $X\setminus F$; h vanishes since it is dominated by p outside a compact set.)

§ 6.3. Thinness and Semi-Polar Sets

A subset A of a harmonic space X is called *thin at a point* $x\in X$ if there exist two open neighbourhoods U, V of x, $V\subset U$, and a positive, hyperharmonic function u on U such that

$${}^U\hat{R}^{A\cap V}_u(x) < u(x).$$

(This notion was introduced in classical potential theory by M. Brelot 1939 [2].) We remark that in this case, if U', V' are any two open neighbourhoods of x such that $V'\subset U'\subset U$ and $V'\subset V$, we have that

$${}^{U'}\hat{R}^{A\cap V'}_{u|_{U'}}(x) < u(x).$$

If $x\notin \bar{A}$, then A is obviously thin at x. A fine open set is not thin at any of its points. If A is thin at a point x, then any subset of A is thin at x. A subset of a harmonic space is called *totally thin* if it is thin at every point of the space. Any subset of a totally thin set is totally thin, and any polar set is obviously totally thin. There exists a Brelot space X and $x\in X$ such that $\{x\}$ is totally thin but not polar (Exercise 6.3.10). A subset of a harmonic space is called *semi-polar* if it is the union of a countable family of totally thin sets. The notion of a semi-polar set was introduced by M. Brelot 1962 [18]. It turned out to be useful (by replacing the notion of polar set) in generalising some results from classical potential theory, where semi-polar and polar sets coincide.

Proposition 6.3.1. *Let A, B be subsets of a harmonic space X, u a positive hyperharmonic function on X, and x a point of X such that*

$$\hat{R}_u^A(x) < u(x) \quad \textit{and} \quad \hat{R}_u^B(x) < u(x).$$

Then there exists a positive, hyperharmonic function v on X, such that

$$\hat{R}_v^{A \cup B}(x) < v(x).$$

We set

$$w := \hat{R}_u^A \wedge \hat{R}_u^B.$$

Let u', v' be positive, hyperharmonic functions on X which are greater than u on A, B respectively. Then we have

$$u + w \leq u' + v'$$

on $A \cup B$. Hence

$$\hat{R}_u^{A \cup B} + \hat{R}_w^{A \cup B} \leq u' + v'$$

(Theorem 4.2.1). But u', v' are arbitrary and thus we obtain

$$\hat{R}_u^{A \cup B}(x) + \hat{R}_w^{A \cup B}(x) \leq \hat{R}_u^A(x) + \hat{R}_u^B(x)$$

(Corollary 4.1.1).

Assume that

$$\hat{R}_u^{A \cup B}(x) = u(x) \quad \text{and} \quad \hat{R}_w^{A \cup B}(x) = w(x).$$

This leads to the contradictory relation

$$\begin{aligned} u(x) + w(x) &\leq \hat{R}_u^A(x) + \hat{R}_u^B(x) \\ &= \sup(\hat{R}_u^A(x), \hat{R}_u^B(x)) + \inf(\hat{R}_u^A(x), \hat{R}_u^B(x)) < u(x) + w(x). \quad \square \end{aligned}$$

Proposition 6.3.2. *Let A be a subset of a $\mathfrak{P}$-harmonic space X and let $x \in A$. The following assertions are equivalent:*

a) A is thin at x;

b) for any positive, hyperharmonic function u, on X, which is finite, continuous and strictly positive at x, there exists a neighbourhood U of x such that

$$\hat{R}_u^{A \cap U}(x) < u(x);$$

c) there exists a finite, continuous potential p on X, harmonic outside a compact set and such that

$$\hat{R}_p^A(x) < p(x).$$

$a \Rightarrow b$. By the definition of thin sets there exist two open neighbourhoods U', V' of x, $V' \subset U'$, and a positive, hyperharmonic function

u' on U' such that

$${}^{U'}\hat{R}_{u'}^{A\cap V'}(x) < u'(x).$$

Let V be an open, relatively compact (and therefore, resolutive) set of X such that

$$x \in V \subset U'.$$

By Proposition 2.3.2, there exists a positive, finite, continuous superharmonic function v on X such that

$$\mu^V v(x) < v(x).$$

Let α be a positive real number such that

$$\alpha\, {}^{U'}\hat{R}_{u'}^{A\cap V'}(x) < v(x) - \mu^V v(x) < \alpha\, u'(x)$$

and let W be a neighbourhood of x contained in $V \cap V'$ such that

$$v - \mu^V v < \alpha\, u'$$

on W. Let v' be a positive, hyperharmonic function on U' which is greater than u' on $A \cap V'$. The function on X which is equal to v on $X \smallsetminus \bar{V}$, $\inf(v, \mu^V v + \alpha v')$ on V, and

$$y \mapsto \inf\big(v(y), \liminf_{x\to y} \mu^V v(x) + \alpha v'(x)\big)$$

on ∂V is hyperharmonic (Proposition 2.1.3) and greater than v on $A \cap W$. Hence

$$\hat{R}_v^{A\cap W} \leq \mu^V v + \alpha v'$$

on V. Since v' is arbitrary, we obtain

$$\hat{R}_v^{A\cap W}(x) \leq \mu^V v(x) + \alpha\, {}^{U'}\hat{R}_{u'}^{A\cap V'}(x) < v(x).$$

Let β be a positive real number such that

$$\beta\, \hat{R}_v^{A\cap W}(x) < u(x) < \beta\, v(x)$$

and let U be a neighbourhood of x contained in W such that $u < \beta v$ on U. Then

$$\hat{R}_u^{A\cap U}(x) < \beta\, \hat{R}_v^{A\cap U}(x) \leq \beta\, \hat{R}_v^{A\cap W}(x) < u(x).$$

$b \Rightarrow c$. Let u be a positive, hyperharmonic function on X which is finite, continuous and strictly positive at x and let U be a relatively compact, open neighbourhood of x such that

$$\hat{R}_u^{A\cap U}(x) < u(x).$$

By Proposition 2.3.2, there exists a finite, continuous, positive, superharmonic function v on X such that

$$\mu^U v(x) < v(x).$$

With the aid of Proposition 5.3.3, we get

$$\hat{R}_v^{A \setminus U}(x) \leq \hat{R}_v^{X \setminus U}(x) = \mu^U v(x) < v(x).$$

Hence

$$\hat{R}_{u+v}^{A \cap U}(x) < (u+v)(x), \qquad \hat{R}_{u+v}^{A \setminus U}(x) < (u+v)(x).$$

By Proposition 6.3.1, there exists a positive, hyperharmonic function w on X such that

$$\hat{R}_w^A(x) < w(x).$$

Let p be a finite, continuous potential on X, harmonic outside a compact set, less than w, and such that

$$\hat{R}_w^A(x) < p(x)$$

(Corollary 2.3.1). We deduce that

$$\hat{R}_p^A(x) \leq \hat{R}_w^A(x) < p(x).$$

$c \Rightarrow a$ is trivial. □

Theorem 6.3.1. *The union of a finite family of sets, each of which are thin at a point x, is thin at x.*

Let A, B be two thin sets at x, let V be a $\mathfrak{P}$-set containing x and let u be a positive, hyperharmonic function on V which is finite, continuous and strictly positive at x. By the preceding proposition there exists a neighbourhood U of x such that

$${}^V\hat{R}_u^{A \cap U}(x) < u(x), \qquad {}^V\hat{R}_u^{B \cap U}(x) < u(x).$$

By Proposition 6.3.1 there exists a positive, hyperharmonic function v on V such that

$${}^V\hat{R}_v^{(A \cup B) \cap U}(x) < v(x).$$

Hence $A \cup B$ is thin at x. The general case follows now by induction. □

Corollary 6.3.1. *Let A, B be two subsets of a harmonic space such that $(A \setminus B) \cup (B \setminus A)$ is totally thin. Then A and B are simultaneously thin at a given point.* □

Corollary 6.3.2. *Let x be a point of a harmonic space X such that $\{x\}$ is thin at x. Then a subset A of X is thin at x if and only if $A \setminus \{x\}$ is thin at x. In particular this is true when $\{x\}$ is polar.* □

Proposition 6.3.3. *Let A be a subset of X, B be the fine closure of A and let $x \in X$.*

a) If A is thin at x, then B is thin at x.

b) If x does belong to B, then A is thin at x.

c) If A is thin at x, $x \in X \setminus A$ and $\{x\}$ is of type G_δ, then $x \in X \setminus B$.

a) Let U, V be open neighbourhoods of x, $V \subset U$, and let u be a positive, hyperharmonic function on U such that

$$ {}^U\hat{R}_u^{A \cap V}(x) < u(x). $$

Let v be a positive, hyperharmonic function on U which is greater than u on $A \cap V$. Then v is greater than u on $B \cap V$. We deduce that

$$ {}^U\hat{R}_u^{B \cap V}(x) = {}^U\hat{R}_u^{A \cap V} < u(x). $$

b) We may assume $x \in \bar{A}$. Then there exists a hyperharmonic function u defined on an open neighbourhood of x such that

$$ u(x) < \liminf_{A \ni y \to x} u(y) $$

(Proposition 5.1.1). Let U be an open, relatively compact neighbourhood of x such that u is defined on $\bar{U}$ and such that there exists a strictly positive, harmonic function h defined on an open neighbourhood of $\bar{U}$. Let α, β be positive real numbers such that $u + \alpha h$ is positive on U and

$$ u(x) + \alpha h(x) < \beta h(x) < \liminf_{A \ni y \to x} \big(u(y) + \alpha h(y)\big). $$

Let V be a neighbourhood of x with $V \subset U$ and such that

$$ \beta h < u + \alpha h $$

on $A \cap V$. Then

$$ {}^U\hat{R}_{\beta h}^{A \cap V}(x) < u(x) + \alpha h(x) < \beta h(x) $$

and A is thin at x.

c) By Proposition 6.3.2, there exist two open neighbourhoods U, V of x, $V \subset U$, and a positive, superharmonic function u on U such that

$$ {}^U\hat{R}_u^{A \cap V}(x) < u(x). $$

From Corollary 5.3.2

$$ {}^U R_u^{A \cap V}(x) = {}^U\hat{R}_u^{A \cap V}(x). $$

Hence there exists a hyperharmonic function v on U which is greater than u on $A \cap V$ and less than u at x; we obtain

$$ v(x) < u(x) \leq \liminf_{A \ni y \to x} u(y) \leq \liminf_{A \ni y \to x} v(y). \quad \square $$

Remark. The assertion *c)* of the proposition is not true if $\{x\}$ is not of type G_δ (Exercise 6.2.15).

Corollary 6.3.3. *The fine closure of a totally thin set is totally thin and fine nowhere dense. A semi-polar set is of the first category in the fine topology.* $\square$

Corollary 6.3.4. *Two hyperharmonic functions on a harmonic space coincide if they coincide outside a semi-polar set.*

The assertion follows from the preceding corollary since the fine topology is a Baire topology (Corollary 5.1.1). □

Corollary 6.3.5. *A totally thin set on a harmonic space with countable base is fine closed.* □

Theorem 6.3.2. *Let $\mathscr{V}$ be a naturally lower bounded set of hyperharmonic functions on a harmonic space. Then*

$$\{x\in X \mid (\wedge\mathscr{V})(x) < \inf_{u\in\mathscr{V}} u(x)\}$$

is a semi-polar set (M. Brelot 1962 [18], H. Bauer 1965 [7]).

We set

$$f := \inf_{u\in\mathscr{V}} u.$$

Then

$$\hat{f} = \wedge\mathscr{V}$$

(Proposition 5.1.4 and Theorem 5.1.1). For any $n\in\mathbb{N}$, let

$$A_n := \left\{x\in X \mid \hat{f}(x) < \inf\left(n, f(x) - \frac{1}{n}\right)\right\}.$$

Obviously

$$\{x\in X \mid (\wedge\mathscr{V})(x) < \inf_{u\in\mathscr{V}} u(x)\} = \bigcup_{n\in\mathbb{N}} A_n.$$

It is therefore sufficient to show that for any $n\in\mathbb{N}$, A_n is totally thin.

Let $x_0\in X$. If $\hat{f}(x_0) > n$, then the set $\{x\in X \mid \hat{f}(x) > n\}$ is a neighbourhood of x_0 which is contained in $X\setminus A_n$. Hence A_n is thin at x_0. Now assume $\hat{f}(x_0)\le n$. Let U be an open, relatively compact neighbourhood of x_0 such that there exists a strictly positive, harmonic function h defined on an open neighbourhood of $\bar{U}$. We may suppose $h(x_0)=1$. Let α be a positive real number such that $\hat{f}|_U + \alpha h|_U$ is positive and let V be an open neighbourhood of x_0 such that

$$\hat{f} > \left(\hat{f}(x_0) - \frac{1}{4n}\right) h, \qquad h < 2$$

on V. Then

$$f + \alpha h > \hat{f} + \frac{1}{2n} h + \alpha h > \left(\hat{f}(x_0) + \frac{1}{4n} + \alpha\right) h$$

on $A_n\cap V$. If we set

$$\beta := \hat{f}(x_0) + \frac{1}{4n} + \alpha$$

then

$${}^U\hat{R}_{\beta h}^{A_n\cap V}(x_0) \le \hat{f}(x_0) + \alpha < \beta = \beta h(x_0). \quad \square$$

In classical potential theory, E. Szpilrajn and T. Rado have shown that the set of the above theorem is of Lebesgue measure zero. M. Brelot 1938 [1] improved this result by proving that all of its compact subsets are polar and H. Cartan 1942 [1] showed that the set itself is polar. In harmonic spaces however, it is no longer true that this exceptional set is polar. In Chapter IX this problem will be discussed in detail.

Corollary 6.3.6. *Let A be a subset of a harmonic space and u be a positive, hyperharmonic function. Then the set*

$$\{x \in A \mid \hat{R}_u^A(x) < u(x)\}$$

is semi-polar (H. Bauer 1965 [7]). □

Theorem 6.3.3. *Let U be an open set of a $\mathfrak{P}$-harmonic space X. A boundary point, x, of U is a regular boundary point* (U is resolutive as an open set of a $\mathfrak{P}$-harmonic space (Theorem 2.4.2)) *if and only if $X \setminus U$ is not thin at x* (M. Brelot 1939 [2], 1960 [15], 1962 [18], M. Brelot-R.-M. Hervé 1958 [1], H. Bauer 1963 [6], 1966 [8].

Assume first that $X \setminus U$ is thin at x. Then by Proposition 6.3.2 there exists a finite, continuous potential p on X such that

$$\hat{R}_p^{X \setminus U}(x) < p(x).$$

Let $f \in \mathscr{K}(\partial U)$ with $f \leq p$ on ∂U and $f(x) = p(x)$. Then, since $R_p^{X \setminus U}$ is equal to p on $X \setminus U$ and equal to $\overline{H}_p^U$ on U (Proposition 5.3.3),

$$\liminf_{y \to x} H_f^U(y) \leq \liminf_{y \to x} \overline{H}_p^U = \hat{R}_p^{X \setminus U}(x) < p(x) = f(x).$$

Hence x is not a regular boundary point of U.

Assume now that $X \setminus U$ is not thin at x and let $f \in \mathscr{K}(\partial U)$. Let p_0 be a finite, continuous potential on X, strictly positive on the carrier of f and let ε be a strictly positive real number. Then there exist two finite, continuous potentials p, q on X such that

$$|p - q - f| < \varepsilon p_0$$

on ∂U and $p - q \in \mathscr{K}(X)$ (Theorem 2.3.1). We have

$$\overline{H}_{|p-q-f|}^U < \varepsilon p_0$$

on U. By Proposition 2.4.3, p and q are resolutive and by Proposition 5.3.3

$$H_p^U = \hat{R}_p^U, \quad H_q^U = \hat{R}_q^U$$

on U. Since $X \setminus U$ is not thin at x, we obtain

$$\lim_{y \to x} H_p^U(y) = \lim_{U \ni y \to x} \hat{R}_p^U(y) = p(x),$$

$$\lim_{y \to x} H_q^U(y) = \lim_{U \ni y \to x} \hat{R}_q^U(y) = q(x).$$

Hence

$$\limsup_{y \to x} H_f^U(y) \le p(x) - q(x) + \varepsilon\, p_0(x) \le f(x) + 2\varepsilon\, p_0(x),$$

$$\liminf_{y \to x} H_f^U(y) \ge p(x) - q(x) - \varepsilon\, p_0(x) \ge f(x) - 2\varepsilon\, p_0(x).$$

Now H_f^U converges to $f(x)$ at x since ε is arbitrary and x is a regular boundary point of U since f is arbitrary. □

Remark. The "if" part of this theorem holds also on $\mathfrak{S}$-harmonic spaces (Exercise 6.3.4). However, the "only if" part is not true even if U is a $\mathfrak{P}$-set and X is an $\mathfrak{S}$-Brelot space (Exercise 6.3.10).

Corollary 6.3.7. *Let U, V be open sets of a $\mathfrak{P}$-harmonic space such that $U \subset V$ and let $x \in \partial U \cap \partial V$. If x is a regular boundary point of V, then X is a regular boundary point of U. Conversely, if x is a regular boundary point of U and if there exists a neighbourhood W of x such that $U \cap W = V \cap W$, then x is a regular boundary point of V* (M. Brelot 1960 [15], H. Bauer 1962 [5]). □

Corollary 6.3.8. *The intersection of two regular sets of a $\mathfrak{P}$-harmonic space is regular* (H. Bauer 1962 [5]). □

Corollary 6.3.9. *If F is an absorbent set of a $\mathfrak{P}$-harmonic space X, then any boundary point of $X \setminus F$ is regular.*

Let x be a boundary point of $X \setminus F$. Then $x \in F$ and F is not thin at x since it is fine open. □

Exercises

6.3.1. An absorbent semi-polar set is empty.

6.3.2. A subset of a harmonic space X is called *subbasic* if it is not thin at any of its points. If A is subbasic, then for any positive, hyperharmonic function u on X we have $\hat{R}_u^A = R_u^A$. Hence if $(u_\iota)_{\iota \in I}$ (resp. $(A_\kappa)_{\kappa \in J}$) is an upper directed family of positive hyperharmonic functions on X (resp. of subbasic sets of X), then

$$R_u^A = \sup_{\substack{\iota \in I \\ \kappa \in J}} R_{u_\iota}^{A_\kappa}$$

where

$$u := \bigvee_{\iota \in I} u_\iota, \qquad A := \bigcup_{\kappa \in J} A_\kappa$$

(M. Brelot 1967 [21]).

6.3.3. Let A be a subset of X and u be a positive, hyperharmonic function on X. If $\hat{R}_u^A = u$ on $X \setminus A$, then $\hat{R}_u^A = u$ everywhere (use Corollary 6.3.6 and Corollary 6.3.4).

6.3.4. Let U be an open set of a harmonic space X and x be a boundary point of U such that $X \setminus U$ is not thin at x. Then for any numerical function f on ∂U such that

$$\limsup_{y \to x} \overline{H}_f^U(y) < \infty,$$

we have

$$\limsup_{y \to x} \overline{H}_f^U(y) \leq \limsup_{y \to x} f(y).$$

Hence if U is resolutive and X is an $\mathfrak{S}$-harmonic space, then x is a regular boundary point (use Exercise 6.1.6 and Theorem 6.3.3).

6.3.5. Assume that X is a $\mathfrak{P}$-harmonic space and let K be a compact set of X. The following assertions are equivalent:

a) $X \setminus K$ is not thin at any point of ∂K.

b) for any continuous real function f on ∂K and any strictly positive real number ε, there exists a harmonic function h, on an open neighbourhood of K, such that $|f - h| < \varepsilon$ on ∂K.

(This is the approximation theorem of Keldych-Brelot) (M. Brelot 1945 [7], 1966 [20], N. Boboc-A. Cornea 1967 [2].)

6.3.6. Assume that X is a $\mathfrak{P}$-harmonic space and let K be a compact set of X. If K is polar, then for any positive, continuous real function f on K and any strictly positive real number ε, there exists a finite, continuous potential p on X such that p is harmonic on a neighbourhood of K and $|f - p| < \varepsilon$ on K. Conversely, if this property holds and every point of K is polar, then K is semi-polar; moreover if X is the smallest absorbent set containing K, then K is polar (Use the preceding exercise) (H. Wallin 1963 [1]).

6.3.7. In the example of Theorem 2.1.2, a subset of X is semi-polar if and only if it is countable.

6.3.8. For any real number α, the set $\mathbb{R}^n \times \{\alpha\}$ is totally thin and not polar for the heat equation on $\mathbb{R}^n \times \mathbb{R}$. If $\mathbb{Q}$ denotes the set of rational numbers, then $\mathbb{R}^n \times \mathbb{Q}$ is a semi-polar, dense set in the fine topology.

6.3.9. Let f be a positive real function on $]0, \infty[$ and let

$$A := \{(x, y, z) \in \mathbb{R}^3 \mid x > 0, \ \sqrt{y^2 + z^2} < f(x)\}.$$

If

$$\limsup_{x\to 0} \frac{f(x)}{x} > 0$$

then A is not thin at $(0,0,0)$ for the Laplace equation in $\mathbb{R}^3$ (Use Proposition 6.3.3 *c)* and Proposition 5.1.3). There exists a strictly monotonous function f, of class $\mathscr{C}^\infty$, converging to 0 at 0 and such that the set

$$\{(x,y,z)\in\mathbb{R}^3 \mid x>0,\ \sqrt{y^2+z^2}\le f(x)\}$$

is thin at $(0,0,0)$ (The spine of Lebesgue) (H. Lebesgue 1913 [2]). (Denote by u, the function $(x,y,z)\mapsto \int_0^1 \frac{t\,dt}{((x-t)^2+y^2+z^2)^{\frac{1}{2}}}$ let $\alpha\in\mathbb{R}$, $\alpha>1$. Choose f so that $u(x,f(x),0)=\alpha$).

6.3.10. Let f be a function possessing the properties of the last proposition of the preceding exercise and let X be the Alexandroff compactification of the subspace Y of $\mathbb{R}^3$ where

$$Y := \{(x,y,z)\in\mathbb{R}^3 \mid 0<x^2+y^2+z^2<1\}\cap(\{(x,y,z)\subset\mathbb{R}^3 \mid x\le 0\}$$
$$\cup\{(x,y,z)\in\mathbb{R}^3 \mid x>0,\ \sqrt{y^2+z^2}>f(x)\}).$$

For any open set U of X, we denote by $\mathscr{H}(U)$ the set of real, continuous functions on U whose restrictions to $U\cap Y$ are solutions of the Laplace equation on $\mathbb{R}^3$. Prove that X endowed with $\mathscr{H}$ is a Brelot space on which the constant functions are harmonic; Y is a regular $\mathfrak{P}$-set of X, but $X\setminus Y$ is totally thin and not polar.

6.3.11. Any nonempty compact set of $\mathbb{R}^n\times\mathbb{R}$ possesses a point at which it is thin for the heat equation.

6.3.12. A subset of the harmonic space defined in Exercise 3.2.7 is thin at a point if and only if it is thin at a point for the harmonic space associated to the Laplace equation (R.-M. Hervé 1962 [4], 1965 [7], N. Boboc-P. Mustaţă 1967 [2], 1968 [5]).

6.3.13. Let F be an absorbent set of X and let A be a subset of F. Then A is thin at a point x of F if and only if it is thin at x in the harmonic space F defined in Exercise 6.1.8. Hence A is totally thin (resp. semi-polar) if and only if it is totally thin (resp. semi-polar) in the harmonic space F.

Chapter 7

Balayage of Measures

Throughout this chapter we shall denote by X a $\mathfrak{P}$-harmonic space.

§ 7.1. General Properties of the Balayage of Measures

Proposition 7.1.1. *Let p be a finite, continuous potential on X and $\mathcal{Q}_p$ be the set of finite, continuous potentials on X which are equal to p outside a compact set. Then $\mathcal{Q}_p$ is lower directed and its infimum is identically* 0.

If $p', p'' \in \mathcal{Q}_p$ then obviously $\inf(p', p'') \in \mathcal{Q}_p$. Hence $\mathcal{Q}_p$ is lower directed. We set

$$\mathcal{F} := \{ f \in \mathcal{K}(X) \mid 0 \leq f \leq 1 \}.$$

For any $f \in \mathcal{F}$ and for any $p' \in \mathcal{Q}_p$ the function $R((1-f)\,p')$ belongs to $\mathcal{Q}'_p$, is less than p' and is harmonic on the interior of the set $\{x \in X \mid f(x) = 1\}$ (Proposition 2.2.3). Hence the infimum of $\mathcal{Q}_p$ is harmonic. Since it is dominated by the potential p, it is identically 0. □

We denote by $\mathcal{P}_c$, the set of finite, continuous potentials on X which are harmonic outside a compact set of X and by Λ, the set of measures on X for which the potentials in $\mathcal{P}_c$ are integrable. Obviously Λ is a convex cone and any measure with compact carrier belongs to Λ. Moreover, any measure μ on X for which there exists a μ-integrable, strictly positive, hyperharmonic function u on X, also belongs to Λ. Indeed, let $p \in \mathcal{P}_c$ and let K be any compact set in X such that p is harmonic on $X \setminus K$. Then there exists an $\alpha \in \mathbb{R}_+$ such that $p < \alpha u$ on K. The function, $\sup(p - \alpha u, 0)$, is obviously a subharmonic minorant of p and therefore negative. Hence $p \leq \alpha u$ on X.

Proposition 7.1.2. *For any $\mu \in \Lambda$ and for any subset A of X there exists a unique measure, $\mu^A \in \Lambda$, such that for every finite, continuous potential p on X, we have*

$$\overset{*}{\int} p\, d\mu^A = \overset{*}{\int} \hat{R}_p^A\, d\mu.$$

We prove first the following lemma.

Lemma. *Let Y be a locally compact space and let $\mathscr{F}$ be a convex cone of continuous, positive, real functions on Y such that for any function f of $\mathscr{K}(Y)$, any strictly positive real number ε and any neighbourhood U of the carrier of f, there exist $f', f'' \in \mathscr{F}$ such that the carrier of $f'-f''$ lies in U and $|f-(f'-f'')|<\varepsilon$. Let φ be a positive, real function on $\mathscr{F}$ such that:*

α) $f, f' \in \mathscr{F} \Rightarrow \varphi(f+f') = \varphi(f)+\varphi(f')$;

β) $f, f' \in \mathscr{F}$, $f \le f' \Rightarrow \varphi(f) \le \varphi(f')$;

γ) for any $f \in \mathscr{F}$ we have

$$\inf_{f' \in \mathscr{F}_f} \varphi(f') = 0,$$

where $\mathscr{F}_f$ denotes the set of functions f', of $\mathscr{F}$, for which $\{y \in Y \mid f'(y) < f(y)\}$ is relatively compact. Then there exists a unique measure ν, on Y, such that for every $f \in \mathscr{F}$

$$\varphi(f) = \int f\, d\nu.$$

The unicity follows immediately from the properties of $\mathscr{F}$. By α) the function ψ

$$f - f' \mapsto \varphi(f) - \varphi(f')$$

is well defined on $\mathscr{K}(Y) \cap (\mathscr{F} - \mathscr{F}) := \{f - f' \in \mathscr{K}(Y) \mid f, f' \in \mathscr{F}\}$ and by α) and β) it is linear and it is positive on the positive functions. Let $f \in \mathscr{K}(X)$. By hypothesis, there exist a sequence $(f_n)_{n \in \mathbb{N}}$ in $\mathscr{K}(Y) \cap (\mathscr{F} - \mathscr{F})$ and a compact set K in Y such that for any $n \in \mathbb{N}$, the carrier of f_n lies in K and

$$\lim_{n \to \infty} \sup_Y |f - f_n| = 0.$$

There also exists $g \in \mathscr{F}$ such that $g \ge 1$ on K. For any $m, n \in \mathbb{N}$ we have

$$|\psi(f_m) - \psi(f_n)| \le (\sup_Y |f - f_m| + \sup_Y |f - f_n|)\, \varphi(g).$$

It follows that $(\psi(f_n))_{n \in \mathbb{N}}$ is a Cauchy sequence. Since the sequence $(f_n)_{n \in \mathbb{N}}$ is arbitrary, its limit depends only on f. We denote it by $\nu(f)$. It is immediate that ν is linear and positive on $\mathscr{K}(Y)$ and is therefore a measure on Y.

Let $f \in \mathscr{F} - \mathscr{F}$, $g \in \mathscr{K}(Y)$ with $g \leq f$. Choose $f_0 \in \mathscr{F}$ such that $f_0 \geq 1$ on the carrier of g and let ε be a strictly positive real number. Then there exists a sequence $(f_n)_{n \in \mathbb{N}}$ in $\mathscr{K}(Y) \cap (\mathscr{F} - \mathscr{F})$ such that

$$\nu(g) = \lim_{n \to \infty} \psi(f_n) \quad \text{and} \quad f_n \leq f + \varepsilon f_0$$

for any $n \in \mathbb{N}$; we obtain

$$\nu(g) \leq \psi(f) + \varepsilon \varphi(f_0).$$

Since g and ε are arbitrary,

$$\int f \, d\nu \leq \psi(f).$$

Now let $f' \in \mathscr{F}_f$. Since $\sup(f - f', 0) \in \mathscr{K}(Y)$ and $f - f' \leq \sup(f - f', 0)$ we deduce from the above considerations

$$\varphi(f) - \varphi(f') \leq \int \sup(f - f', 0) \, d\nu \leq \int f \, d\nu.$$

We get

$$\varphi(f) \leq \varphi(f') + \int f \, d\nu$$

and by $\gamma)$

$$\varphi(f) \leq \int f \, d\nu. \quad \square$$

We now prove the proposition. In the preceding lemma take Y equal to X and $\mathscr{F}$ equal to $\mathscr{P}_c$. By Theorem 2.3.1, $\mathscr{P}_c$ satisfies the hypothesis of the lemma.

Let φ be the function on $\mathscr{P}_c$

$$p \mapsto \int \hat{R}_p^A \, d\mu.$$

Now $\alpha)$ and $\beta)$ of the lemma follow immediately from the properties of balayage (Theorem 4.2.1). $\gamma)$ follows from the inequality

$$\varphi(p) \leq \int p \, d\mu$$

for every $p \in \mathscr{P}_c$ and from Proposition 7.1.1. By the lemma, there exists a unique measure μ^A on X such that for any $p \in \mathscr{P}_c$

$$\int p \, d\mu^A = \varphi(p) = \int \hat{R}_p^A \, d\mu.$$

Let now p be a finite, continuous potential on X. We denote by $\mathscr{Q}$ the set of potentials of $\mathscr{P}_c$ which are less than p on X and let $\mathscr{Q}_p$ be the set defined in Proposition 7.1.1. For any $q \in \mathscr{Q}_p$, $R(p-q) \in \mathscr{Q}$ and

$$p \leq q + R(p-q).$$

Hence

$$\hat{R}_p^A \leq \hat{R}_q^A + \hat{R}_{R(p-q)}^A \leq \hat{R}_q^A + \sup_{q' \in \mathscr{Q}} \hat{R}_{q'}^A.$$

Since

$$\inf_{q\in\mathcal{Q}_p} \hat{R}_q^A = 0$$

(Proposition 7.1.1), we get

$$R_p^A = \sup_{q\in\mathcal{Q}} \hat{R}_q^A .$$

$\mathcal{Q}$ is upper directed since for any two functions q_1, q_2 in $\mathcal{Q}$ the function $R(\sup(q_1, q_2))$ belongs to $\mathcal{Q}$ (Proposition 2.2.3). Since p is the supremum of $\mathcal{Q}$, we see, by the above considerations,

$$\int^* p\, d\mu^A = \sup_{q\in\mathcal{Q}} \int q\, d\mu^A = \sup_{q\in\mathcal{Q}} \int \hat{R}_q^A\, d\mu = \int^* \hat{R}_p^A\, d\mu . \quad \square$$

For any $\mu\in\Lambda$ and any subset A of X, the measure μ^A defined in the preceding proposition is called *the balayaged measure of* μ *on* A. Obviously $\mu^X = \mu$ and $(\mu+\nu)^A = \mu^A + \nu^A$ for every $\mu, \nu\in\Lambda$ and $A\subset X$.

Corollary 7.1.1. *If* $A\subset B\subset X$, *then for any* $\mu\in\Lambda$ *and any positive, hyperharmonic function* u *on* X, *we have*

$$\int^* u\, d\mu^A \le \int^* u\, d\mu^B .$$

Let $\mathcal{V}$ be an upper directed set of finite, continuous potentials on X whose supremum is u (Corollary 2.3.1). Then

$$\begin{aligned}\int^* u\, d\mu^A &= \sup_{p\in\mathcal{V}} \int^* p\, d\mu^A = \sup_{p\in\mathcal{V}} \int^* \hat{R}_p^A\, d\mu \le \sup_{p\in\mathcal{V}} \int^* \hat{R}_p^B\, d\mu \\ &= \sup_{p\in\mathcal{V}} \int^* p\, d\mu^B = \int^* u\, d\mu^B . \quad \square\end{aligned}$$

Corollary 7.1.2. *Let* u *be a positive, hyperharmonic function on* X, $\mu\in\Lambda$ *and* $A\subset X$. *Then*

$$\int^* u\, d\mu^A \le \int^* \hat{R}_u^A\, d\mu .$$

If X *has a countable base, then*

$$\int^* u\, d\mu^A = \int^* \hat{R}_u^A\, d\mu$$

(M. Brelot 1945 [8], R.-M. Hervé 1959 [1], N. Boboc-C. Constantinescu-A. Cornea 1965 [4]).

By Corollary 2.3.1, there exists an upper directed set $\mathcal{Q}$ of finite, continuous potentials on X such that

$$u = \sup_{q\in\mathcal{Q}} q .$$

Then

$$\int^* u\,d\mu^A = \sup_{q\in\mathcal{Q}} \int^* q\,d\mu^A = \sup_{q\in\mathcal{Q}} \int^* \hat{R}_q^A\,d\mu \leq \int^* \hat{R}_u^A\,d\mu.$$

If X has a countable base, we may take $\mathcal{Q}$ countable (Corollary 2.3.1) and by Corollary 4.2.1, the above inequality becomes an equality. □

Remark. The above equality does not hold always if X does not have a countable base (Exercise 5.3.5 *d)* and *e)*).

Corollary 7.1.3. *Assume that X has a countable base and let $A\subset X$, $x\in X$. Every polar set of $X\setminus\{x\}$ is of ε_x^A-measure zero* (M. Brelot 1948 [9], N. Boboc-A. Cornea 1966 [1]).

Let B be a polar set of $X\setminus\{x\}$. Since

$$\hat{R}_\infty^B = 0$$

(Corollary 6.2.2), there exists by Corollary 5.3.2 a positive, hyperharmonic function u on X which is finite at x and infinite on B. From

$$\int u\,d\varepsilon_x^A = \hat{R}_u^A(x) \leq u(x) < \infty,$$

it follows that B is of ε_x^A-measure zero. □

Proposition 7.1.3. *Let A be a subset of X, let U be the interior of A and let $\mu\in\Lambda$. If $\mu(X\setminus U)=0$, then $\mu^A=\mu$. If $\mu(U)=0$, then the carrier of μ^A is contained in ∂A. Hence for any $\mu\in\Lambda$, the carrier of μ^A is contained in $\bar{A}$.*

Assume first that $\mu(X\setminus U)=0$. Then for any $p, q\in\mathcal{P}_c$ such that $p-q\in\mathcal{K}(X)$, we have

$$\int (p-q)\,d\mu^A = \int (\hat{R}_p^A - \hat{R}_q^B)\,d\mu = \int (p-q)\,d\mu$$

since $p=\hat{R}_p^A$ and $q=\hat{R}_q^A$ on U. Hence $\mu^A=\mu$ (Theorem 2.3.1).

Now assume that $\mu(U)=0$. Choose $p, q\in\mathcal{P}_c$ such that $q\leq p$, $p-q\in\mathcal{K}(X)$ and such that the carrier, K, of $p-q$ does not meet ∂A. Let u be a positive, hyperharmonic function on X which is greater than q on A. Then the function on X which is equal to $\inf(p,u)$ on $X\setminus U\cap K$ and p on U is hyperharmonic and positive and it is equal to p on A. It follows that

$$\hat{R}_p^A \leq u$$

on $X\setminus U$. Since u is arbitrary,

$$\hat{R}_p^A = \hat{R}_q^A$$

on $X \setminus U$. Hence

$$\int (p-q)\, d\mu^A = \int (\hat{R}_p^A - \hat{R}_q^A)\, d\mu = 0.$$

The assertion follows now from Theorem 2.3.1. □

Proposition 7.1.4. *Let A be a subset of X. For any $f \in \mathscr{K}(X)$, $x \mapsto \varepsilon_x^A(f)$ is a fine continuous Borel function on X, harmonic on $X \setminus \bar{A}$ and for any $\mu \in \Lambda$*

$$\mu^A(f) = \int \varepsilon_x^A(f)\, d\mu(x)$$

(M. Brelot 1945 [8]).

Let p_0 be a potential of $\mathscr{P}_c$ which is strictly positive on the carrier of f. For any $n \in \mathbb{N}$, there exist $p_n, q_n \in \mathscr{P}_c$ such that

$$|f - (p_n - q_n)| < \frac{1}{n} p_0, \quad p_n - q_n \in \mathscr{K}(X)$$

(Theorem 2.3.1). We get for any $x \in X$,

$$|\varepsilon_x^A(f) - (\hat{R}_{p_n}^A(x) - \hat{R}_{q_n}^A(x))| \leq \int |f - (p_n - q_n)|\, d\varepsilon_x^A < \frac{1}{n} p_0(x).$$

Hence $(\hat{R}_{p_n}^A - \hat{R}_{q_n}^A)_{n \in \mathbb{N}}$ converges uniformly, on any compact set of X, to the function $x \mapsto \varepsilon_x^A(f)$. Since the functions $\hat{R}_{p_n}^A - \hat{R}_{q_n}^A$ are fine continuous Borel functions which are harmonic on $X \setminus \bar{A}$ (Proposition 5.3.1), the function $x \mapsto \varepsilon_x^A(f)$ is a fine continuous Borel function which is harmonic on $X \setminus \bar{A}$. By Lebesgue's theorem, this function is μ-integrable and

$$\int \varepsilon_x^A(f)\, d\mu(x) = \lim_{n \to \infty} \int (\hat{R}_{p_n}^A - \hat{R}_{q_n}^A)\, d\mu = \lim_{n \to \infty} \int (p_n - q_n)\, d\mu^A = \int f\, d\mu^A. \quad \square$$

Proposition 7.1.5. *A subset A of X is polar if and only if $\mu^A = 0$ for every $\mu \in \Lambda$* (R.-M. Hervé 1962 [4], H. Bauer 1963 [6]).

If A is polar then $\hat{R}_p^A = 0$ for all $p \in \mathscr{P}_c$ (Proposition 6.2.4). Hence $\mu^A = 0$ for every $\mu \in \Lambda$.

Assume now that $\mu^A = 0$ for every $\mu \in \Lambda$. Let W be a relatively compact, open set of X and let p be a potential in $\mathscr{P}_c$ which is strictly positive on W. Then, for any $x \in W$,

$${}^W\hat{R}_{p|W}^{A \cap W}(x) \leq \hat{R}_p^A(x) = \int p\, d\varepsilon_x^A = 0.$$

Since W is arbitrary, A is polar. □

Theorem 7.1.1. *A subset A of X is thin at a point $x \in X$ if only if $\varepsilon_x^A \neq \varepsilon_x$* (M. Brelot 1945 [8], C. Constantinescu 1967 [6]).

Assume first that A is thin at x. Then there exists $p \in \mathscr{P}_c$ such that

$$\hat{R}_p^A(x) < p(x)$$

(Proposition 6.3.2). We obtain

$$\int p\, d\varepsilon_x^A = \hat{R}_p^A(x) \neq p(x) = \int p\, d\varepsilon_x, \qquad \varepsilon_x^A \neq \varepsilon_x$$

(Proposition 7.1.2).

Conversely, if $\varepsilon_x^A \neq \varepsilon_x$ there exists $p \in \mathscr{P}_c$ such that

$$\hat{R}_p^A(x) = \int p\, d\varepsilon_x^A \neq \int p\, d\varepsilon_x = p(x)$$

and A is thin at x. □

Corollary 7.1.4. *Let U be an open set of X and $x \in \partial U$. Then x is a regular boundary point of U if and only if*

$$\varepsilon_x^{X \setminus U} = \varepsilon_x$$

(O. Frostman 1940 [3]).

The corollary follows immediately from the theorem and Theorem 6.3.3. □

Theorem 7.1.2. *For any open set U of X and any $x \in U$ we have,*

$$\varepsilon_x^{X \setminus U} = \mu_x^U$$

(Ch. de la Vallée Poussin 1931 [1], 1932 [2], O. Frostman 1935 [1]).

By Theorem 2.4.2, U is resolutive and by Proposition 2.4.3 and Propositions 5.3.3 and 5.3.1 for any finite, continuous potential p on X we have

$$\hat{R}_p^{X \setminus U}(x) = R_p^{X \setminus U}(x) = H_p^U(x).$$

Hence for any $p, q \in \mathscr{P}_c$ such that $p - q \in \mathscr{K}(X)$, we have

$$\varepsilon_x^{X \setminus U}(p-q) = \hat{R}_p^{X \setminus U}(x) - \hat{R}_q^{X \setminus U}(x) = H_p^U(x) - H_q^U(x) = \mu_x^U(p-q), \quad \varepsilon_x^{X \setminus U} = \mu_x^U$$

(Theorem 2.3.1). □

Proposition 7.1.6. *Let A, B be subsets of X. For every $\mu \in \Lambda$ we have*

$$\mu^{A \cup B} \leq \mu^A + \mu^B$$

(C. Constantinescu 1967 [6]).

Choose $p, q \in \mathscr{P}_c$ such that $q \leq p$ and $p - q \in \mathscr{K}(X)$. Then (Proposition 5.3.4 *b)*)

$$\hat{R}_q^{A, B} \leq \hat{R}_p^{A, B}$$

and thus (Proposition 5.3.4 *a)*)

$$\begin{aligned}\mu^{A\cup B}(p-q) &= \int (\hat{R}_p^{A\cup B} - \hat{R}_q^{A\cup B})\, d\mu \\ &= \int ((\hat{R}_p^A - \hat{R}_q^A) + (\hat{R}_p^B - \hat{R}_q^B) - (\hat{R}_p^{A,B} - \hat{R}_q^{A,B}))\, d\mu \\ &\le \mu^A(p-q) + \mu^B(p-q).\end{aligned}$$

The proposition follows now from Theorem 2.3.1. □

Exercises

7.1.1. The second assertion of Corollary 7.1.2 is no longer true if X does not have a countable base (C. Constantinescu 1966 [3]).

7.1.2. Let A be a subbasic (Exercise 6.3.2) set of X. Then for any $\mu \in \Lambda$ and for any positive, hyperharmonic function u on X, we have

$$\int u\, d\mu^A = \int \hat{R}_u^A\, d\mu.$$

7.1.3. Let u be the positive, hyperharmonic function on X

$$x \mapsto \int^* u_y(x)\, d\mu(y)$$

defined in Exercise 2.1.5. If X has a countable base, then for any $A \subset X$ and any $x \in A$ we have

$$\hat{R}_u^A(x) = \int^* \hat{R}_{u_y}^A(x)\, d\mu(y).$$

(Use Corollary 7.1.2.)

§ 7.2. Fine Properties of the Balayage of Measures

Throughout this section it will be assumed that X is a $\mathfrak{P}$-harmonic space with a countable base.

A potential p on a harmonic space X is called *strict* provided any two measures μ, ν on X coincide if:

a) $\int p\, d\mu = \int p\, d\nu < \infty$,

b) for every positive, hyperharmonic function u on X, we have

$$\int^* u\, d\mu \le \int^* u\, d\nu.$$

From *a)* it follows that $\mu, \nu \in \Lambda$ (see p. 159). Indeed, a strict potential p cannot vanish anywhere; if there exists $x \in X$ such that $p(x) = 0$ then taking $\mu = 0$ and $\nu = \varepsilon_x$ we get

$$\int p\, d\mu = \int p\, d\nu = 0,$$

which leads to the contradictory relation

$$\varepsilon_x = 0.$$

A potential p on X is called *hyperstrict* if for any positive function $f \in \mathscr{K}(X)$ and any strictly positive real number ε there exist two finite, continuous potentials p', p'' on X satisfying:

a) there exists $\alpha \in \mathbb{R}_+$ such that $p' + p'' \leqslant \alpha p$;

b) p' and p'' are harmonic on the complement of the carrier of f;

c) $0 \leq p'' - p' \leq f \leq p'' - p' + \varepsilon$.

Any hyperstrict potential is obviously strict. We do not know if the converse is also true.

Proposition 7.2.1. *Let $\mu \in \Lambda$ and let u be a strictly positive, hyperharmonic function on X. There exists a μ-integrable continuous finite hyperstrict (and therefore strict) potential on X which is less than u* (R.-M. Hervé 1962 [4], H. Bauer 1963 [6]).

We prove first the following lemma.

Lemma. *There exists a countable set $\mathscr{F}$, in $\mathscr{K}(X)$, such that for any $g \in \mathscr{K}(X)$ and for any strictly positive real number ε, there exists a function $f \in \mathscr{F}$ whose carrier is contained in the carrier of g and such that $|f - g| < \varepsilon$ on X.*

Let $\mathfrak{B}$ be a countable base of X and for any $U, V \in \mathfrak{B}$ with $\overline{U}$ compact and contained in V, let $f_{U,V}$ be a positive function of $\mathscr{K}(X)$ whose carrier lies in V, is less than 1 on X and equal to 1 on U. Let $\mathscr{F}'$ be the set of functions of the form

$$\sup_{i \in I} \alpha_i f_{U_i, V_i},$$

where I is a finite set, α_i are positive, rational numbers and f_{U_i, V_i} are the functions defined above. We set $\mathscr{F} := \mathscr{F}' - \mathscr{F}'$. $\mathscr{F}'$ and therefore $\mathscr{F}$ is countable.

Let $g \in \mathscr{K}(X)$. It is sufficient to prove the lemma for g positive. We set

$$\mathscr{F}_g := \{ f \in \mathscr{F}' \mid f \leq g \}.$$

Since $f', f'' \in \mathscr{F}_g$ implies $\sup(f', f'') \in \mathscr{F}_g$, we only have to show that g is the least upper bound of $\mathscr{F}_g$. Let $x \in X$. We may assume $g(x) > 0$. Let α be any positive, rational number, such that $\alpha < g(x)$. There exist $U, V \in \mathfrak{B}$ such that $x \in U$, $\overline{U}$ is compact and contained in V and such that g is greater than α on V. The function $\alpha f_{U,V} \in \mathscr{F}_g$ and therefore

$$\left(\sup_{f \in \mathscr{F}_g} f\right)(x) \geq \alpha.$$

The assertion now follows from the fact that α is arbitrary. □

In order to prove the proposition, let $(f_n)_{n\in\mathbb{N}}$ be a sequence in $\mathcal{K}(X)$ having the properties stated in the lemma for $\mathcal{F}$. For any $n\in\mathbb{N}$, there exists by Theorem 2.3.1, two finite, continuous potentials p_n, q_n on X such that p_n and q_n are harmonic outside the carrier of f_n and such that

$$0\leq p_n-q_n\leq f_n\leq p_n-q_n+n^{-1}.$$

Let $(U_n)_{n\in\mathbb{N}}$ be an increasing sequence of relatively compact, open sets of X such that for any $n\in\mathbb{N}$

$$\overline{U}_n\subset U_{n+1}, \qquad \bigcup_{n\in\mathbb{N}} U_n=X.$$

We set for any $n\in\mathbb{N}$

$$\alpha_n:=\sup_{U_n}(p_n+q_n), \quad \beta_n:=\int(p_n+q_n)\,d\mu, \quad \gamma_n:=\sup_X\frac{p_n+q_n}{u}.$$

All these numbers are finite. We set also

$$p:=\sum_{n\in\mathbb{N}}\frac{p_n+q_n}{2^{n+1}(\alpha_n+\beta_n+\gamma_n+1)}.$$

Then p is a μ-integrable finite, continuous potential on X which is less than u.

We want to show that it is a hyperstrict potential. Let f be a positive function of $\mathcal{K}(X)$ and let ε be a strictly positive real number. There exists $n\in\mathbb{N}$ such that

$$2<\varepsilon n, \qquad f_n\leq f\leq f_n+\frac{\varepsilon}{2}.$$

The potentials p_n, q_n possess then, the following properties:

a) $p_n+q_n\preccurlyeq 2^{n+1}(\alpha_n+\beta_n+\gamma_n+1)\,p$;

b) p_n and q_n are harmonic outside the carrier of f;

c) $0\leq p_n-q_n\leq f\leq p_n-q_n+\varepsilon$.

Hence p is a hyperstrict potential. □

For any subset A, of X, we denote by $b(A)$, the set of points of X where A is not thin. The fine closure of A is exactly the set $A\cup b(A)$ (Proposition 6.3.3). We have (Theorem 6.3.1)

$$A\subset B\Rightarrow b(A)\subset b(B), \qquad b(A\cup B)=b(A)\cup b(B).$$

Proposition 7.2.2. *For any finite, strict potential p on X and for any subset A of X we have*

$$b(A)=\{x\in X\,|\,\hat{R}_p^A(x)=p(x)\}$$

(R.-M. Hervé 1959 [1], C. Constantinescu 1967 [6]).

Let $x\in X$. If $x\in b(A)$, then $\varepsilon_x^A=\varepsilon_x$ (Theorem 7.1.1) and therefore

$$\hat{R}_p^A(x)=\int p\,d\varepsilon_x^A=p(x).$$

Conversely, if

$$\hat{R}_p^A(x) = p(x)$$

then $\varepsilon_x^A = \varepsilon_x$ by the definition of a strict potential. Hence A is not thin at x (Theorem 7.1.1). □

Corollary 7.2.1. *For any subset A of X, $b(A)$ is of type G_δ and fine closed and $A \setminus b(A)$ is semi-polar. The fine closure of a Borel set is a Borel set* (H. Bauer 1965 [7], C. Constantinescu 1967 [6]).

Let p be a finite, continuous, strict potential an X (Proposition 7.2.1). Since $\hat{R}_p^A$ is lower semi-continuous and fine continuous, it follows from the proposition that $b(A)$ is of type G_δ and fine closed. Since

$$A \setminus b(A) = \{x \in A \mid \hat{R}_p^A(x) < p(x)\},$$

it follows from Corollary 6.3.6 that $A \setminus b(A)$ is semi-polar. □

Corollary 7.2.2. *For any open set of X the set of non-regular boundary points is semi-polar and of type K_σ* (H. Bauer 1965 [7], 1966 [8]).

Let U be an open set of X. By the preceding corollary the set $(X \setminus U) \setminus b(X \setminus U)$ is semi-polar and of type K_σ. By Theorem 6.3.3 this set is exactly the set of non-regular boundary points of U. □

Remark. This theorem remains true if X is an $\mathfrak{S}$-harmonic space with a countable base (Exercise 7.2.4).

Proposition 7.2.3. *For any subset A of X, there exists a fine closed set A', of type G_δ, containing A and such that*

$$\mu^A = \mu^{A'}$$

for any $\mu \in \Lambda$ (C. Constantinescu 1967 [6]).

We prove first the following lemma:

Lemma (G. Choquet). *For any set $\mathscr{F}$ of lower semi-continuous, numerical functions on a topological space with countable base, there exists a countable subset $\mathscr{F}'$ of $\mathscr{F}$ such that*

$$\widehat{\inf_{f \in \mathscr{F}} f} = \widehat{\inf_{f \in \mathscr{F}'} f}.$$

Let $\mathfrak{B}$ be a countable base of the space. For any $U \in \mathfrak{B}$, there exists a countable subset $\mathscr{F}_U$ of $\mathscr{F}$ such that

$$\inf_{(x,f) \in U \times \mathscr{F}} f(x) = \inf_{(x,f) \in U \times \mathscr{F}_U} f(x).$$

We may take

$$\mathscr{F}' := \bigcup_{U \in \mathfrak{B}} \mathscr{F}_U. \quad \square$$

Let p be a finite continuous, strict potential on X. By the lemma there exists a decreasing sequence $(p_n)_{n\in\mathbb{N}}$ of potentials on X such that p_n is equal to p on A and

$$\widehat{\lim_{n\to\infty} p_n} = \hat{R}_p^A.$$

We set

$$A' := \bigcap_{n\in\mathbb{N}} \left\{ x\in X \,\middle|\, \frac{n}{n+1}\, p(x) < p_n(x) \right\}.$$

It is obvious that

$$A' = \bigcap_{n\in\mathbb{N}} \left\{ x\in X \,\middle|\, \frac{n}{n+1}\, p(x) \le p_n(x) \right\}.$$

Hence A' is of type G_δ and fine closed. On the other hand, $A\subset A'$ and

$$R_p^{A'} \le \frac{n+1}{n}\, p_n$$

for any $n\in\mathbb{N}$. Hence

$$\hat{R}_p^A \le \hat{R}_p^{A'} \le \widehat{\lim_{n\to\infty} p_n} = \hat{R}_p^A.$$

Let $x\in X$. We get

$$\int p\, d\varepsilon_x^A = \hat{R}_p^A(x) = \hat{R}_p^{A'}(x) = \int p\, d\varepsilon_x^{A'} < \infty.$$

Since

$$\int^* u\, d\varepsilon_x^A \le \int^* u\, d\varepsilon_x^{A'}$$

for any positive, hyperharmonic function u on X (Corollary 7.1.1), it follows that

$$\varepsilon_x^A = \varepsilon_x^{A'}.$$

Let $\mu\in\Lambda$. For any $f\in\mathscr{K}(X)$, we have

$$\mu^A(f) = \int \varepsilon_x^A(f)\, d\mu(x) = \int \varepsilon_x^{A'}(f)\, d\mu(x) = \mu^{A'}(f)$$

(Proposition 7.1.4). Hence $\mu^A = \mu^{A'}$. □

Corollary 7.2.3. *Every totally thin set of X is contained in a totally thin set of X of type G_δ.*

Let A be a totally thin set and let A' be the set of the proposition. For any $x\in X$, we have

$$\varepsilon_x^{A'} = \varepsilon_x^A \neq \varepsilon_x.$$

Hence A' is thin at x (Theorem 7.1.1). □

Theorem 7.2.1. *For any subset A of X and any $\mu\in\Lambda$, the fine interior of $X\setminus A$ is of inner μ^A-measure zero* (M. Brelot 1945 [8], R.-M. Hervé 1959 [1], C. Constantinescu 1967 [6]).

Let K be a compact set contained in the fine interior of $X \setminus A$ and let p be a μ-integrable, finite, strict potential (Proposition 7.2.1). Let ε be any strictly positive real number. By Corollary 5.3.2 there exists a positive, superharmonic function v which is less than p on X equal to p on A, and less than $\hat{R}_p^A + \varepsilon$ on K. We have

$$\int v\, d\mu^A = \int \hat{R}_v^A\, d\mu = \int \hat{R}_p^A\, d\mu = \int p\, d\mu^A \leq \int p\, d\mu < \infty .$$

Hence

$$0 \leq \int_K (p-v)\, d\mu^A \leq \int p\, d\mu^A - \int v\, d\mu^A = 0.$$

We get

$$\int_K p\, d\mu^A = \int_K v\, d\mu^A \leq \int_K \hat{R}_p^A\, d\mu^A + \varepsilon\, \mu^A(K).$$

Since ε being arbitrary, we deduce that

$$\int_K (p - \hat{R}_p^A)\, d\mu^A = 0 .$$

By Proposition 7.2.2, the function $p - \hat{R}_p^A$ is strictly positive on K. Hence $\mu^A(K) = 0$. □

Remark. If A is a Borel set, then the fine interior of $X \setminus A$ is a Borel set (Corollary 7.2.1) and therefore, it is of μ^A-measure zero.

Corollary 7.2.4. *Let A be a subset of X and x be a point of $X \setminus A$ such that A is thin at x. Then*

$$\varepsilon_x^A(\{x\}) = 0 .$$

Since A is thin at x, $\{x\}$ is contained in the fine interior of $X \setminus A$. □

Proposition 7.2.4. *Let A be a subset of X, $(A_n)_{n \in \mathbb{N}}$ be a sequence of subsets of X and x be a point of X. Then $(\varepsilon_x^{A_n})_{n \in \mathbb{N}}$ converges vaguely to ε_x^A if one of the following conditions are satisfied:*

a) $(A_n)_{n \in \mathbb{N}}$ is increasing and $A = \bigcup_{n \in \mathbb{N}} A_n$;

b) $(A_n)_{n \in \mathbb{N}}$ is decreasing, $A = \bigcap_{n \in \mathbb{N}} A_n$ and for any neighbourhood U of A, there exists $n \in \mathbb{N}$ such that $A_n \subset U$, and either $x \in b(A) \cup (X \setminus A)$ or there exists a neighbourhood V of x such that $A \cap V = A_n \cap V$ for any $n \in \mathbb{N}$.

By Theorem 2.3.1 it is sufficient to prove the relation

$$\lim_{n \to \infty} \hat{R}_p^{A_n}(x) = \hat{R}_p^A(x)$$

for any $p \in \mathscr{P}_c$. This relation follows in the case *a)* from Corollary 4.2.2. Let us now prove *b)*. If $x \in b(A)$, the above relation is trivial since

$$\hat{R}_p^{A_n}(x) = \hat{R}_p^A(x) = p(x)$$

for any $n \in \mathbb{N}$ and for any $p \in \mathscr{P}_c$. Assume now that $x \in X \setminus A$ and let $p \in \mathscr{P}_c$. Let ε be a strictly positive real number and p_0 be a strictly positive potential on X. There exists a positive, hyperharmonic function v on X which is greater than p on A and such that

$$v(x) \leq \hat{R}_p^A(x) + \varepsilon$$

(Corollary 5.3.2). Since the set $\{y \in X \mid p(y) < v(y) + \varepsilon p_0(y)\}$ is open and contains A, there exists $m \in \mathbb{N}$ such that it contains A_m. We obtain

$$\lim_{n \to \infty} \hat{R}_p^{A_n}(x) \leq \hat{R}_p^{A_n}(x) \leq v(x) + \varepsilon p_0(x) \leq \hat{R}_p^A(x) + \varepsilon p_0(x) + \varepsilon$$

and the relation

$$\lim_{n \to \infty} \hat{R}_p^{A_n}(x) = \hat{R}_p^A(x)$$

follows since ε is arbitrary.

Assume now that there exists a neighbourhood V of x such that $A \cap V = A_n \cap V$ for any $n \in \mathbb{N}$. We set

$$A'_n := A_n \setminus V, \qquad A' := A \setminus V, \qquad A'' := A \cap V$$

for any $n \in \mathbb{N}$. Let $p \in \mathscr{P}_c$. By Proposition 5.3.4, we have, for any $n \in \mathbb{N}$,

$$\hat{R}_p^{A_n}(x) + \hat{R}_p^{A'_n, A''}(x) = R_p^{A'_n}(x) + R_p^{A''}(x),$$

$$\hat{R}_p^{A}(x) + \hat{R}_p^{A', A''}(x) = \hat{R}_p^{A'}(x) + \hat{R}_p^{A''}(x),$$

$$\hat{R}_p^{A', A''}(x) \leq \hat{R}_p^{A'_{n+1}, A''}(x) \leq \hat{R}_p^{A'_n, A''}(x).$$

From the above proof,

$$\lim_{n \to \infty} R_p^{A'_n}(x) = \hat{R}_p^{A'}(x)$$

and hence,

$$\begin{aligned} \hat{R}_p^{A}(x) + \hat{R}_p^{A', A''}(x) &\leq \lim_{n \to \infty} \hat{R}_p^{A_n}(x) + \lim_{n \to \infty} \hat{R}_p^{A'_n, A''}(x) \\ &= \lim_{n \to \infty} \hat{R}_p^{A'_n}(x) + \hat{R}_p^{A''}(x) = \hat{R}_p^{A'}(x) + \hat{R}_p^{A''}(x) \\ &= \hat{R}_p^{A}(x) + \hat{R}_p^{A', A'''}(x), \end{aligned}$$

$$\lim_{n \to \infty} \hat{R}_p^{A_n}(x) = \hat{R}_p^{A}(x). \quad \square$$

Proposition 7.2.5. *Let A, B be subsets of X, such that A is a Borel set contained in B, and let $\tilde{A}$ be the fine closure of A. Then for any $x \in X \setminus (b(B) \setminus b(A)) \cap A$, the restriction of ε_x^B to $\tilde{A}$ is less than the restriction of ε_x^A to $\tilde{A}$* (C. Constantinescu 1967 [6]).

We show first that the restriction of ε_x^B to $\tilde{A} \setminus \{x\}$ is less than the restriction of ε_x^A to $\tilde{A} \setminus \{x\}$. Let K be a compact subset of $\tilde{A} \setminus \{x\}$ and let $(K_n)_{n\in\mathbb{N}}$ be a decreasing sequence of compact neighbourhoods of K such that

$$K=\bigcap_{n\in\mathbb{N}} K_n.$$

We set, for any $n\in\mathbb{N}$,

$$A_n := (A\cup K_n)\cap B, \qquad B_n := B\smallsetminus K_n.$$

We have

$$\varepsilon_x^B(K)\leq \varepsilon_x^{A_n}(K)+\varepsilon_x^{B_n}(K)$$

(Proposition 7.1.6). Since

$$\varepsilon_x^{B_n}(K)=0$$

(Proposition 7.1.3) and $(\varepsilon_x^{A_n})_{n\in\mathbb{N}}$ converges vaguely to ε_x^A (Proposition 7.2.4), we get

$$\varepsilon_x^B(K)\leq \limsup_{n\to\infty} \varepsilon_x^{A_n}(K)\leq \varepsilon_x^A(K).$$

If $x\notin\tilde{A}$, the proposition is proved. Assume now that $x\in\tilde{A}$. If $x\in b(A)$, then the proposition is trivial since

$$\varepsilon_x^A=\varepsilon_x^B=\varepsilon_x.$$

It remains therefore, to show only that if $x\in A\smallsetminus b(A)$, then

$$\varepsilon_x^B(\{x\})\leq \varepsilon_x^A(\{x\}).$$

Since $x\notin B\smallsetminus A$ and since $x\notin b(B)$, we have

$$\varepsilon_x^{B\smallsetminus A}(\{x\})=0$$

(Corollary 7.2.4), and from Proposition 7.1.6 we have

$$\varepsilon_x^B(\{x\})\leq \varepsilon_x^A(\{x\})+\varepsilon_x^{B\smallsetminus A}(\{x\})=\varepsilon_x^A(\{x\}). \quad \square$$

Theorem 7.2.2. *For any subsets A, B of X and any $\mu\in\Lambda$ we have*

$$\mu^{A\cup B}\leq \sup(\mu^A,\mu^B)$$

(*where* $\sup(\mu^A,\mu^B)$ *denotes the smallest measure on X that is greater than μ^A and μ^B*) (C. Constantinescu 1967 [6]).

By Proposition 7.2.3, we may assume that A and B are fine closed Borel sets.

Let K be a compact subset of $b(A)$ (resp. $A\smallsetminus b(A\cup B)$). Then

$$\begin{aligned}\mu^{A\cup B}(K)&=\int \varepsilon_x^{A\cup B}(K)\,d\mu(x)\\ &=\int_{b(A\cup B)\smallsetminus b(A)} \varepsilon_x^{A\cup B}(K)\,d\mu(x)+\int_{X\smallsetminus(b(A\cup B)\smallsetminus b(A))} \varepsilon_x^{A\cup B}(K)\,d\mu(x)\end{aligned}$$

(Proposition 7.1.4). For any $x \in b(A \cup B) \setminus b(A)$, we have

$$\varepsilon_x^{A \cup B}(K) = \varepsilon_x(K) = 0$$

(Theorem 7.1.1). By the preceding proposition, $x \in X \setminus (b(A \cup B) \setminus b(A))$ implies that

$$\varepsilon_x^{A \cup B}(K) \leq \varepsilon_x^A(K).$$

Hence

$$\begin{aligned} \mu^{A \cup B}(K) &\leq \int_{X \setminus (b(A \cup B) \setminus b(A))} \varepsilon_x^A(K)\, d\mu \leq \int \varepsilon_x^A(K)\, d\mu \\ &= \mu^A(K) \leq (\sup(\mu^A, \mu^B))(K). \end{aligned}$$

It follows that the restriction of $\mu^{A \cup B}$ to $b(A)$ (resp. $A \setminus b(A \cup B)$) is less than the restriction of $\sup(\mu^A, \mu^B)$ to the same set. Analogously, the restriction of $\mu^{A \cup B}$ to $b(B)$ (resp. $B \setminus b(A \cup B)$) is less than the restriction of $\sup(\mu^A, \mu^B)$ to the same set. The theorem follows now from the relations

$$A \cup B = b(A) \cup b(B) \cup (A \setminus b(A \cup B)) \cup (B \setminus b(A \cup B))$$

and from the fact that $X \setminus (A \cup B)$ is of $\mu^{A \cup B}$ measure zero (Theorem 7.2.1). □

Corollary 7.2.5. *Let $(A_n)_{n \in \mathbb{N}}$ be a sequence of subsets of X, let A be its union and let $\mu \in \Lambda$. If $\sup\limits_{n \in \mathbb{N}} \mu^{A_n}$ exists, then it is greater than μ^A.*

By using the theorem, one can easily prove the assertion for finite sequences by induction. The general result follows from Proposition 7.2.4 *a)*. □

Theorem 7.2.3. *Let A be a subset of X, x be a point of X and $\mathfrak{U}$ be an ultrafilter on X that converges to x. There exists then, a measure λ on X such that:*

a) every finite, continuous potential p on X is μ-integrable and

$$\lim_{\mathfrak{U}} \hat{R}_p^A = \int p\, d\lambda;$$

in particular $\lambda \in \Lambda$ and $y \mapsto \varepsilon_y^A$ converges vaguely to λ along $\mathfrak{U}$;

b) $\lambda(\{x\}) \in [\varepsilon_x^A(\{x\}), 1]$;

c) $\lambda = \lambda(\{x\})\, \varepsilon_x + (1 - \lambda(\{x\}))\, \varepsilon_x^{A \setminus \{x\}}$

(N. Boboc-A. Cornea 1967 [3]).

We denote by $\mathscr{F}$, the set of finite, continuous potentials on X and by φ, the real function on $\mathscr{F}$

$$p \mapsto \lim_{\mathfrak{U}} \hat{R}_p^A.$$

By Theorem 2.3.1, $\mathscr{F}$ satisfies the hypothesis of the lemma of Proposition 7.1.2. Properties $\alpha)$ and $\beta)$ for the function φ follow immediately from its definition and from the properties of balayage. Property $\gamma)$ is a consequence of the obvious inequality

$$\varphi(p) \leq p(x)$$

for any $p \in \mathscr{F}$ and of Proposition 7.1.1. Hence by the quoted lemma, there exists a measure λ, on X, such that

$$\varphi(p) = \int p\, d\lambda$$

for any finite, continuous potential p on X and this proves the first assertion of *a)*. The second assertion follows from Theorem 2.3.1.

c) We set

$$\alpha := \lambda(\{x\}).$$

For any $p \in \mathscr{P}_c$, we have

$$\alpha\, p(x) \leq \int p\, d\lambda \leq p(x).$$

Hence $\alpha \leq 1$.

Assume first that $\alpha = 1$. Then we get from the above inequalities

$$\int p\, d\lambda = p(x)$$

for any $p \in \mathscr{P}_c$. It follows that $\lambda = \varepsilon_x$ (Theorem 2.3.1) which proves the assertion.

Assume now that $\alpha < 1$. We set

$$\mu := \frac{1}{1-\alpha}(\lambda - \alpha\, \varepsilon_x).$$

For any $p \in \mathscr{P}_c$, we have

$$\int p\, d\mu = \frac{1}{1-\alpha} \int p\, d\lambda - \frac{\alpha}{1-\alpha}\, p(x) \leq \frac{1}{1-\alpha}\, p(x) - \frac{\alpha}{1-\alpha}\, p(x) = p(x).$$

Hence for any positive, hyperharmonic function u, on X, we obtain

$$\int^{*} u\, d\mu \leq u(x)$$

(Corollary 2.3.1).

Let $p \in \mathscr{P}_c$. We want to show that

$$\int p\, d\mu = \int p\, d\varepsilon_x^{A \setminus \{x\}}.$$

This will prove that $\mu = \varepsilon_x^{A \setminus \{x\}}$ (Theorem 2.3.1).

Let U be an open neighbourhood of x. For any $y \in X$, we have

$$\varepsilon_y^A \leq \varepsilon_y^{A \cap U} + \varepsilon_y^{A \setminus U}$$

(Proposition 7.1.6). For any positive function $f \in \mathcal{K}(X)$ which is equal to 0 on a neighbourhood of $\bar{U}$, we get

$$\varepsilon_y^A(f) \leq \varepsilon_y^{A \setminus U}(f),$$

and hence

$$(1-\alpha)\,\mu(f) = \lambda(f) = \lim_{y,\mathfrak{U}} \varepsilon_y^A(f) \leq \lim_{y,\mathfrak{U}} \varepsilon_y^{A \setminus U}(f) = \varepsilon_x^{A \setminus U}(f)$$

(Proposition 7.1.3 and Proposition 7.1.4). Let u be a positive, hyperharmonic function on X which is less than p on X and equal to p on $A \setminus U$ and let f be a positive function of $\mathcal{K}(X)$, with $f \leq 1$, and equal to 0 on a neighbourhood of $\bar{U}$. We get

$$\begin{aligned} 0 \leq (1-\alpha) \int f(p-u)\, d\mu \leq \int f(p-u)\, d\varepsilon_x^{A \setminus U} \leq \int (p-u)\, d\varepsilon_x^{A \setminus U} \\ = \hat{R}_p^{A \setminus U}(x) - \hat{R}_u^{A \setminus U}(x) = 0 \end{aligned}$$

(Corollary 7.1.2). Hence

$$\int f p\, d\mu = \int f u\, d\mu \leq \int u\, d\mu \leq u(x),$$

and u being arbitrary implies that

$$\int f p\, d\mu \leq R_p^{A \setminus U}(x) = \hat{R}_p^{A \setminus U}(x) \leq R_p^{A \setminus \{x\}}(x) = \int p\, d\varepsilon_x^{A \setminus \{x\}}$$

(Proposition 7.1.2). Since U and f are arbitrary and $\mu(\{x\}) = 0$, it follows that

$$\int p\, d\mu \leq \int p\, d\varepsilon_x^{A \setminus \{x\}}.$$

Again let U be an open neighbourhood of x. Let ε be a strictly positive real number and let u be a positive hyperharmonic function on X which is less than p on X, equal to p on $A \setminus U$ and less than $\hat{R}_p^{A \setminus U} + \varepsilon$ on a neighbourhood of x. If $\bar{u}$ is the function on X

$$y \mapsto \limsup_{z \to y} u(z),$$

we have

$$\int (p-\bar{u})\, d\lambda \leq \lim_{y,\mathfrak{U}} \int (p-\bar{u})\, d\varepsilon_y^A.$$

Hence by *a)*

$$\begin{aligned} \lim_{y,\mathfrak{U}} \int \bar{u}\, d\varepsilon_y^A = \lim_{y,\mathfrak{U}} \int p\, d\varepsilon_y^A - \lim_{y,\mathfrak{U}} \int (p-\bar{u})\, d\varepsilon_y^A \\ \leq \int p\, d\lambda - \int (p-\bar{u})\, d\lambda = \int \bar{u}\, d\lambda. \end{aligned}$$

For any $y \in X$,

$$\hat{R}_p^{A \setminus U}(y) = \hat{R}_u^{A \setminus U}(y) \leq \hat{R}_u^A(y) = \int u\, d\varepsilon_y^A \leq \int \bar{u}\, d\varepsilon_y^A$$

(Corollary 7.1.2). Then

$$\begin{aligned}\hat{R}_p^{A \setminus U}(x) &= \lim_{y, \mathfrak{U}} \hat{R}_p^{A \setminus U}(y) \le \lim_{y, \mathfrak{U}} \int \bar{u}\, d\varepsilon_y^A \\ &\le \int \bar{u}\, d\lambda = \alpha\, \bar{u}(x) + (1-\alpha) \int \bar{u}\, d\mu \\ &\le \alpha\, \hat{R}_p^{A \setminus U}(x) + \alpha\, \varepsilon + (1-\alpha) \int p\, d\mu.\end{aligned}$$

Since ε is arbitrary

$$\hat{R}_p^{A \setminus U}(x) \le \int p\, d\mu.$$

But U is arbitrary and X has a countable base. Therefore, by Corollary 4.2.2 and Proposition 7.1.2, we obtain

$$\int p\, d\varepsilon_x^{A \setminus \{x\}} = \hat{R}_p^{A \setminus \{x\}}(x) \le \int p\, d\mu.$$

b) The inequality

$$\lambda(\{x\}) \le 1$$

was already proved. If A is not thin at x, then obviously $\lambda = \varepsilon_x$, $\varepsilon_x^A = \varepsilon_x$ and *b)* is proved.

Assume now that A is thin at x. Let $\mathfrak{U}_0$ be the altrafilter containing $\{x\}$. Then $y \mapsto \varepsilon_y^A$ converges vaguely along $\mathfrak{U}_0$ to ε_x^A. Hence by *c)*

$$\varepsilon_x^A = \varepsilon_x^A(\{x\})\, \varepsilon_x + \big(1 - \varepsilon_x^A(\{x\})\big)\, \varepsilon_x^{A \setminus \{x\}}.$$

Since A is thin at x, $A \setminus \{x\}$ is also thin at x and there exists $p \in \mathscr{P}_c$ such that

$$\int p\, d\varepsilon_x^{A \setminus \{x\}} = \hat{R}_p^{A \setminus \{x\}}(x) < p(x)$$

(Proposition 7.1.2, Proposition 6.3.2). We get

$$\begin{aligned}\varepsilon_x^A(\{x\})\big(p(x) - \int p\, d\varepsilon_x^{A \setminus \{x\}}\big) &+ \int p\, d\varepsilon_x^{A \setminus \{x\}} \\ &= \int p\, d\varepsilon_x^A = \hat{R}_p^A(x) \le \lim\nolimits_{\mathfrak{U}_0} \hat{R}_p^A = \int p\, d\lambda \\ &= \lambda(\{x\})\big(p(x) - \int p\, d\varepsilon_x^{A \setminus \{x\}}\big) + \int p\, d\varepsilon_x^{A \setminus \{x\}}, \\ \varepsilon_x^A(\{x\}) &\le \lambda(\{x\}). \quad \square\end{aligned}$$

Corollary 7.2.6. *Let U be an open set of X, x be a boundary point of U and $\mathfrak{U}$ be an ultrafilter on U converging to x. Then there exists a real number $\alpha \in [0, 1]$ such that for any continuous real function f on ∂U with $|f|$ dominated by a potential, we have*

$$\lim\nolimits_{\mathfrak{U}} H_f^U = \alpha f(x) + (1-\alpha) \int f\, d\varepsilon_x^{X \setminus (U \cup \{x\})}$$

(U is resolutive by Theorem 2.4.2 and f is resolutive by Proposition 2.4.1 and Corollary 2.4.1) (O. Frostman 1939 [2]).

Let λ be the measure from the theorem and $\alpha := \lambda(\{x\})$. By Theorem 7.1.2 $\varepsilon_x^{X \setminus U} = \mu_x^U$. Hence, if f has a compact carrier, the corollary follows immediately from the theorem.

Let $g \in \mathscr{K}(X)$, $0 \leq g \leq 1$, $g(x) = 1$. Then

$$H^U_{|f|(1-g)} \leq R(|f|(1-g)) \quad \text{on } U,$$
$$\int |f|(1-g)\, d\varepsilon_x^{X \setminus (U \cup \{x\})} \leq R(|f|(1-g))(x).$$

We have that

$$\begin{aligned}
&|\lim_{\mathfrak{U}} H^U_f - \alpha f(x) - (1-\alpha) \textstyle\int f\, d\varepsilon_x^{X \setminus (U \cup \{x\})}| \\
&\quad \leq |\lim_{\mathfrak{U}} H^U_{fg} - \alpha (fg)(x) + (1-\alpha) \textstyle\int fg\, d\varepsilon_x^{X \setminus (U \cup \{x\})}| \\
&\qquad + \lim_{\mathfrak{U}} H^U_{|f|(1-g)} + (1-\alpha) \textstyle\int |f|(1-g)\, d\varepsilon_x^{X \setminus (U \cup \{x\})} \\
&\quad \leq 2R(|f|(1-g))(x).
\end{aligned}$$

Since $|f|\, g$ is dominated by a potential,

$$\inf_g R(|f|(1-g))(x) = 0. \quad \square$$

Exercises

7.2.1. Let A be a subset of X and $x \in X$. If x is a fine adherent point of $A \setminus \{x\}$ then $x \in b(A)$. The converse is also true if $\{x\}$ is semi-polar. If X is the harmonic space associated with Laplace equation on $\mathbb{R}^1$ and $A = \{0\}$, then 0 belongs to $b(A)$ and is fine isolated.

7.2.2. Let A be a subset of X, and $\mu \in \Lambda$. We denote by μ' (resp. μ''), the measure on X

$$f \mapsto \int_{b(A)} f\, d\mu \quad \left(\text{resp. } f \mapsto \int_{X \setminus b(A)} f\, d\mu\right) \quad (f \in \mathscr{K}(X)).$$

a) $\mu'^A = \mu'$.

b) Every polar set of X is of μ''^A measure zero.

(For *b)* assume that the polar set B, of X, is a Borel set (Exercise 6.2.1). Let $x \in X \setminus b(A)$. If $x \notin B$, then $\varepsilon_x^A(B) = 0$ (Corollary 7.1.3). If $x \in B$, then $\{x\}$ is a polar set and $\varepsilon_x^A = \varepsilon_x^{A \setminus \{x\}}$ (Corollary 6.2.1). Since x does not belong to the fine closure of $A \setminus \{x\}$, we have $\varepsilon_x^{A \setminus \{x\}}(\{x\}) = 0$ (Theorem 7.2.1). On the other hand, $\varepsilon_x^{A \setminus \{x\}}(B \setminus \{x\}) = 0$ (Corollary 7.1.3), and hence $\varepsilon_x^A(B) = 0$. The assertion follows now with the aid of Proposition 7.1.4.)

7.2.3. The fine closure of a Borel set is a Borel set.

7.2.4. Let U be an open set of an $\mathfrak{S}$-harmonic space with a countable base. The set of non-regular boundary points of U is semi-polar and of type K_σ. (Use Corollary 7.2.2, Theorem 2.3.3 and Exercise 2.4.6.)

7.2.5. Let $\mathfrak{B}$ be the smallest set of subsets of X such that:

a) $\mathfrak{B}$ contains every fine closed set;

b) $A \in \mathfrak{B} \Rightarrow X \setminus A \in \mathfrak{B}$;

c) if $(A_n)_{n\in\mathbb{N}}$ is a sequence in $\mathfrak{B}$ then $\bigcup_{n\in\mathbb{N}} A_n \in \mathfrak{B}$.

For any $A \in \mathfrak{B}$ there exists a Borel set B of X such that

$$(A-B)\cup(B-A)$$

is semi-polar.

7.2.6. Any family of fine open sets of X contains a countable subfamily whose union differs from the union of the whole family by a semi-polar set (J.L. Doob 1966 [2]). (Use Choquet's lemma (p. 169), Proposition 7.2.2 and Theorem 6.3.2.)

7.2.7. Let μ be a mesure on X such that any compact total thin set is μ-negligible. Then: *a)* any semi-polar set is μ-negligible; *b)* any set of $\mathfrak{B}$ from Exercise 7.2.5 is μ-measurable; *c)* there exists a fine closed set of type G_δ which is not thin at any of its points and which is the minimal fine closed set with μ-negligible complement (G. Choquet 1965 [3], C. Constantinescu 1967 [6]). (For *a)* use Corollary 7.2.3; for *b)* use Exercise 7.2.5; for *c)* use Exercise 7.2.6, Proposition 6.3.3 and Corollary 7.2.1.)

7.2.8. Let $f \in \mathscr{K}_+(X)$. For any $\alpha \in \mathbb{R}_+$ we set

$$U_\alpha := \{x \in X \mid f(x) > \alpha\}.$$

For any $x \in X$ and $g \in \mathscr{K}(X)$, the function

$$\alpha \mapsto \mu_x^{U_\alpha}(g):\ [0, f(x)[\to \mathbb{R}$$

is continuous from the left (use Theorem 5.3.1). Let μ be the measure on X

$$g \mapsto \int_0^{f(x)} \mu_x^{U_\alpha}(g)\, d\alpha.$$

Every semi-polar set of X is of μ-measure zero. (By Corollary 7.2.3 it is sufficient to prove the assertion for compact totally thin sets of X. Let K be such a set and let p be a finite, continuous, strict potential on X. Then for any $\alpha \in [0, f(x)[$

$$\int \hat{R}_p^K\, d\mu_x^{U_\alpha} \le \int R_p^K\, d\mu_x^{U_\alpha} \le \lim_{\substack{\beta\to\alpha \\ \beta>\alpha}} \int \hat{R}_p^K\, d\mu_x^{U_\beta}.$$

Hence K is of $\mu_x^{U_\alpha}$ measure zero with the exception of a countable number of α.)

7.2.9. Let A be a subset of X and B be a neighbourhood of A. Then for any positive, hyperharmonic function u, we have

$$\hat{R}^B_{\hat{R}^A_u} = \hat{R}^A_u .$$

The assertion is not true if B is only a fine neighbourhood of A. (It is sufficient to prove the assertion for B open and for u finite and continuous (Corollary 4.2.1). If A is compact, then $X \setminus A$ is an MP-set and the assertion follows from Proposition 5.3.1, Proposition 5.3.3 and Proposition 2.4.4. By Corollary 5.3.3 and Corollary 5.2.2, we may now assume that A is a Borel set. The general case follows then from Proposition 7.2.3. The last assertion may be proved by using the harmonic space associated with the heat equation.)

7.2.10. If A, B are subsets of X, then for every positive hyperharmonic function u on X we have

$$\hat{R}^A_u \vee \hat{R}^B_u = \hat{R}^{A \cup B}_u$$

(N. Boboc-A. Cornea 1968 [4]). (First prove the assertion when $A \cup B$ is compact by using Proposition 5.3.3. The general case then follows from Corollary 5.2.2, Corollary 5.3.3, Proposition 7.2.3.)

7.2.11. A closed, fine open set of a harmonic space with a countable base is absorbent (C. Berg 1971 [1]). (Let A be such a set. We may assume that the space is a $\mathfrak{P}$-harmonic space. For any $x \in A$, we have $\varepsilon_x^{X \setminus A}(A) = 0$ (Theorem 7.2.1) and $\varepsilon_x^{X \setminus A}(X \setminus A) = 0$ (Proposition 7.1.3). Hence for any strictly positive potential p on X, we have

$$A = \{x \in X \mid \hat{R}_p^{X \setminus A}(x) = 0\}.)$$

7.2.12. Let A be a subset of X. There exists a K_σ-set $B \subset A$ such that for any positive, hyperharmonic function u on X

$$\hat{R}^B_u = \sup_K \hat{R}^K_u ,$$

where K runs through the set of compact subsets of A.

7.2.13. With the notation of Theorem 7.2.3, we have

$$\hat{R}^A_u(x) \leq \int^* u \, d\lambda$$

for any positive, hyperharmonic function u on X. For any positive, finite, continuous, superharmonic function u on X, we have

$$\lim_{\mathfrak{U}} \hat{R}^A_u = \int u \, d\lambda .$$

(Assume first that u is a harmonic function such that $\hat{R}_u^A = u$ and use Theorem 7.2.3 *c)* and the relation $\hat{R}_u^{A \setminus \{x\}} = u$. Assume then that $\hat{R}_u^A$ is a potential, use Exercise 7.2.16, Exercise 5.3.2 *d)* and observe that the assertion is trivial if A is relatively compact. For the general case, use Exercise 5.3.2 *a)* and *b)*.)

In exercises 7.2.14, 7.2.15, 7.2.16 *it will be not assumed that X has a countable base.*

7.2.14. Let A be a subset of X and $x \in X$. Then there exists $\varepsilon_{A/x} \in \Lambda$ such that for any finite, continuous potential p on X we have

$$\int p \, d\varepsilon_{A/x} = \sup_V \hat{R}_p^{A \setminus V}(x),$$

where V runs through the set of neighbourhoods of x. If $\{x\}$ is of type G_δ (resp. polar), then $\varepsilon_{A/x} = \varepsilon_x^{A \setminus \{x\}}$ (resp. $\varepsilon_{A/x} = \varepsilon_x^A$). (Use the lemma of Proposition 7.1.2.)

7.2.15. Let $A \subset X$ and $x \in X$. Assume that for any $p \in \mathcal{P}_c$ there exists a fundamental system of neighbourhoods $\mathfrak{B}$ of x such that for any $V \in \mathfrak{B}$, for any $y \in V$ and for any positive, hyperharmonic function u on X such that $u = p$ on $A \setminus V$, we have

$$\hat{R}_u^{A \setminus V}(y) = \int u \, d\varepsilon_y^{A \setminus V}.$$

In this case prove that Theorem 7.2.3 holds with $\varepsilon_x^{A \setminus \{x\}}$ replaced by $\varepsilon_{A/x}$ (Exercise 7.2.14). Show that if A is subbasic (Exercise 6.3.2) or closed, the above equation is valid. Show that Corollary 7.2.6 holds with $\varepsilon_x^{X \setminus (U \cup \{x\})}$ replaced by $\varepsilon_{X \setminus U/x}$ and the hypothesis concerning the countable base deleted.

7.2.16. Let A, A' be subsets of X, x be a point of X, $\mathfrak{U}$ be an ultrafilter on X which converges to x and λ (resp. λ') be the measure associated with A (resp. A') in Theorem 7.2.3. If $A \subset A'$, then

$$\int^* u \, d\lambda \leq \int^* u \, d\lambda'$$

for every positive, hyperharmonic function u on X.

7.2.17. The following assertions are equivalent:

a) every semi-polar set is countable;

b) the fine topology is Lindelöf;

c) the fine topology is paracompact;

d) the fine topology is normal

(C. Berg 1971 [1]).

7.2.18. Given any subset A of X, there exists a fine closed set B such that: *a)* B is contained in the fine closure of A; *b)* $A \setminus B$ is semi-polar; *c)* for any semi-polar set C of X, $B \setminus C$ is not thin at any point of B. (Let p be a strict potential on X and let $\mathscr{F}$ be the set of positive, hyperharmonic functions on X such that $\{x \in A \mid u(x) < p(x)\}$ is semi-polar. Then $q := \bigwedge \mathscr{F} \in \mathscr{F}$ (Theorem 6.3.2 and lemma of Proposition 7.2.3). Let $B := \{x \in X \mid p(x) = q(x)\}$. Then *a)* and *b)* follow immediately. For *c)* note that $\hat{R}_p^{B \setminus C} \in \mathscr{F}$ and use the fact that p is strict.)

7.2.19. For any non-semi-polar set A, there exists a potential p on X, not-identically 0, such that, with the notation of the preceding exercise,

$$p = \hat{R}_p^B$$

(B is non-empty).

Chapter 8

Positive Superharmonic Functions. Specific Order

Throughout this chapter X denotes a harmonic space.

§ 8.1. Abstract Carriers

This section is devoted to an abstract presentation of specific multiplication which was introduced by R.-M. Hervé 1962 [4]. All of the results obtained here are slightly modified versions of those from N. Boboc-C. Constantinescu-A. Cornea 1965 [3].

A *convex cone* is a set C endowed with an "addition", denoted by $+$, and an exterior "multiplication" by positive real numbers such that the addition is associative, commutative and possesses a null-element denoted by 0, and such that

$$\alpha \in \mathbb{R}_+, \ u, v \in C \Rightarrow \alpha(u+v) = \alpha u + \alpha v;$$

$$\alpha, \beta \in \mathbb{R}_+, \ u \in C \Rightarrow (\alpha+\beta) u = \alpha u + \beta u, \ \alpha(\beta u) = (\alpha \beta) u;$$

$$u \in C \Rightarrow 1 \cdot u = u.$$

Assume that the following axioms are satisfied:

$$a) \quad u, v, w \in C, \ u+v = u+w \Rightarrow v = w;$$

$$b) \quad u, v \in C, \ u+v = 0 \Rightarrow u = v = 0.$$

Let $(u, v) \in C^2$. We denote $u \preccurlyeq v$ the relation: there exists $w \in C$ such that $u+w=v$. This relation is an order relation on C, for which 0 is the smallest element. If $u \preccurlyeq v$, then there exists a unique element of C which will be denoted by $(v-u)$ such that

$$v = u + (v-u).$$

A convex cone C, is called a *prevector lattice* if it satisfies axioms *a)* and *b)* and if it is a lattice with respect to the order relation $\preccurlyeq$. We denote $\curlyvee$ (resp. $\curlywedge$) the supremum (resp. infimum) in C with respect to this order relation.

A subset B of a lower complete, prevector lattice C, is called a *band* if:

a) $u, v \in B \Rightarrow u+v \in B$;

b) $u \preccurlyeq v,\ v \in B \Rightarrow u \in B$;

c) the least upper bound of any upper directed subset of B belongs to B if it exists.

From *a)* and *b)* it follows that if u belongs to a band B, then for any $\alpha \in \mathbb{R}_+$, αu belongs also to B. The least upper bound of any subset of a band B belongs to B, if it exists.

The set of positive elements of a vector lattice is a prevector lattice. It is easily proved that any prevector lattice is the set of positive elements of a vector lattice (with the same addition, multiplication and order relation) (see for instance Proposition 11.2.1).

Proposition 8.1.1. *Let C be a lower complete, prevector lattice.*

a) If $(u_\iota)_{\iota\in I}$ is a family (resp. an upper bounded family) in C, then for any $u \in C$ and $\alpha \in \mathbb{R}_+$, we have

$$u + \bigcurlywedge_{\iota\in I} u_\iota = \bigcurlywedge_{\iota\in I}(u+u_\iota) \qquad \text{(resp. } u + \bigcurlyvee_{\iota\in I} u_\iota = \bigcurlyvee_{\iota\in I}(u+u_\iota)\text{)},$$

$$\bigcurlywedge_{\iota\in I} \alpha u_\iota = \alpha \bigcurlywedge_{\iota\in I} u_\iota \qquad \text{(resp. } \alpha \bigcurlyvee_{\iota\in I} u_\iota = \bigcurlyvee_{\iota\in I} \alpha u_\iota\text{)};$$

$$u \curlyvee \Big(\bigcurlywedge_{\iota\in I} u_\iota\Big) = \bigcurlywedge_{\iota\in I}(u \curlyvee u_\iota) \qquad \Big(\text{resp. } u \curlywedge \Big(\bigcurlyvee_{\iota\in I} u_\iota\Big) = \bigcurlyvee_{\iota\in I}(u \curlywedge u_\iota)\Big).$$

b) $u \in C \Rightarrow \bigcurlywedge_{\substack{\alpha\in\mathbb{R}_+ \\ \alpha>0}} \alpha u = 0.$

c) $u, v, w \in C \Rightarrow (u+v) \curlywedge w \preccurlyeq u \curlywedge v + v \curlywedge w.$

d) if $(u_i)_{i\in I}$, $(v_j)_{j\in J}$ are two finite families in C such that

$$\sum_{i\in I} u_i = \sum_{j\in J} v_j,$$

then there exists a family $(w_{ij})_{(i,j)\in I\times J}$ in C such that for any $i \in I$ and any $j \in J$ we have

$$u_i = \sum_{k\in J} w_{ik}, \qquad v_j = \sum_{k\in I} w_{kj}.$$

e) Let B be a band of C, let $u \in C$, and let

$$u_B := \bigcurlyvee_{\substack{v\in B \\ v \preccurlyeq u}} v.$$

For any $w \in B$, we have

$$(u - u_B) \curlywedge w = 0.$$

a) We set

$$v := \bigwedge_{\iota\in I} u_\iota, \qquad w := \bigwedge_{\iota\in I}(u+u_\iota) \qquad \left(\text{resp. } v := \bigvee_{\iota\in I} u_\iota,\ w := \bigvee_{\iota\in I}(u+u_\iota)\right).$$

Obviously

$$u+v\leqslant w \qquad (\text{resp. } w\leqslant u+v).$$

From

$$w\leqslant u+u_\iota \qquad (\text{resp. } u+u_\iota\leqslant w)$$

we get

$$w-u\leqslant u_\iota \qquad (\text{resp. } u_\iota\leqslant w-u)$$

for any $\iota\in I$. Hence

$$w-u\leqslant v \qquad (\text{resp. } v\leqslant w-u),$$

$$w\leqslant u+v \qquad (\text{resp. } u+v\leqslant w).$$

The second assertion is trivial.

For the last assertion, we set

$$v := \bigwedge_{\iota\in I} u_\iota \qquad \left(\text{resp. } v := \bigvee_{\iota\in I} u_\iota\right),$$

$$v_\iota := u_\iota - v \qquad (\text{resp. } v_\iota := v - u_\iota)$$

for any $\iota\in I$. From the first assertion, we immediately obtain

$$\bigwedge_{\iota\in I} v_\iota = 0.$$

We have, again by the first part of the proof,

$$u\curlyvee u_\iota\leqslant(u+v_\iota)\curlyvee(v+v_\iota)=u\curlyvee v+v_\iota,$$

$$\left(\text{resp. } u\curlywedge v\leqslant(u+v_\iota)\curlywedge(u_\iota+v_\iota)=u\curlywedge u_\iota+v_\iota\leqslant \bigvee_{\iota\in I}(u\curlywedge v_\iota)+v_\iota\right),$$

$$\bigwedge_{\iota\in I}(u\curlyvee u_\iota)\leqslant \bigwedge_{\iota\in I}(u\curlyvee v+v_\iota)=u\curlyvee v$$

$$\left(\text{resp. } u\curlywedge v\leqslant \bigvee_{\iota\in I}(u\curlywedge u_\iota)\right).$$

The converse inequality is trivial.

b) follows from

$$\bigwedge_{\alpha>0}\alpha u = \bigwedge_{\alpha>0} 2\alpha u = 2\bigwedge_{\alpha>0}\alpha u.$$

c) $u\curlywedge w+v\curlywedge w=(u\curlywedge w+v)\curlywedge(u\curlywedge w+w)$

$$=(u+v)\curlywedge(w+v)\curlywedge(u+w)\curlywedge(w+w)\geqslant(u+v)\curlywedge w.$$

d) Assume first that $I=J=\{1,2\}$. We set

$$w_{11}:=u_1\curlywedge v_1, \quad w_{12}:=u_1-w_{11}, \quad w_{21}:=v_1-w_{11}.$$

By *a)* we have

$$w_{11}+w_{12}\curlywedge w_{21}=(w_{11}+w_{12})\curlywedge(w_{11}+w_{21})=u_1\curlywedge v_1=w_{11},$$

$$w_{12}\curlywedge w_{21}=0.$$

We have also

$$w_{11}+w_{12}+u_2=w_{11}+(u_1-w_{11})+u_2=u_1+u_2=v_1+v_2=w_{11}+w_{21}+v_2,$$

$$w_{12}\preccurlyeq w_{12}+u_2=w_{21}+v_2.$$

By *c)* we get

$$w_{12}\preccurlyeq w_{12}\curlywedge w_{21}+w_{12}\curlywedge v_2=w_{12}\curlywedge v_2\preccurlyeq v_2.$$

Let

$$w_{22}=v_2-w_{12}.$$

Then

$$w_{12}+u_2=w_{21}+w_{22}+w_{12},$$

and finally

$$u_2=w_{21}+w_{22}.$$

The general case now follows by induction on the sum of the cardinals of I and J.

e) We set

$$v:=(u-u_B)\curlywedge w.$$

We have $v\in B$, $v\preccurlyeq u-u_B$. Hence $v+u_B\in B$, $v+u_B\preccurlyeq u$ and thus

$$v+u_B\preccurlyeq u_B, \quad v=0. \quad \square$$

Let Y be a regular topological space and C be a lower complete, prevector lattice. An *abstract carrier* on (C, Y) is a map $u\mapsto S(u)$ of C into the set of closed sets of Y, which satisfies the following axioms:

a) $u=0\Leftrightarrow S(u)=\varnothing$;

b) $u\preccurlyeq v\Rightarrow S(u)\subset S(v)$;

c) for any $u\in C$ and for any two closed sets F_1, F_2 of Y such that $F_1\cup F_2=Y$, there exists $u_1, u_2\in C$ such that

$$u=u_1+u_2, \quad S(u_i)\subset F_i \quad (i=1,2).$$

$S(u)$ will be called *the abstract carrier of u.*

In the remainder of this section, we shall let Y (resp. C, resp. S) denote a regular topological space (resp. a lower complete, prevector lattice, resp. an abstract carrier on (C, Y)).

Proposition 8.1.2. *For any* $u, v \in C$ *we have:*

a) $S(u \wedge v) \subset S(u) \cap S(v)$;

b) $S(u) \cap S(v) = \varnothing \Rightarrow u \wedge v = 0$;

c) $S(u) \cup S(v) = S(u \vee v) = S(u+v)$.

a) and *b)* follow immediately from the above properties *a)* and *b)* of S. We have also the relations

$$S(u) \cup S(v) \subset S(u \vee v) \subset S(u+v).$$

Let U be an open neighbourhood of $S(u) \cup S(v)$. By property *c)* of an abstract carrier, there exist $u', v' \in C$ such that

$$u+v=u'+v', \qquad S(u') \subset \overline{U}, \qquad S(v') \subset X \setminus U.$$

Hence we have

$$u \wedge v' = v \wedge v' = 0,$$

$$v' = (u+v) \wedge v' \leqslant u \wedge v' + v \wedge v' = 0,$$

$$S(u+v) = S(u') \subset \overline{U}$$

(Proposition 8.1.1 *c)*). Since U is arbitrary and Y regular, we have

$$S(u+v) \subset S(u) \cup S(v). \quad \square$$

Let F be a closed set of Y and let $u \in C$. We set

$$C_F := \{v \in C \mid S(v) \subset F\}, \qquad u_F := \bigvee_{\substack{v \in C_F \\ v \leqslant u}} v.$$

Proposition 8.1.3. *For any closed set* F, *the set* C_F *is a band. For any* $u \in C$ *we have*

$$S((u-u_F)) \subset \overline{S(u) \setminus F}.$$

That C_F satisfies conditions *a)* and *b)* of the definition of a band follows immediately from the definition of an abstract carrier and the preceding proposition.

Let A be a subset of C_F and let v be its least upper bound in C. We want to show that $v \in C_F$. Let U be an open neighbourhood of F. There exist $v', v'' \in C$ such that

$$v = v' + v'', \qquad S(v') \subset \overline{U}, \qquad S(v'') \subset Y \setminus U.$$

Then for any $u \in A$, we have

$$S(u \wedge v'') \subset S(u) \cap S(v'') = \varnothing.$$

Hence

$$u \wedge v'' = 0, \qquad u = u \wedge v \leqslant u \wedge v' + u \wedge v'' \leqslant v'.$$

(Proposition 8.1.1 *c)*). Since u is arbitrary,

$$v \preccurlyeq v', \qquad S(v) \subset S(v') \subset \overline{U}.$$

Since U is arbitrary, it follows that $v \in C_F$. Hence C_F is a band.

In order to prove the second assertion, let U be an open neighbourhood of $\overline{S(u) \setminus F}$ and let u', u'' be elements of C such that

$$(u - u_F) = u' + u'', \qquad S(u') \subset \overline{U}, \qquad S(u'') \subset Y \setminus U.$$

With the aid of Proposition 8.1.1 *e)*, we get

$$S(u'') \subset F, \qquad u'' \in C_F,$$

$$u'' \preccurlyeq u_F \curlywedge (u - u_F) = 0, \qquad (u - u_F) = u', \qquad S((u - u_F)) \subset \overline{U}.$$

The assertion now follows from the fact that U is arbitrary and Y is regular. □

Proposition 8.1.4. *Let $u \in C$ and let $(F_i)_{1 \le i \le n}$ be a finite sequence of closed sets such that*

$$S(u) \subset \bigcup_{i=1}^{n} F_i,$$

Then there exists a finite sequence $(u_i)_{1 \le i \le n}$ in C such that

$$u = \sum_{i=1}^{n} u_i, \qquad S(u_i) \subset F_i$$

for any i.

We prove the assertion by induction on n. Assume that the assertion is true for $n-1$. By the preceding proposition,

$$S((u - u_{F_n})) \subset \overline{S(u) \setminus F_n} \subset \bigcup_{i=1}^{n-1} F_i.$$

Hence, there exists a finite sequence $(u_i)_{1 \le i \le n-1}$, in C such that

$$(u - u_{F_n}) = \sum_{i=1}^{n-1} u_i, \qquad S(u_i) \subset F_i$$

for any i. The assertion follows by setting $u_n := u_{F_n}$. □

Let $u \in C$. A *partition* of u is a finite family from C whose sum is equal to u. We denote by Δ_u, the set of all partitions of u. A partition, $(u_i)_{i \in I}$, of u is called *finer* than a partition, $(v_j)_{j \in J}$, of u if there exists a decomposition $(I_j)_{j \in J}$ of I (i.e. $j \neq j' \Rightarrow I_j \cap I_{j'} = \emptyset$ and $I = \bigcup_{j \in I} I_j$) such that

$$v_j = \sum_{i \in I_j} u_i$$

for all $j \in J$. This relation is an upper directed, preorder relation on Δ_u (Proposition 8.1.1 *d*)). For any positive, bounded, continuous real function f on Y and any $\delta \in \Delta_u$, $\delta := (u_i)_{i \in I}$, we set

$$\delta^*(f) := \sum_{i \in I} (\sup_{S(u_i)} f)\, u_i,$$

$$\delta_*(f) := \sum_{i \in I} (\inf_{S(u_i)} f)\, u_i,$$

where we make the convention

$$\sup_{\emptyset} f = \inf_{\emptyset} f = 0.$$

Proposition 8.1.5. *Let $u \in C$ and let f be a positive, bounded, continuous real function on Y. If $\delta, \delta' \in \Delta_u$ and δ' is finer than δ, then*

$$\delta_*(f) \leqslant \delta'_*(f) \leqslant \delta'^*(f) \leqslant \delta^*(f)$$

and

$$\bigvee_{\delta \in \Delta_u} \delta_*(f) = \bigwedge_{\delta \in \Delta_u} \delta^*(f).$$

The first assertion is obvious. Since Δ_u is upper directed and since C is conditionally complete, we get

$$\bigvee_{\delta \in \Delta_u} \delta_*(f) \leqslant \bigwedge_{\delta \in \Delta_u} \delta^*(f).$$

In order to prove the converse inequality, we remark that

$$\delta^*(f) \leqslant \delta_*(f) + \big(\sup_{i \in I} (\sup_{S(u_i)} f - \inf_{S(u_i)} f)\big)\, u.$$

Let ε be a strictly positive real number. There exists a finite, closed covering $(F_i)_{i \in I}$ of Y such that

$$\sup_{i \in I} (\sup_{F_i} f - \inf_{F_i} f) < \varepsilon$$

By the preceding proposition, there exists a partition $\delta := (u_i)_{i \in I}$ of u such that

$$S(u_i) \subset F_i$$

for any $i \in I$. We get

$$\delta^*(f) \leqslant \delta_*(f) + \varepsilon u.$$

The assertion follows now from the fact that ε is arbitrary and Proposition 8.1.1 *b*). $\square$

Let $u \in C$ and f be a positive, bounded, continuous real function on Y. We set

$$f \cdot u := \bigwedge_{\delta \in \Delta_u} \delta^*(f) = \bigvee_{\delta \in \Delta_u} \delta_*(f).$$

Obviously

$$f \leq g \Rightarrow f \cdot u \leqslant g \cdot u,$$

and

$$\alpha \cdot u = \alpha u$$

for any $\alpha \in \mathbb{R}_+$.

Theorem 8.1.1. *Let l be a map of the set of positive, bounded, continuous, real functions on Y into C, such that for any positive, bounded, continuous, real functions f, g on Y,*

$$l(f+g) = l(f) + l(g)$$

$$S(l(f)) \subset \operatorname{Supp} f.$$

Then for every positive, bounded, continuous, real function f on Y, we have

$$l(f) = f \cdot l(1)$$

(R.-M. Hervé 1962 [4]).

Notice that

$$f \leq g \Rightarrow l(f) \preccurlyeq l(g)$$

and that

$$l(\alpha f) = \alpha\, l(f)$$

for any $\alpha \in \mathbb{N}$. This relation may be immediately extended to any positive rational number α. Let $\alpha \in \mathbb{R}_+$ and let α', α'' be positive rational numbers such that $\alpha' \leq \alpha \leq \alpha''$. Then

$$\alpha'\, l(f) = l(\alpha' f) \preccurlyeq l(\alpha f) \preccurlyeq l(\alpha'' f) = \alpha''\, l(f).$$

But α' and α'' are arbitrary which implies that

$$\alpha\, l(f) \preccurlyeq l(\alpha f) \preccurlyeq \alpha\, l(f)$$

and therefore

$$l(\alpha f) = \alpha\, l(f).$$

Let f be a positive, bounded, continuous, real function on Y and let ε be a strictly positive real number. For any $n \in \mathbb{N}$, we denote by g_n, the real function on $\mathbb{R}_+$ which is equal to 0 on $(\mathbb{R}_+ \cap [0, (n-1)\varepsilon]) \cup [(n+1)\varepsilon, \infty[$, equal to 1 at $n\varepsilon$ and linear on $](n-1)\varepsilon, n\varepsilon[$ and $]n\varepsilon, (n+1)\varepsilon[$. We set, for any $n \in \mathbb{N}$,

$$f_n := g_n \circ f,$$

$$I := \{n \in \mathbb{N} \mid f_n \not\equiv 0\}.$$

Since

$$(n-1)\,\varepsilon\, l(f_n) \preccurlyeq l(f_n f) \preccurlyeq (n+1)\,\varepsilon\, l(f_n), \qquad \sum_{n \in I} f_n f = f,$$

we obtain

$$\sum_{n \in I} (n-1)\,\varepsilon\, l(f_n) \preccurlyeq l(f) \preccurlyeq \sum_{n \in I} (n+1)\,\varepsilon\, l(f_n).$$

Since $\delta := (l(f_n))_{n \in I}$ is a partition of $l(1)$ and since

$$S(l(f_n)) \subset \operatorname{Supp} f_n \subset \{x \in Y \mid (n-1)\,\varepsilon \leq f(x) \leq (n+1)\,\varepsilon\}$$

we get

$$\sum_{n \in I} (n-1)\,\varepsilon\, l(f_n) \preccurlyeq \delta_*(f) \preccurlyeq f \cdot l(1) \preccurlyeq \delta^*(f) \preccurlyeq \sum_{n \in I} (n+1)\,\varepsilon\, l(f_n).$$

Hence

$$f \cdot l(1) \preccurlyeq l(f) + 2\varepsilon\, l(1), \qquad l(f) \preccurlyeq f \cdot l(1) + 2\varepsilon\, l(1).$$

Since ε is arbitrary, we get

$$f \cdot l(1) = l(f)$$

(Proposition 8.1.1 *b)*). □

Proposition 8.1.6. *Let* $u \in C$ *and let* f *be a positive, bounded, real, continuous function on* Y. *Then*

$$S(f \cdot u) \subset S(u) \cap \operatorname{Supp} f.$$

Let U be a neighbourhood of $S(u) \cap \operatorname{Supp} f$. There exists a partition $\delta := (u_1, u_2)$ of u such that

$$S(u_1) \subset \bar{U}, \qquad S(u_2) \subset S(u) \setminus U$$

(Proposition 8.1.4). Then

$$f \cdot u \preccurlyeq \delta^*(f) \preccurlyeq (\sup_Y f)\, u_1.$$

Hence

$$S(f \cdot u) \subset S(u_1) \subset \bar{U}.$$

The assertion follows now from the fact that U is arbitrary and Y regular. □

Theorem 8.1.2. *Let* f, g *be positive, bounded, continuous, real functions on* Y *and* $u, v \in C$. *Then*

a) $(f+g) \cdot u = f \cdot u + g \cdot u$;

b) $f \cdot (u+v) = f \cdot u + f \cdot v$;

c) $(f\,g) \cdot u = f \cdot (g \cdot u)$

(N. Boboc-C. Constantinescu-A. Cornea 1965 [3]).

a) For any $\delta \in \Delta_u$, we have

$$\delta_*(f) + \delta_*(g) \preccurlyeq \delta_*(f+g) \preccurlyeq \delta^*(f+g) \preccurlyeq \delta^*(f) + \delta^*(g).$$

Hence

$$(f+g) \cdot u = f \cdot u + g \cdot u.$$

b) is trivial.

c) Let l be the map $f \mapsto (fg) \cdot u$. By *a)* and the preceding proposition, l satisfies the hypothesis of Theorem 8.1.1. Hence

$$(fg) \cdot u = l(f) = f \cdot l(1) = f \cdot (g \cdot u). \quad \square$$

Theorem 8.1.3.

a) Let $(f_\iota)_{\iota \in I}$ be a family of positive, bounded, continuous, real functions on Y. We set

$$f := \inf_{\iota \in I} f_\iota \quad (\text{resp. } f := \sup_{\iota \in I} f_\iota).$$

If f is continuous (resp. continuous and bounded), then for any $u \in C$, we have

$$\curlywedge_{\iota \in I} f_\iota \cdot u = f \cdot u \quad (\text{resp. } \curlyvee_{\iota \in I} f_\iota \cdot u = f \cdot u).$$

b) Let $(u_\iota)_{\iota \in I}$ be a family (resp. an upper bounded family) in C. Then for any positive, bounded, continuous, real function f on Y, we have

$$\curlywedge_{\iota \in I} f \cdot u_\iota = f \cdot \curlywedge_{\iota \in I} u_\iota \quad (\text{resp. } \curlyvee_{\iota \in I} f \cdot u_\iota = f \cdot \curlyvee_{\iota \in I} u_\iota)$$

(N. Boboc-C. Constantinescu-A. Cornea 1965 [3]).

a) Assume first that $f = 0$ and let ε be a strictly positive real number. For any $\iota \in I$, we set

$$F_\iota := \{x \in Y \mid f_\iota(x) \geq \varepsilon\}, \quad g_\iota := \sup(f_\iota - \varepsilon, 0), \quad v := \curlywedge_{\iota \in I} (f_\iota \cdot u).$$

Then, for any $\iota \in I$

$$v \preccurlyeq f_\iota \cdot u \preccurlyeq g_\iota \cdot u + \varepsilon u, \quad (v \curlyvee \varepsilon u - \varepsilon u) \preccurlyeq g_\iota \cdot u, \quad S((v \curlyvee \varepsilon u - \varepsilon u)) \subset F_\iota$$

(Proposition 8.1.6). Hence

$$S((v \curlyvee \varepsilon u - \varepsilon u)) \subset \bigcap_{\iota \in I} F_\iota = \varnothing, \quad (v \curlyvee \varepsilon u - \varepsilon u) = 0, \quad v \preccurlyeq \varepsilon u.$$

Since ε is arbitrary, $v = 0$ (Proposition 8.1.1 *b)*).

Now let f be arbitrary. We set, for any $\iota \in I$,

$$g_\iota := f_\iota - f \quad (\text{resp. } g_\iota := f - f_\iota).$$

Since

$$\inf_{\iota \in I} g_\iota = 0$$

we have

$$\curlywedge_{\iota \in I} g_\iota \cdot u = 0.$$

Therefore

$$\bigwedge_{\iota\in I} f_\iota \cdot u = \bigwedge_{\iota\in I}(f\cdot u + g_\iota \cdot u) = f\cdot u + \bigwedge_{\iota\in I} g_\iota \cdot u = f\cdot u$$

$$\left(\text{resp. } f\cdot u = f_\iota \cdot u + g_\iota \cdot u \leqslant \bigvee_{\iota\in I} f_\iota \cdot u + g_\iota \cdot u,\right.$$

$$\left.\bigvee_{\iota\in I} f_\iota \cdot u \leqslant f\cdot u \leqslant \bigvee_{\iota\in I} f_\iota \cdot u + \bigwedge_{\iota\in I} g_\iota \cdot u = \bigvee_{\iota\in I} f_\iota \cdot u\right)$$

(Theorem 8.1.2 and Proposition 8.1.1 *a)*).

b) We set

$$u := \bigwedge_{\iota\in I} u_\iota \qquad \left(\text{resp. } u = \bigvee_{\iota\in I} u_\iota\right).$$

If $u=0$, then

$$\bigwedge_{\iota\in I} f\cdot u_\iota \leqslant (\sup_Y f)\cdot \bigwedge_{\iota\in I} u_\iota = 0.$$

(Proposition 8.1.1 *a)*). We set, for any $\iota\in I$,

$$v_\iota := (u_\iota - u) \qquad (\text{resp. } v_\iota := (u - u_\iota)).$$

Then

$$\bigwedge_{\iota\in I} f\cdot u_\iota = \bigwedge_{\iota\in I}(f\cdot u + f\cdot v_\iota) = f\cdot u + \bigwedge_{\iota\in I} f\cdot v_\iota = f\cdot u$$

$$\left(\text{resp. } f\cdot u = f\cdot u_\iota + f\cdot v_\iota \leqslant \bigvee_{\iota\in I} f\cdot u_\iota + f\cdot v_\iota,\right.$$

$$\left.\bigvee_{\iota\in I} f\cdot u_\iota \leqslant f\cdot u \leqslant \bigvee_{\iota\in I} f\cdot u_\iota + \bigwedge_{\iota\in I} f\cdot v_\iota = \bigvee_{\iota\in I} f\cdot u_\iota\right)$$

(Theorem 8.1.2 and Proposition 8.1.1 *a)*). □

Exercises

8.1.1. Let S be an abstract carrier on (C, Y). Let Y' be a regular topological space and $\varphi\colon Y\to Y'$ be a continuous map. For any $u\in C$, we set

$$S'(u) := \overline{\varphi(S(u))}.$$

Then S' is an abstract carrier on (C, Y') and for any positive, bounded, continuous, real function f' on Y', we have

$$f'\cdot u = (f'\circ\varphi)\cdot u.$$

8.1.2. Let S be an abstract carrier on (C, Y) and let A be a subset of Y. We denote by C_A, the band in C generated by the set $\{u\in C \mid S(u)\subset A\}$. Then $u\mapsto S(u)\cap A$ is an abstract carrier on (C_A, A). (Use Proposition 8.1.4.)

§ 8.2. Sets of Nonharmonicity

The aim of this section is: *a)* to show that every potential can be associated with a sort of kernel (Theorem 8.2.1). This association is the key to the construction of a suitable Markov process on a harmonic space and to integral representations; *b)* to prove the Theorem 8.2.2, which will play an important role in the study of the ideal $\mathcal{M}$, quasi-continuity, and the axioms of polarity and domination.

We denote by $\mathcal{P}$ (resp. $\mathcal{S}$), the convex cone of potentials (resp. of positive, superharmonic functions) on X. If $u, v, w \in \mathcal{S}$ are such that

$$u + w = v + w$$

then $u = v$ since they coincide outside the polar set $\{x \in X \mid w(x) = \infty\}$ (Corollary 6.3.4). It follows that $\mathcal{P}$ and $\mathcal{S}$ are lower complete, prevector lattices (Theorem 4.1.1 and Theorem 5.1.1). One can also prove with the aid of Riesz' proposition and of Proposition 8.1.1 *a)*, that $\mathcal{P}$ and $\mathcal{H}_+(X)$ are bands of $\mathcal{S}$.

Let p be a potential on X and U be the largest open set in X on which p is harmonic. We denote by $S(p)$ the set $X \setminus U$.

Let X_0 be the Alexandroff compactification of X [1]. Let u be a positive, superharmonic function on X and let h be its greatest minorant. We set

$$S_0(u) := \begin{cases} \overline{S(u-h)} & \text{if } h \equiv 0 \\ \overline{S(u-h)} \cup (X_0 \setminus X) & \text{if } h \not\equiv 0, \end{cases}$$

where the closure is taken in X_0.

Proposition 8.2.1. *S (resp. S_0) is an abstract carrier on $(\mathcal{P}, X)$ (resp. $(\mathcal{S}, X_0)$)* (R.-M. Hervé 1959 [1], N. Boboc-C. Constantinescu-A. Cornea 1965 [3]).

The assertion about S_0 follows immediately from the assertion about S. Conditions *a), b)* of the definition of an abstract carrier are trivially satisfied by S. Let F_1, F_2 be two closed sets such that $F_1 \cup F_2 = X$ and let p be a potential on X. The set of specific minorants q, of p, for which $S(q) \subset F_1$ is obviously specifically upper directed. Let p_1 be its specific supremum and let p_2 be the potential on X for which

$$p = p_1 + p_2 .$$

Then $S(p_1) \subset F_1$.

We want to show that $S(p_2) \subset F_2$. We denote by f, the function on X which is equal to $p_2 - \hat{R}_{p_2}^{F_2}$ wherever $\hat{R}_{p_2}^{F_2}$ is finite, and 0 elsewhere. Let $\mathcal{F}$ be the set of hyperharmonic functions which dominate f. The

[1] If X is compact, then X_0 is obtained by adding an isolated point to X.

set of the functions $\mathscr{F}$, restricted to the interior of F_2, is a Perron set and therefore Rf is harmonic on the interior of F_2. Hence $S(\widehat{Rf})\subset F_1$. By Proposition 4.1.5 and Theorem 5.1.1,

$$\widehat{Rf}\preccurlyeq p_2.$$

We get

$$p_1+\widehat{Rf}\preccurlyeq p, \qquad S(p_1+\widehat{Rf})\subset F_1,$$

and these relations yield

$$p_1+\widehat{Rf}\preccurlyeq p_1, \qquad \widehat{Rf}=0.$$

Since

$$p_2\leq \hat{R}_{p_2}^{F_2}+\widehat{Rf},$$

we deduce that

$$p_2\leq \hat{R}_{p_2}^{F_2}+\widehat{Rf}=\hat{R}_{p_2}^{F_2}, \qquad p_2=\hat{R}_{p_2}^{F_2}$$

(Corollary 4.1.1, Proposition 5.1.4 and Theorem 5.1.1). The last equality shows that $S(p_2)\subset F_2$. □

This theorem enables us to apply to the theory of harmonic spaces, all of the results from the preceding section. In particular, *for any potential* p (resp. *for any positive, superharmonic function* u) *on* X *and for any positive, bounded, continuous, real function* f *on* X (resp. X_0) *we shall use the notation* $f\cdot p$ (resp. $f\cdot u$) *corresponding to the definition given in the preceding section.*

Proposition 8.2.2. *For any potential* p *on* X *and any neighbourhood* U *of* $S(p)$, *we have*

$$\hat{R}_p^U=p.$$

The function on X which is equal to $p-\hat{R}_p^U$ on $X\setminus S(p)$ and 0 on $S(p)$ is obviously a subharmonic minorant of p. Hence it vanishes identically. □

Let p be a potential on X and let x be a point of X where p is finite. By Theorem 8.1.2 *a)*, there exists a (unique) measure $\mathsf{V}_{p,x}$ on X such that for every $f\in\mathscr{K}_+(X)$,

$$\mathsf{V}_{p,x}(f)=f\cdot p(x).$$

Also every positive bounded continuous real function f on X is $\mathsf{V}_{p,x}$ integrable and thus we have,

$$f\cdot p(x)=\int f\,d\mathsf{V}_{p,x}$$

(Theorem 8.1.3 *a)* and Proposition 8.2.1). A numerical function on X is called *p-measurable* if, for any $x\in X$ at which p is finite, it is $\mathsf{V}_{p,x}$-measurable. A subset A, of X, is called *p-measurable* if its characteristic function is p-measurable.

Theorem 8.2.1. *Let p be a potential on X. For any positive, bounded, p-measurable function f on X, there exists a unique potential on X, which will be denoted by $f \cdot p$, such that for any x at which p is finite, we have*

$$f \cdot p(x) = \int f \, d\mathsf{V}_{p,x}$$

(R.-M. Hervé 1960 [3]).

The unicity follows immediately from Corollary 6.3.4.

Assume first that f is lower semi-continuous. We set

$$f \cdot p := \curlyvee_{\substack{g \in \mathscr{K}_+(X) \\ g \leq f}} g \cdot p.$$

By Proposition 4.1.4 and Theorem 5.1.1, we have

$$f \cdot p(x) = \sup_{\substack{g \in \mathscr{K}_+(X) \\ g \leq f}} g \cdot p(x) = \sup_{\substack{g \in \mathscr{K}_+(X) \\ g \leq f}} \mathsf{V}_{p,x}(g) = \int f \, d\mathsf{V}_{p,x}.$$

If f is not lower semi-continuous we set

$$f \cdot p := \curlywedge_{g} g \cdot p,$$

where g runs through the set of bounded, lower semi-continuous, real functions g on X which are greater than f. Again by Proposition 4.1.4 and Theorem 5.1.1, we obtain

$$f \cdot p = \inf_g g \cdot p(x) = \inf_g \int g \, d\mathsf{V}_{p,x} = \int f \, d\mathsf{V}_{p,x}. \quad \square$$

Corollary 8.2.1. *Let p be a potential on X and let f and g be positive, bounded, p-measurable functions on X. Then*

$$(f+g) \cdot p = f \cdot p + g \cdot p, \qquad S(f \cdot p) \subset (\operatorname{Supp} f) \cap S(p). \quad \square$$

Let p be a potential on X and A be a p-measurable subset of X. We shall denote by p_A, the potential $f \cdot p$, where f is the characteristic function of A. Obviously

$$p_A = \curlyvee_K p_K = \bigvee_K p_K,$$

where K runs through the set of compact subsets of A.

Proposition 8.2.3. *For any locally bounded potential p on X and any p-measurable, polar set A, we have*

$$p_A = 0.$$

Let K be a compact subset of A. Then K is a polar set and p_K is a locally bounded potential with

$$S(p_K) \subset K$$

(Corollary 8.2.1). By Corollary 6.2.5, there exists a harmonic function h on X which is equal to p_K on $X \setminus K$. Hence $h = p_K$ on X (Corollary 6.3.4) and we deduce that $p_K = 0$. □

Theorem 8.2.2. *Every finite potential p on X is the sum of a family $(p_\iota)_{\iota \in I}$ of potentials on X which has the property that for any $\iota \in I$, $S(p_\iota)$ is compact and the restriction of p_ι to $S(p_\iota)$ is continuous* (N. Boboc-C. Constantinescu-A. Cornea 1965 [3]).

Let x be a point of X such that $p(x) \neq 0$. We want to show that there exists a potential q on X with

$$q \leqslant p, \quad q(x) \neq 0,$$

and such that its restriction to $S(q)$ is continuous. By Theorem 8.2.1, we have

$$\mathsf{V}_{p,x}(X) = 1 \cdot p(x) = p(x) \neq 0.$$

Hence there exists a compact set K of X such that $\mathsf{V}_{p,x}(K) \neq 0$. Since p is lower semi-continuous, there exists (by Lusin's theorem, N. Bourbaki [5], Chap. IV, §5 Corollaire de la Proposition 8) a compact subset L of K such that $\mathsf{V}_{p,x}(L) \neq 0$ and such that the restriction of p to L is continuous. Since the restrictions of p_L and $p_{X \setminus L}$ to L are lower semi-continuous and their sum is finite and continuous, it follows that the restriction of p_L to L is continuous. Obviously

$$p_L \leqslant p, \quad p_L(x) = \mathsf{V}_{p,x}(L) \neq 0,$$
$$S(p_L) \subset K$$

(Corollary 8.2.1).

Let $\mathfrak{C}$ be the set of sets $\mathscr{F}$ of non-identically zero potentials on X satisfying

$$\sum_{q \in \mathscr{F}} q \leqslant p$$

and such that for any $q \in \mathscr{F}$, the restriction of q to $S(q)$ is continuous. The set $\mathfrak{C}$ ordered by the inclusion relation is obviously inductive. By Zorn lemma, there exists a maximal element $\mathscr{F}_0$ of $\mathfrak{C}$. Let q_0 be a potential on X such that

$$p = q_0 + \sum_{q \in \mathscr{F}_0} q$$

and choose any $x \in X$. If $q_0(x) \neq 0$, there exists by the first part of the proof, a potential q' on X such that $q' \leqslant q_0$ with $q'(x) \neq 0$, and such that the restriction of q' to $S(q')$ is continuous. For any $\alpha \in [0, 1]$ the set $\mathscr{F}_0 \cup \{\alpha q'\}$ belongs to $\mathfrak{C}$ and this contradicts the maximality of $\mathscr{F}_0$. Hence q_0 is identically 0 and

$$p = \sum_{q \in \mathscr{F}_0} q. \quad \square$$

Exercises

8.2.1. Let p be a potential which is an extremal superharmonic function. Then $S(p)$ is a singleton (R-M. Hervé 1959 [1]).

8.2.2. Assume that X is a $\mathfrak{P}$-harmonic space with a countable base and let x be a point of X such that either $\{x\}$ is polar or every potential p on X, with $S(p)=\{x\}$, is continuous. If a subset A of X is thin at x and if $x\in\bar{A}\setminus A$, then there exists a potential p on X, finite at x and such that

$$\lim_{\substack{y\to x\\ y\in A}} p(y)=\infty$$

(M. Brelot 1944 [6], R.-M. Hervé 1962 [4]). (Let q be a locally bounded potential on X such that

$$q(x)<\liminf_{\substack{y\to x\\ y\in A}} q(y).$$

Choose a decreasing sequence $(U_n)_{n\in\mathbb{N}}$ of open neighbourhoods of x such that

$$\sum_{n\in\mathbb{N}} q_{U_n}$$

is a potential finite at x.)

8.2.3. A *kernel on* X is an additive, positively homogeneous map, V, of $\mathscr{K}_+(X)$ into $\mathscr{P}$ such that for any $f\in\mathscr{K}_+(X)$,

$$S(\mathsf{V}f)\subset \operatorname{Supp} f.$$

For every potential p on X, $f\mapsto f\cdot p\colon \mathscr{K}_+(X)\to\mathscr{P}$ is a kernel which will be denoted by V_p. Let A_{V} be the set of points $x\in X$ for which there exists $f\in\mathscr{K}_+(X)$ such that $\mathsf{V}f(x)=\infty$. Prove that A_{V} is polar. For any $x\in X\setminus A_{\mathsf{V}}$, we denote by V_x, the measure on X for which

$$\mathsf{V}_x(f)=\mathsf{V}f(x)$$

for all $f\in\mathscr{K}_+(X)$. A positive, numerical function f on X is called V-integrable if it is V_x-integrable for every $x\in X\setminus A_{\mathsf{V}}$ and if there exists a superharmonic function $\mathsf{V}f$ on X such that

$$\mathsf{V}f(x)=\int f\,d\mathsf{V}_x.$$

The function $\mathsf{V}f$ possessing the above property is unique and a potential. We denote by $\mathscr{L}^1_+(\mathsf{V})$, the set of positive V-integrable functions on X. Any universally measurable function that is bounded and whose carrier is compact, is V-integrable. $\mathscr{L}^1_+(\mathsf{V})$ is a convex cone, the map

$$f\mapsto \mathsf{V}f\colon\ \mathscr{L}^1_+(\mathsf{V})\to\mathscr{P}$$

is additive and positively homogeneous and we have

$$S(\mathsf{V}f)\subset \operatorname{Supp} f.$$

If A is a set whose characteristic functions, f, is V-integrable, we say that A is V-integrable and set $\mathsf{V}A := \mathsf{V}f$. If X is V-integrable, then $\mathsf{V} = \mathsf{V}_p$ where $p := \mathsf{V}X$. A kernel, V, is called strict if, for any two measures μ, ν on X such that

$$\int^* u\, d\mu \leq \int^* u\, d\nu$$

for all positive hyperharmonic functions u on X, we have:

$$\int \mathsf{V}f\, d\mu = \int \mathsf{V}f\, d\nu < \infty$$

for all $f \in \mathscr{K}_+(X)$ implies $\mu = \nu$. A potential p is strict if and only if V_p is strict.

8.2.4. We denote by $\mathscr{V}$, the set of kernels on X (see the preceding exercise). If addition and multiplication are defined by

$$(\mathsf{V} + \mathsf{W}) f := \mathsf{V}f + \mathsf{W}f \qquad (f \in \mathscr{K}_+(X))$$

$$(\alpha \mathsf{V}) f := \alpha \mathsf{V} f \qquad (f \in \mathscr{K}_+(X),\ \alpha \in \mathbb{R}_+),$$

then $\mathscr{V}$ is a convex cone. The null-element of $\mathscr{V}$ is the kernel

$$f \mapsto 0\colon \mathscr{K}_+(X) \to \mathscr{P}.$$

We have ($\mathsf{V}, \mathsf{W}, \mathsf{V}' \in \mathscr{V}$)

$$\mathsf{V}' + \mathsf{V} = \mathsf{V}' + \mathsf{W} \Rightarrow \mathsf{V} = \mathsf{W},$$

$$\mathsf{V} + \mathsf{W} = 0 \Rightarrow \mathsf{V} = \mathsf{W} = 0.$$

$\mathscr{V}$ is a lower complete, prevector lattice. Let $(\mathsf{V}_\iota)_{\iota \in I}$ be a family of $\mathscr{V}$. If

$$\mathsf{V} := \curlyvee_{\iota \in I} \mathsf{V}_\iota$$

exists and $f \in \mathscr{L}^1_+(\mathsf{V})$, then $f \in \mathscr{L}^1_+(\mathsf{V}_\iota)$ for any $\iota \in I$ and we have

$$\mathsf{V}f = \curlyvee_{\iota \in I} \mathsf{V}_\iota f.$$

If $f \in \mathscr{L}^1_+(\mathsf{V}_\iota)$ for every $\iota \in I$, then $f \in \mathscr{L}^1_+(\curlywedge_{\iota \in I} \mathsf{V}_\iota)$ and

$$(\curlywedge_{\iota \in I} \mathsf{V}_\iota)(f) = \curlywedge_{\iota \in I} (\mathsf{V}_\iota f).$$

For any $\mathsf{V} \in \mathscr{V}$, we denote by $S(\mathsf{V})$, the smallest closed set F, of X, for which

$$\mathsf{V}f = 0$$

for all $f \in \mathscr{K}_+(X)$ and such that

$$F \cap \operatorname{Supp} f = \varnothing.$$

The map $\mathsf{V} \mapsto S(\mathsf{V})$ is an abstract carrier on $(\mathscr{V}, X)$.

The map
$$p \mapsto \mathsf{V}_p \colon \mathscr{P} \to \mathscr{V}$$
is an injection that preserves the algebraic and lattice operations and the band in $\mathscr{V}$ generated by $\{\mathsf{V}_p \mid p \in \mathscr{P}\}$ is equal to $\mathscr{V}$.

8.2.5. Assume that X has a countable base and let p be a potential (resp. a finite potential) on X. Let A be a Borel set such that $p = p_A$. Then there exists a sequence, $(K_n)_{n \in \mathbb{N}}$, of pairwise disjoint, compact subsets of A such that
$$p = \sum_{n \in \mathbb{N}} p_{K_n}$$
and such that, for any $n \in \mathbb{N}$, the restriction of p (resp. p_{K_n}) to K_n is continuous. (The proof is similar to the proof of Theorem 8.2.2.) Obtain a similar result with p replaced by a kernel.

8.2.6. The band $\mathscr{P}_f$, of $\mathscr{P}$, generated by the finite potentials and the band of $\mathscr{P}$, generated by the potentials, p, having the property that $S(p)$ is compact and $p|_{S(p)}$ is finite and continuous, coincide (use Theorem 8.2.2). Any $p \in \mathscr{P}$ may be uniquely written in the form
$$p =: p' + p'',$$
where $p' \in \mathscr{P}_f$ and p'' possesses the property that for any $q \in \mathscr{P}$, $q \preccurlyeq p''$,
$$q_A = 0,$$
where $A := \{x \in X \mid q(x) < \infty\}$.

8.2.7. Let p, q be two potentials on X and let U be an open set for which there exists a harmonic function h, on U, such that $p|_U = q|_U + h$. Then, for every Borel set $A \subset U$, we have $p_A = q_A$.

8.2.8. Let $(X_\iota)_{\iota \in I}$ be a family of pairwise disjoint, open and closed sets of X whose union is equal to X. For any $\iota \in I$, we set
$$\mathscr{S}_\iota := \{u \in \mathscr{S} \mid u|_{X \setminus X_\iota} = 0\}.$$

a) for any $\iota \in I$, $\mathscr{S}_\iota$ is a band of $\mathscr{S}$;

b) for any $\iota, \kappa \in I$, $\iota \neq \kappa$, $\mathscr{S}_\iota \cap \mathscr{S}_\kappa = \{0\}$;

c) for any $u \in \mathscr{S}$, there exists a unique family $(u_\iota)_{\iota \in I}$ in $\mathscr{S}$ such that for any $\iota \in I$, $u_\iota \in \mathscr{S}_\iota$ and
$$u = \sum_{\iota \in I} u_\iota.$$

Let $\iota \in I$ and let $\mathscr{S}(X_\iota)$ be the prevector lattice of positive, superharmonic functions on X_ι. The map
$$u \mapsto u|_{X_\iota} \colon \mathscr{S}_\iota \to \mathscr{S}(X_\iota)$$
is an isomorphism with respect to the algebraic and order structures.

§ 8.3. The Band $\mathcal{M}$

In classical potential theory, every locally bounded potential can be written as the sum of a family of finite, continuous potentials. This property is no longer true on harmonic spaces. Nevertheless, classical results connected with the above property, for instance, Frostman's domination principle or the Evans-Vasilescu continuity principle, still hold for the set of potentials which have the above property. In this section we shall study this set of potentials.

We denote by $\mathcal{M}$, the set of potentials on X which are the specific supremum of their finite, continuous minorants (N. Boboc-C. Constantinescu-A. Cornea 1965 [3]).

Proposition 8.3.1. *Every potential in $\mathcal{M}$ is the sum of a set of finite, continuous potentials on X.*

Let $p \in \mathcal{M}$ and let $\mathfrak{C}$ be the set of sets, $\mathcal{F}$, of non-identically zero, finite, continuous potentials on X such that

$$\sum_{q \in \mathcal{F}} q \preccurlyeq p.$$

Since $\mathfrak{C}$ ordered by the inclusion relation is inductive it has a maximal element $\mathcal{F}_0$. Let p_0 be a potential on X such that

$$p = \sum_{q \in \mathcal{F}_0} q + p_0.$$

Suppose q_0 is a finite, continuous potential on X and a specific minorant of p. Then for any $\alpha \in [0, 1]$, $\alpha(p_0 \curlywedge q_0)$ is a finite, continuous specific minorant of p and $\mathcal{F}_0 \cup \{\alpha(p_0 \curlywedge q_0)\} \in \mathfrak{C}$. From the maximality of $\mathcal{F}_0$ it follows

$$p_0 \curlywedge q_0 = 0.$$

Hence

$$q_0 \preccurlyeq \sum_{q \in \mathcal{F}_0} q$$

(Proposition 8.1.1 *c)*). Since q_0 is arbitrary, we have that

$$p = \sum_{q \in \mathcal{F}_0} q. \quad \square$$

Theorem 8.3.1. *$\mathcal{M}$ is a band of $\mathcal{P}$.*

Conditions *a)* and *c)* of the definition of a band are trivially satisfied. Let $p \in \mathcal{M}$, $q \in \mathcal{P}$ with $q \preccurlyeq p$. We set

$$\mathcal{Q} = \{p' \in \mathcal{P} \mid p' \preccurlyeq p,\ p' \text{ finite and continuous}\}.$$

For any $p' \in \mathscr{Q}$, $p' \curlywedge q$ is finite and continuous, since the sum of two lower semi-continuous real functions is finite and continuous only if both are continuous. From

$$q = q \curlywedge p = q \curlywedge \big(\curlyvee_{p' \in \mathscr{Q}} p'\big) = \curlyvee_{p' \in \mathscr{Q}} (q \curlywedge p')$$

(Proposition 8.1.1 *a*)) it follows that $q \in \mathscr{M}$. □

Proposition 8.3.2. *Let u be a superharmonic function on X and U be an MP-set of X. If $\overline{H}_u^U$ is the greatest harmonic minorant of $u|_U$, then for any superharmonic function v on X such that $v \preccurlyeq u$, the function $\overline{H}_v^U$ is the greatest harmonic minorant of $v|_U$.*

Let w be a positive, superharmonic function on X such that

$$v + w := u.$$

and let h be a harmonic minorant of $v|_U$. Then $h + \overline{H}_w^U$ is a harmonic minorant of $u|_U$. Hence

$$h + \overline{H}_w^U \leq \overline{H}_u^U \leq \overline{H}_v^U + \overline{H}_w^U, \qquad h \leq \overline{H}_v^U. \quad □$$

Proposition 8.3.3. *Let p be a finite (resp. locally bounded) potential on X such that there exists an open covering $\mathfrak{W}$ of X with the property that for any relatively compact MP-set U of X contained in a set of $\mathfrak{W}$, $\overline{H}_p^U$ is the greatest harmonic minorant of $p|_U$ (resp. for any $x \in U$, the set $\{y \in \partial U \mid p(y) < \liminf_{U \ni z \to y} p(z)\}$ is of μ_x^U-measure zero). Then $p \in \mathscr{M}$. Moreover if $p|_{S(p)}$ is continuous, then p is continuous.*

With the aid of the previous proposition, we see that any potential on X which is a specific minorant of p possesses the same property. The first assertion now follows from the second one, by applying Theorem 8.2.2.

Assume that $p|_{S(p)}$ is continuous. Let $x \in X$ and K be a compact neighbourhood of x contained in a relatively compact $\mathfrak{P}$-set V, which in turn is contained in a set of $\mathfrak{W}$. We set

$$q := p_K.$$

We have

$$S(q) \subset V.$$

As a specific minorant of p, $q|_{S(q)}$ is obviously continuous. Let f be a positive, continuous, real function on X, which is less than q on X and equal to q on $S(q) \cup (X \setminus V)$. Then Rf is a finite, continuous potential on X (Proposition 2.2.3). Let q' be a finite, continuous potential on V which is strictly positive on $S(q)$. We set

$$U := \{x \in V \mid Rf(x) + q'(x) < q(x)\}.$$

Obviously U is a relatively compact MP-set of X (Corollary 2.3.3) contained in a set of $\mathfrak{W}$ and $U \cap S(q) = \varnothing$. We want to show that U is empty.

Assume first that $\overline{H}_q^U$ is the greatest harmonic minorant of $q|_U$. Since $(Rf + q')|_U \in \overline{\mathscr{U}}_q^U$, we have

$$\overline{H}_q^U \leq Rf + q'.$$

Since q is harmonic on U, we have

$$q \leq \overline{H}_q^U$$

on U. Hence

$$q \leq Rf + q'$$

and U is empty.

Assume now that for any $x \in U$,

$$\mu_x^U(\{y \in \partial U \mid q(y) < \liminf_{U \ni z \to y} q(z)\}) = 0.$$

We have

$$\{y \in \partial U \mid q(y) < \liminf_{U \ni z \to y} q(z)\} = S(q) \cap \overline{U}.$$

Let g be the characteristic function of the set $S(p) \cap \overline{U}$. Since this set is compact, it follows

$$\overline{H}_g^U = 0$$

(Theorem 1.2.1). Choose any $v \in \overline{\mathscr{U}}_g^U$. Then

$$Rf + q' + (\sup_{\overline{U}} q)\, v - q$$

is positive since it is hyperharmonic and its lower limit at every boundary point of U is positive. Hence, v being arbitrary,

$$q \leq Rf + q'$$

and as before U is empty.

It follows that for either hypothesis of the proposition

$$q \leq Rf + q'$$

on V. Since q' is arbitrary,

$$q \leq Rf.$$

on X. The converse inequality is trivial and therefore q is equal to Rf. Hence q is continuous. Since

$$p - q = p_{X \setminus K}$$

and since $p_{X \setminus K}$ is harmonic on a neighbourhood of x, it follows that p is continuous at x and therefore on X, since x was arbitrarily chosen. $\square$

Remark. The last assertion is true in classical potential theory without any restriction and is known as the Evans-Vasilescu continuity principle (F. Vasilesco 1930 [1]).

Theorem 8.3.2. *Let $p \in \mathscr{P}$. The following assertions are equivalent:*

a) $p \in \mathscr{M}$;

b) for any compact set K of X, we have

$$\hat{R}_p^K = \bigwedge_U \hat{R}_p^U,$$

where U runs through the set of neighbourhoods of K;

c) for any $q \in \mathscr{P}$, $q \preccurlyeq p$, we have

$$\hat{R}_q^{S(q)} = q;$$

d) for any MP-set U of X, $\overline{H}_p^U$ is the greatest harmonic minorant of the restriction of p to U;

e) there exists an open covering $\mathfrak{W}$ of X such that for any MP-set U of X contained in a set of $\mathfrak{W}$, $\overline{H}_p^U$ is the greatest harmonic minorant of $p|_U$ (N. Boboc-C. Constantinescu-A. Cornea 1965 [3], C. Constantinescu 1968 [7], N. Boboc-A. Cornea 1968 [5]).

$a \Rightarrow b$ follows from Theorem 5.3.1 *a)* and Proposition 8.3.1.

$b \Rightarrow c$. Let $q' \in \mathscr{P}$ be such that

$$q + q' := p.$$

For any compact set K of X, we have

$$\bigwedge_U \hat{R}_q^U + \bigwedge_U \hat{R}_{q'}^U = \bigwedge_U \hat{R}_p^U = \hat{R}_q^K + \hat{R}_{q'}^K,$$

where U runs through the set of neighbourhoods of K (Theorem 5.1.1, Corollary 4.1.1, Theorem 4.2.1).

Since

$$\hat{R}_q^K \le \bigwedge_U \hat{R}_q^U, \qquad \hat{R}_{q'}^K \le \bigwedge_U \hat{R}_{q'}^U$$

we get

$$\hat{R}_q^K = \bigwedge_U \hat{R}_q^U.$$

If $S(q)$ is compact, then the assertion follows from the above considerations by using Proposition 8.2.2. For a general q,

$$\hat{R}_q^{S(q)} \ge \bigvee_{\substack{f \in \mathscr{K}_+(X) \\ f \le 1}} \hat{R}_{f \cdot q}^{S(f \cdot q)} = \bigvee_{\substack{f \in \mathscr{K}_+(X) \\ f \le 1}} f \cdot q \ge q$$

(Theorem 8.2.1, Corollary 8.2.1).

$c \Rightarrow d$. Let K be a compact subset of U. We set

$$p_1 := p_{X \smallsetminus U}, \qquad p_2 := p_K, \qquad p_3 := p_{U \smallsetminus K}.$$

Then

$$p = p_1 + p_2 + p_3$$

(Corollary 8.2.1). Let h be the greatest harmonic minorant of the restriction of p to U. Since the axiom of natural decomposition is satisfied by the convex cone of hyperharmonic functions on U (Theorem 5.1.1), there exist positive, hyperharmonic functions h_i on U such that

$$h = h_1 + h_2 + h_3, \qquad h_i \leq p_i$$

for $i = 1, 2, 3$. The functions h_i are harmonic being specific minorants of a harmonic function. Since

$$S(p_1) \subset X \smallsetminus U$$

(Corollary 8.2.1), we have by hypothesis

$$h_1 \leq p_1 = \hat{R}_{p_1}^{X \smallsetminus U}.$$

Let q be an Evans function of p. Then $h_2 - q|_U \in \underline{\mathscr{U}}_{p_2}^U$ since p_2 is continuous at every point of ∂U. Hence

$$h_2 - q \leq \underline{H}_{p_2}^U \leq \overline{H}_{p_2}^U = \hat{R}_{p_2}^{X \smallsetminus U}$$

(Proposition 5.3.3) and

$$h_2 \leq \hat{R}_{p_2}^{X \smallsetminus U} + \bigwedge_q q = \hat{R}_{p_2}^{X \smallsetminus U}$$

(Proposition 2.2.4) on U. We deduce

$$h = h_1 + h_2 + h_3 \leq \hat{R}_{p_1}^{X \smallsetminus U} + \hat{R}_{p_2}^{X \smallsetminus U} + p_3 \leq \hat{R}_p^{X \smallsetminus U} + p_3 = \overline{H}_p^U + p_{U \smallsetminus K}$$

on U (Theorem 4.2.1, Theorem 5.1.1, Proposition 5.3.3). Since K is arbitrary, it follows that

$$h \leq \overline{H}_p^U + \bigwedge_K p_{U \smallsetminus K} = \overline{H}_p^U$$

(Theorem 8.2.1).

$d \Rightarrow e$ is trivial.

$e \Rightarrow a$. By Proposition 8.3.2, any specific minorant of p possesses the same property with the same $\mathfrak{W}$. Since $\mathscr{M}$ is a band (Theorem 8.3.1), we may assume that 0 is the only specific minorant of p belonging to $\mathscr{M}$ (Proposition 8.1.1 *e*)). Let K be a compact set of X. We want to show that if K is contained either in $\{x \in X \mid p(x) = \infty\}$ or in $\{x \in X \mid p(x) < \infty\}$, then $p_K = 0$.

Assume first that

$$K \subset \{x \in X \mid p(x) = \infty\}$$

Since K is contained in the union of a finite family of compact sets, each of which is contained in an MP-set and in a set of $\mathfrak{W}$, we may assume by Corollary 8.2.1, that K is contained in an MP-set U which is contained in a set of $\mathfrak{W}$. For any $u \in \overline{\mathscr{U}}_{p_K}^U$, and for any strictly positive real number ε, we have $u|_{U \setminus K} + \varepsilon\, p|_{U \setminus K} \in \overline{\mathscr{U}}_{p_K}^{U \setminus K}$. Hence

$$p_K = \overline{H}_{p_K}^{U \setminus K} \leq u + \varepsilon\, p$$

on $\overline{U} \setminus K$. But u and ε are arbitrary which implies that

$$p_K \leq \overline{H}_{p_K}^U$$

on $U \setminus K$. By Corollary 6.3.4, we deduce that p_K and $\overline{H}_{p_K}^U$ coincide on U. Hence p_K is harmonic on U and therefore on X (Corollary 8.2.1). Since it is a potential, it vanishes identically.

Assume now that

$$K \subset \{x = X \mid p(x) < \infty\}$$

Then p_K is a finite potential on X. By Proposition 8.3.3, it belongs to $\mathscr{M}$ and being a specific minorant of p, it vanishes identically.

Let A be the set $\{x \in X \mid p(x) = \infty\}$. From the above considerations and Theorem 8.2.1, we have

$$p_A = 0, \qquad p_{X \setminus A} = 0.$$

Hence (Corollary 8.2.1)

$$p = p_A + p_{X \setminus A} = 0. \quad \square$$

Corollary 8.3.1. *If the restriction of a potential p of $\mathscr{M}$ to $S(p)$ is finite and continuous, then p is finite and continuous.*

The corollary follows from the last assertion of Proposition 8.3.3 by using the implication, $a \Rightarrow d$, of the theorem. $\square$

Proposition 8.3.4. *Let $p \in \mathscr{M}$. For every totaly thin compact set K, we have $p_K = 0$.*

We may assume that p is finite and continuous. Since K is contained in the union of a family of $\mathfrak{P}$-sets of X, it is the union of a finite family of compact sets, each of which is contained in a $\mathfrak{P}$-set. Hence we may assume that K is contained in a $\mathfrak{P}$-set U. Let

$$p_K|_U =: p' + h$$

be the Riesz decomposition of $p_K|_U$. Let $\mathcal{F}$ be the set of functions of the form $(q+\alpha p')|_K$, where q is a finite, continuous potential on U and $\alpha \in \mathbb{R}$. If p' does not vanish identically on K, then there exists a point $x \in K$ and two functions $f, g \in \mathcal{F}$ such that $f(x)<0$, $g(x)<\infty$ and $g(y)>0$ for all $y \in K$. By lemma 2, from the proof of Theorem 1.3.1, there exists a point $x_0 \in K$ such that ε_{x_0} is the only measure μ on K satisfying the relation

$$\int u\, d\mu \le u(x_0)$$

for every $u \in \mathcal{F}$. We have

$$\int p'\, d^U\varepsilon_{x_0}^K = {}^U\hat{R}_{p'}^K(x_0) = p'(x_0),$$

where ${}^U\varepsilon_{x_0}^K$ denotes the balayage of ε_{x_0} on K, the balayage operation being performed on the harmonic space U (Theorem 8.3.2 $a \Leftrightarrow c$). Hence for any $u \in \mathcal{F}$,

$$\int u\, d^U\varepsilon_{x_0}^K \le u(x_0).$$

It follows that ${}^U\varepsilon_{x_0}^K = \varepsilon_{x_0}$ which contradicts the fact that K is thin at x_0 (Theorem 7.1.1). Hence $p'=0$ on K and therefore on U (Theorem 8.3.2 $a \Rightarrow c$). We deduce that p_K is harmonic on U and therefore on X. Since it is a potential, it vanishes identically. □

Corollary 8.3.2. *Assume that X has a countable base and let $p \in \mathcal{M}$. Then every semi-polar set A is p-measurable and* $p_A=0$ (C. Constantinescu 1968 [7]).

It is sufficient to prove the assertion for a totally thin set A contained in a $\mathfrak{P}$-set. By Corollary 7.2.3, we may assume that A is a Borel set. Now the assertion follows from the proposition and Theorem 8.2.1. □

Exercises

8.3.1. Assume that X is a $\mathfrak{P}$-harmonic space and let $\mu \in \Lambda$. For any $p \in \mathcal{M}$ and for any $A \subset X$, we have

$$\int p\, d\mu^A = \int \hat{R}_p^A\, d\mu.$$

8.3.2. Assume that X has a base of regular sets and let p be locally bounded potential. The following assertions are equivalent:

a) $p \in \mathcal{M}$;

b) for any relatively compact, MP-set U, of X and any $f \in \mathscr{C}(\overline{U})$ such that $f|_U$ is a harmonic minorant of $p|_U$, we have

$$f|_U \le \overline{H}_p^U;$$

c) there exists an open covering $\mathfrak{W}$ of X which satisfies the following condition: for any regular set U contained in a set of $\mathfrak{W}$ and any finite, continuous potential q on X with $q|_U$ a harmonic minorant of $p|_U$, we have

$$q|_U \leq H_p^U$$

(N. Boboc-A. Cornea 1968 [5]).

8.3.3. Let f be a positive, numerical function on X, K be a compact set of X and u be a positive, superharmonic function on X such that

$$\widehat{Rf} \leq u \quad \text{on } X, \qquad f > u \quad \text{on } K.$$

If there exists a family $(u_\iota)_{\iota \in I}$ of positive, hyperharmonic functions on X such that for any $\iota \in I$, u_ι is continuous at every point of K and

$$u = \sum_{\iota \in I} u_\iota$$

on K, then K is polar. In particular, if $\widehat{Rf} \in \mathcal{M}$, then any compact subset of $\{x \in X | \widehat{Rf}(x) < f(x)\}$ is polar (N. Boboc-A. Cornea 1968 [5]).

8.3.4. Let $\mathscr{E}(U)$ be the set of positive, superharmonic functions, u, on U such that for any positive superharmonic function v on U we have

$$u \leq v \Rightarrow u \preccurlyeq v.$$

The set $\mathscr{E}(X)$ is a band of $\mathscr{S}$ containing $\mathscr{H}_+(X)$. If p is a potential on X for which there exists a p-measurable, polar set A of X such that $p = p_A$, then $p \in \mathscr{E}(X)$. $\mathcal{M} \cap \mathscr{E}(X) = \{0\}$. If X is an $\mathfrak{S}$-harmonic space, then every locally bounded function in $\mathscr{E}(X)$ is harmonic. If X is a $\mathfrak{P}$-harmonic space, then $U \mapsto \mathscr{E}(U)$ is a sheaf. It is not known whether a positive, superharmonic function on X whose locally bounded specific minorants are identically 0 belongs to $\mathscr{E}(X)$ (N. Boboc-C. Constantinescu-A. Cornea 1965 [3]).

8.3.5. Let V be a kernel on X (Exercises 8.2.3 and 8.2.4). The following assertions are equivalent:

a) $f \in \mathscr{L}_+^1(V) \Rightarrow \mathsf{V}f \in \mathcal{M}$;

b) $f \in \mathscr{K}_+(X) \Rightarrow \mathsf{V}f \in \mathcal{M}$;

c) V is the least upper bound in $\mathscr{V}$ of the set

$$\{\mathsf{V}_p | \mathsf{V}_p \preccurlyeq \mathsf{V},\ p \text{ finite continuous potential on } X\};$$

d) $f \in \mathscr{L}_+^1(\mathsf{V}) \Rightarrow \hat{R}_{\mathsf{V}f}^{\{f>0\}} = \mathsf{V}f$;

e) if f is the characteristic function of a compact set K, then

$$\hat{R}_{\mathsf{V}f}^K = \mathsf{V}f.$$

The set, $\mathscr{V}_{\mathcal{M}}$, of kernels satisfying one (and hence, all) of the above conditions is the band of $\mathscr{V}$ generated by $\{\mathsf{V}_p | p \in \mathcal{M}\}$ (C. Constantinescu 1968 [7]).

8.3.6. Let V be a kernel on X and let A_{V} be the set of points $x \in X$ for which there exists $f \in \mathscr{K}_+(X)$ such that $\mathsf{V}f(X)=\infty$. We assume: *a)* for any compact subset K of A_{V}, we have $\mathsf{V}K=0$; *b)* there exists a strictly positive, finite, continuous, superharmonic function u on X such that for any compact set K of X and any $\alpha \in \mathbb{R}_+$, we have

$$\mathsf{V}K \leq \alpha u \text{ on } K \Rightarrow \mathsf{V}K \leq \alpha u \text{ on } X.$$

Then $\mathsf{V} \in \mathscr{V}_{\mathscr{M}}$ (see the preceding exercise).

(We may assume that there exists $p \in \mathscr{P}$ such that $\mathsf{V}=\mathsf{V}_p$. Moreover, by Exercise 8.2.6, we may assume that p is finite. Let A be a Borel set and choose $\alpha \in \mathbb{R}_+$ such that $p_A \leq \alpha u$ on A. Show that $p_A \leq \alpha u$ on X. Let K be a compact set such that the restriction of p to K is continuous and let $x \in K$. Show first that $p_{\{x\}}$ is continuous. Hence for any open neighbourhood U, of x, one has

$$\limsup_{y \to x} p_{K \cap (U \setminus \{x\})}(y) - p_{K \cap (U \setminus \{x\})}(x) = \limsup_{y \to x} p_K(y) - p_K(x).$$

The family $(p_{K \cap (U \setminus \{x\})}|_K)_U$, where U is an arbitrary open neighbourhood of x, is a lower directed family of continuous functions whose infimum is 0. Hence for any strictly positive real number ε, there exists an open neighbourhood U of x such that $p_{K \cap (U \setminus \{x\})} \leq \varepsilon u$ on K and therefore on X. By the above remark

$$\limsup_{y \to x} p_K(y) - p_K(x) \leq \varepsilon u(x),$$

and since ε and x are arbitrary, it follows that p_K is continuous.

Let q be the greatest specific minorant of p which belongs to $\mathscr{M}$ and let $x \in X$. By Lusin's theorem, there exists a sequence $(K_n)_{n \in \mathbb{N}}$ of pairwise disjoint, compact sets in X such that for any $n \in \mathbb{N}$, $p|_{K_n}$ is continuous and

$$p(x) = \sum_{n \in \mathbb{N}} p_{K_n}(x).$$

By the above proof $\sum_{n \in \mathbb{N}} p_{K_n} \in \mathscr{M}$. Since $\sum_{n \in \mathbb{N}} p_{K_n} \preccurlyeq p$, we see that $p(x)=q(x)$. Since x is arbitrary, $p \in \mathscr{M}$.)

8.3.7. Assume that X has a countable base and let V be a kernel of $\mathscr{V}_{\mathscr{M}}$ (Exercise 8.3.5). Then:

a) Every semi-polar set, A, is V-integrable and $\mathsf{V}A=0$.

b) Every relatively compact set belonging to $\mathfrak{B}$, defined in Exercise 7.2.5, is V-integrable.

c) Let $\mathfrak{A}$ be the set of fine closed sets A, of X, such that $\mathsf{V}(X \setminus A)=0$; then the set $\bigcap_{A \in \mathfrak{A}} A$ belongs to $\mathfrak{A}$, is of type G_δ and is not thin at any of its points (this set is called *the fine carrier of* V).

d) Let $\mathscr{U}_{\mathsf{V}}$ be the set of hyperharmonic functions u on X for which there exists a sequence $(f_n)_{n\in\mathbb{N}}$ in $\mathscr{L}_+^1(\mathsf{V})$ such that $(\mathsf{V}f_n)_{n\in\mathbb{N}}$ is increasing (in the natural order) and converges to u. Let A be a V-integrable subset of X. If $\mathsf{V}A=0$, then

$$\hat{R}_u^{X\setminus A}=u$$

for any $u\in\mathscr{U}_{\mathsf{V}}$. If A is fine open and

$$\hat{R}_u^{X\setminus A}=u$$

for every $u\in\mathscr{U}_{\mathsf{V}}$, then $\mathsf{V}A=0$.

e) A fine closed set A, of X, belongs to $\mathfrak{A}$ (defined in *c)*) if and only if

$$\hat{R}_u^{X\setminus A}=u$$

for every $u\in\mathscr{U}_{\mathsf{V}}$ (C. Constantinescu 1968 [7]).

8.3.8. If X has a countable base, then for any potential p of $\mathscr{M}$, there exists a sequence $(p_n)_{n\in\mathbb{N}}$ of finite continuous potentials on X such that for any $m,n\in\mathbb{N}$, $S(p_n)$ and $S(p_m)$ are compact and disjoint and such that $p=\sum_{n\in\mathbb{N}}p_n$ (use the preceding exercise *a)*, Exercise 8.2.5 and Corollary 8.3.1).

8.3.9. Assume that X has a countable base and let p be a potential on X. Let $(x_n)_{n\in\mathbb{N}}$ be a dense sequence in X such that p is finite at every x_n. Then

$$\mu:=\sum_{n\in\mathbb{N}}\frac{1}{2^n(1+p(x_n))}\mathsf{V}_{p,x_n}$$

is a measure on X and for every $x\in X$, $\mathsf{V}_{p,x}$ is absolutely continuous with respect to μ. If p is a strict potential of $\mathscr{M}$, then every semi-polar set of X is of μ-measure zero and every fine open set is μ-measurable and of positive μ-measure (use Corollary 8.3.2 and Exercise 8.3.7).

8.3.10. If X is σ-compact and $p\in\mathscr{M}$, then there exists an Evans function of p which belongs to $\mathscr{M}$. (If p is finite and continuous use Proposition 2.2.4. For a general p, let $(p_\iota)_{\iota\in I}$ be a family of finite, continuous potentials on X such that

$$p=\sum_{\iota\in I}p_\iota.$$

For any compact set K and any strictly positive real number ε, there exists $f\in\mathscr{K}_+(X)$, $f\leq 1$, such that

$$\sum_{\iota\in I}R((1-f)\,p_\iota)<\varepsilon$$

on K. Construct an Evans function of the form

$$\sum_{n\in\mathbb{N}}\sum_{\iota\in I}R((1-f_n)p_\iota).)$$

§ 8.4. Quasi-Continuity

H. Cartan 1945 [2] proved that every superharmonic function on R^n is quasi-continuous i.e. there exists an open set, having an arbitrarily small capacity, such that the restriction of the given function to the complementary set is continuous. This property of superharmonic functions is not true in general harmonic spaces. In this section, we introduce the set of quasi-continuous, hyperharmonic functions on harmonic spaces and study their properties.

Let u be a hyperharmonic function on X and let $\mathfrak{K}(u)$ be the set of compact sets K of X for which the function $u|_K$ is continuous. We say that u is *quasi-continuous* if, for any open set U of X and any locally bounded potential p, on U, we have

$$\bigwedge_{\substack{K \in \mathfrak{K}(u) \\ K \subset U}} {}^U R_p^{U \setminus K} = 0.$$

If u is a quasi-continuous hyperharmonic function on X and if U is an open set of X, then clearly $u|_U$ is quasi-continuous on the harmonic subspace U.

Proposition 8.4.1. *Let u be a hyperharmonic function on X and let $\mathfrak{W}$ be a covering of X by $\mathfrak{P}$-sets, such that for any $W \in \mathfrak{W}$ and any finite, continuous potential p on W, we have*

$$\bigwedge_{\substack{K \in \mathfrak{K}(u) \\ K \subset W}} {}^W R_p^{W \setminus K} = 0.$$

Then u is quasi-continuous.

Let U be an open set of X and let p be a locally bounded potential on U. We set

$$p_0 := \bigwedge_{\substack{K \in \mathfrak{K}(u) \\ K \subset U}} {}^U R_p^{U \setminus K}.$$

In order to show that p_0 vanishes identically, it is sufficient to show that it is harmonic.

Let $W \in \mathfrak{M}$. Choose $K \in \mathfrak{K}(u)$ with $K \subset U$, and let h_K be the greatest harmonic minorant of the restriction of ${}^U R_p^{U \setminus K}$ to $U \cap W$. We set

$$q := ({}^U R_p^{U \setminus K})|_{U \cap W} - h_K.$$

Now q is a locally bounded potential on $U \cap W$. Let L be a compact subset of $U \cap W$ and choose $K' \in \mathfrak{K}(u)$ with $K' \subset W$. Let q' be a finite, continuous potential on W which is greater than q on L. The function on U which is equal to ${}^U R_p^{U \setminus K}$ on $U \setminus W$ and

$$\inf({}^U R_p^{U \setminus K}, h_K + {}^{U \cap W} R_p^{U \cap W \setminus L} + {}^W R_{q'}^{W \setminus K'})$$

on $U \cap W$ is hyperharmonic and greater than p on $U \setminus K \cup (K' \cap L)$. Since $K \cup (K' \cap L)$ is contained in U and belongs to $\mathfrak{K}(u)$ we have

$$p_0 \leq h_K + {}^{U \cap W}R_p^{U \cap W \setminus L} + {}^W R_{q'}^{W \setminus K'}$$

on $U \cap W$. Since K' is arbitrary and q' is a finite, continuous potential on W, we have

$$p_0 \leq h_K + {}^{U \cap W}R_q^{U \cap W \setminus L} + \bigwedge_{\substack{K' \in \mathfrak{K}(u) \\ K' \subset W}} {}^W R_{q'}^{W \setminus K} = h_K + {}^{U \cap W}R_q^{U \cap W \setminus L}$$

on $U \cap W$. But L is arbitrary and hence

$$p_0 \leq h_K + \bigwedge_{L \subset U \cap W} {}^{U \cap W}R_q^{U \cap W \setminus L} = h_K$$

on $U \cap W$ (Corollary 5.3.1). Since K is arbitrary, we deduce that

$$p_0 \leq \bigwedge_{\substack{K \in \mathfrak{K}(u) \\ K \subset U}} h_K \leq \bigwedge_{\substack{K \in \mathfrak{K}(u) \\ K \subset U}} {}^U R_p^{U \setminus K} = p_0$$

on $U \cap W$. Now $(h_K)_{K \in \mathfrak{K}(u), K \subset U}$ is a lower directed family of positive harmonic functions on $U \cap W$ and it follows that p_0 is harmonic on $U \cap W$. Finally W is arbitrary and thus p_0 is harmonic on U. □

Remark. From the proposition, it follows that any hyperharmonic function is quasi-continuous if it is locally quasi-continuous and that the potentials involved in the definition of quasi-continuity may be taken to be finite and continuous.

Proposition 8.4.2. *If u and v are two quasi-continuous, hyperharmonic functions on X, then $u+v$ and $u \wedge v$ are quasi-continuous.*

We set

$$w := u + v \qquad (\text{resp. } w := u \wedge v).$$

Let U be an open set of X and p be a locally bounded potential on U. Choose $K' \in \mathfrak{K}(u)$ and $K'' \in \mathfrak{K}(v)$, $K' \cup K'' \subset U$. Then $K' \cap K'' \in \mathfrak{K}(w)$ and

$${}^U R_p^{X \setminus (K' \cap K'')} \leq {}^U R_p^{X \setminus K'} + {}^U R_p^{X \setminus K''}.$$

Hence

$$\bigwedge_{\substack{K \in \mathfrak{K}(w) \\ K \subset U}} {}^U R_p^{X \setminus K} \leq \bigwedge_{\substack{K \in \mathfrak{K}(u) \\ K \subset U}} {}^U R_p^{X \setminus K} + \bigwedge_{\substack{K \in \mathfrak{K}(v) \\ K \subset U}} {}^U R_p^{X \setminus K} = 0$$

(Corollary 4.1.1). □

Proposition 8.4.3. *Let $\mathscr{F}$ be an upper directed family of hyperharmonic functions on X such that*

$$\bigvee \mathscr{F} = \infty$$

on X. If u is a hyperharmonic function on X such that $u \wedge v$ is quasi-continuous for any $v \in \mathscr{F}$, then u is quasi-continuous.

Let U be an open set of X and let K be a compact subset of U. By Dini's theorem, there exists a sequence, $(v_n)_{n\in\mathbb{N}}$, in $\mathscr{F}$ such that $v_n \geq n$ on K. Let $(K_n)_{n\in\mathbb{N}}$ be a sequence of compact set of U such that $K_n \in \mathfrak{K}(u \wedge v_n)$ for any $n \in \mathbb{N}$. Then

$$K' := \bigcap_{n\in\mathbb{N}} (K \cap K_n) \in \mathfrak{K}(u).$$

Let p be a locally bounded potential on U. We have

$${}^U R_p^{U\setminus K'} \leq {}^U R_p^{U\setminus K} + \sum_{n\in\mathbb{N}} {}^U R_p^{U\setminus K_n}.$$

Hence

$$\bigwedge_{\substack{L\in\mathfrak{K}(u)\\ L\subset U}} {}^U R_p^{U\setminus L} \leq {}^U R_p^{U\setminus K} + \sum_{n\in\mathbb{N}} \bigwedge_{\substack{L\in\mathfrak{K}(u\wedge v_n)\\ L\subset U}} {}^U R_p^{U\setminus L} = {}^U R_p^{U\setminus K}$$

(Proposition 4.1.3). Since K is arbitrary,

$$\bigwedge_{\substack{L\in\mathfrak{K}(u)\\ L\subset U}} {}^U R_p^{U\setminus L} = 0$$

(Corollary 5.3.1). $\square$

We denote by $\mathscr{Q}$, the set of positive, quasi-continuous potentials on X.

Theorem 8.4.1. *$\mathscr{Q}$ is a band of $\mathscr{P}$* (N. Boboc-C. Constantinescu-A. Cornea 1965 [3]).

If $p, q \in \mathscr{Q}$, then $p + q \in \mathscr{Q}$ (Proposition 8.4.2).

Now let $p \in \mathscr{Q}$, $q \in \mathscr{P}$ with $q \preccurlyeq p$. Let U be an open set of X and let p_0 be a locally bounded potential on U. Let $K \in \mathfrak{K}(p)$, $K \subset U$, and let $\alpha \in]0, \infty[$. We set

$$K_\alpha := \{x \in K \mid p(x) \leq \alpha\},$$

$$\beta := \sup_K p_0.$$

Then $K_\alpha \in \mathfrak{K}(q)$, $K_\alpha \subset U$ and

$${}^U R_{p_0}^{U\setminus K_\alpha} \leq {}^U R_{p_0}^{U\setminus K} + \frac{\beta}{\alpha} p,$$

$$\bigwedge_{\substack{K\in\mathfrak{K}(q)\\ K\subset U}} {}^U R_{p_0}^{U\setminus K} \leq \bigwedge_{\substack{K\in\mathfrak{K}(p)\\ K\subset U}} {}^U R_{p_0}^{U\setminus K} + \bigwedge_{\alpha\in]0,\infty[} \frac{\beta}{\alpha} p = 0.$$

Hence $q \in \mathscr{Q}$.

Let $\mathscr{F}$ be a specifically upper directed subset of $\mathscr{Q}$ such that

$$q := \curlyvee \mathscr{F} \in \mathscr{P}.$$

Let p_0 be a finite, continuous potential on an open set U, of X and let x be a point of U at which q is finite. There exists, then, a sequence

$(q_n)_{n\in\mathbb{N}}$ in $\mathscr{F}$ such that

$$\sum_{n\in\mathbb{N}} (q(x)-q_n(x)) < \infty.$$

For any $n\in\mathbb{N}$, we denote by q'_n, the potential on X such that

$$q'_n + q_n := q.$$

Let $(K_n)_{n\in\mathbb{N}}$ be a sequence of compact sets of U such that for any $n\in\mathbb{N}$, $K_n\in\mathfrak{K}(q_n)$. Then for any $\alpha\in]0,\infty[$

$$L_\alpha := \{y\in \bigcap_{n\in\mathbb{N}} K_n \mid \sum_{n\in\mathbb{N}} q'_n(y) \le \alpha\, p_0\} \in \mathfrak{K}(q).$$

Indeed, let $y\in L_\alpha$ and $\mathfrak{U}$ be an ultrafilter on L_α converging to y. If

$$\lim_{\mathfrak{U}} q > q(y),$$

then for any $n\in\mathbb{N}$, we have

$$\lim_{\mathfrak{U}} q'_n = q'_n(y) + (\lim_{\mathfrak{U}} q - q(y)) \ge \lim_{\mathfrak{U}} q - q(y).$$

We deduce that

$$\lim_{\mathfrak{U}} \sum_{n\in\mathbb{N}} q'_n = \infty$$

and this contradicts the relation

$$\sum_{n\in\mathbb{N}} q'_n \le \alpha\, p_0$$

on L_α. Hence

$${}^U R_{p_0}^{U\setminus L} \le \frac{1}{\alpha} \sum_{n\in\mathbb{N}} q'_n + \sum_{n\in\mathbb{N}} {}^U R_{p_0}^{U\setminus K_n},$$

$$\bigwedge_{\substack{K\in\mathfrak{K}(q)\\ K\subset U}} {}^U R_{p_0}^{U\setminus K}(x) \le \bigwedge_{\alpha\in]0,\infty[} \left(\frac{1}{\alpha}\sum_{n\in\mathbb{N}} q'_n\right)(x) + \sum_{n\in\mathbb{N}} \bigwedge_{\substack{K\in\mathfrak{K}(q_n)\\ K\subset U}} {}^U R_{p_0}^{U\setminus K}(x) = 0$$

(Proposition 4.1.3). Since x is arbitrary,

$$\bigwedge_{\substack{K\in\mathfrak{K}(p)\\ K\subset U}} {}^U R_{p_0}^{U\setminus K}$$

is equal to 0 outside a polar set and therefore on U (Corollary 6.3.4). □

Theorem 8.4.2. *Let u be a hyperharmonic function on a $\mathfrak{P}$-harmonic space X. The following assertions are equivalent:*

a) u is quasi-continuous;

b) for any subset, A, of X and any locally bounded potential p on X we have

$$\bigwedge_U R_p^{A\cap U} = 0,$$

where U runs through the set of neighbourhoods of the set

$$\{x \in \bar{A} \setminus A \mid \liminf_{A \ni y \to x} u(y) > u(x)\};$$

c) for any relatively compact, open set V and any $x \in V$, the set

$$\{y \in \partial V \mid \liminf_{V \ni z \to y} u(z) > u(y)\}$$

is of μ_x^V measure zero.

d) there exists an open covering $\mathfrak{W}$ of X such that for any relatively compact, open set V contained in a set of $\mathfrak{W}$ and for any $x \in W$, the set

$$\{y \in \partial V \mid \liminf_{V \ni z \to y} u(z) > u(y)\}$$

is of μ_x^V-measure zero

(N. Boboc-A. Cornea 1968 [5]).

a ⇒ *b*. Let $K \in \mathfrak{K}(u)$. For any $x \in \bar{A} \setminus A$ such that

$$\liminf_{A \ni y \to x} u(y) > u(x),$$

there exists a neighbourhood U_x of x for which

$$A \cap U_x \subset X \setminus K.$$

The set

$$U := \bigcup_{x \in \bar{A} \setminus A} U_x$$

is an open neighbourhood of the set $\{x \in \bar{A} \setminus A \mid \liminf_{A \ni y \to x} u(y) > u(x)\}$ such that

$$A \cap U \subset X \setminus K.$$

Hence

$$\bigwedge_U \hat{R}_p^{A \cap U} \le \bigwedge_{K \in \mathfrak{K}(u)} R_p^{X \setminus K} = 0.$$

b ⇒ *c*. Let f be the characteristic function of the set

$$B := \{y \in \partial V \mid \liminf_{V \ni z \to y} u(z) > u(y)\}$$

and let p be a locally bounded potential on X which is greater than 1 on $\bar{V}$. For any neighbourhood U of the set B, we have

$$\hat{R}_p^{V \cap U}|_V \in \bar{\mathscr{U}}_f^V.$$

Hence

$$\bar{H}_f^V \le \hat{R}_p^{V \cap U}.$$

Since U is arbitrary, we obtain by taking A equal to V in *b)*

$$\bar{H}_f^V \le \bigwedge_U \hat{R}_p^{V \cap U} = 0.$$

c ⇒ *d* is trivial.

$d \Rightarrow a$. Assume first that u is positive and let p be a finite, continuous potential on X. The function $u \wedge p$ has property *d)* and is a locally bounded potential. By Proposition 8.3.3, $u \wedge p$ belongs to $\mathscr{M}$ and therefore by the definition of $\mathscr{M}$, it belongs also to $\mathscr{Q}$ (Theorem 8.4.1). From Proposition 8.4.3, we deduce that u is quasi-continuous.

For the general case let U be a relatively compact, open set of X and let p be a finite, continuous potential on X such that $u+p$ is positive on U. It is immediate that $(u+p)|_U$ has property *d)* on the harmonic subspace U. Hence, $(u+p)|_U$ is quasi-continuous. Since $\mathfrak{K}(u+p)=\mathfrak{K}(u)$ it follows that $u|_U$ is quasi-continuous and U being arbitrary implies that u is quasi-continuous (Proposition 8.4.1). □

Corollary 8.4.1. *If $\mathscr{P}_b$ denotes the band of $\mathscr{P}$ which is generated by the set of locally bounded potentials on X, then*

$$\mathscr{M}=\mathscr{Q}\cap\mathscr{P}_b.$$

The inclusion

$$\mathscr{M}\subset\mathscr{Q}\cap\mathscr{P}_b$$

follows from the definition of $\mathscr{M}$ and from Theorem 8.4.1.

Let $p\in\mathscr{Q}\cap\mathscr{P}_b$ and let q be a locally bounded potential on X which is a specific minorant of p. By Theorem 8.4.1, q is quasi-continuous. From $a \Rightarrow d$ we get, using Proposition 8.3.3, $q\in\mathscr{M}$. But q is arbitrary and $p\in\mathscr{P}_b$, and thus $p\in\mathscr{M}$. □

Exercises

8.4.1. Assume that X is a $\mathfrak{P}$-harmonic space, let $\mu\in\Lambda$ and let p be a locally bounded potential on X.

a) For any positive, quasi-continuous, hyperharmonic function u, on X,

$$\inf_{K\in\mathfrak{K}(u)} \int^* R_p^{X\setminus K}\, d\mu=0.$$

b) Let $\mathsf{V}\in\mathscr{V}_{\mathscr{M}}$ (defined in Exercise 8.3.5) and let $\mathfrak{K}(\mathsf{V})$ be the set of compact sets K in X such that for any $f\in\mathscr{L}^1_+(\mathsf{V})$, $\mathsf{V}f|_K$ is continuous; then

$$\inf_{K\in\mathfrak{K}(\mathsf{V})} \int^* R_p^{X\setminus K}\, d\mu=0$$

(C. Constantinescu 1968 [7]).

8.4.2. Let u be a locally bounded, quasi-continuous, hyperharmonic function on an $\mathfrak{S}$-harmonic space X. For any relatively compact MP-set U, of X, the function $\overline{H}_u^U$ is the greatest harmonic minorant of $u|_U$.

8.4.3. Let φ be a map of X into a topological space and let $\mathfrak{K}(\varphi)$ be the set of compact sets K in X such that $\varphi|_K$ is continuous. φ is called *quasi-continuous* if, for any open set U of X and any locally bounded potential p on U, we have
$$\bigwedge_{\substack{K \in \mathfrak{K}(\varphi) \\ K \subset U}} {}^U R_p^{U \setminus K} = 0.$$

a) If φ is quasi-continuous, then the set of points of X at which φ is not fine continuous is a polar set (B. Fuglede 1965 [1]).

b) If $(\varphi_n\colon X \to Y_n)_{n \in \mathbb{N}}$ is a sequence of quasi-continuous maps, then the map
$$x \mapsto (\varphi_n(x))_{n \in \mathbb{N}} : X \to \prod_{n \in \mathbb{N}} Y_n$$
is quasi-continuous (use Exercise 6.2.14*b)*).

8.4.4. Let $\mathscr{F}$ be a finite family in $\mathscr{Q}$ (resp. $\mathscr{M}$). Then $\bigvee \mathscr{F}$ and $\bigwedge \mathscr{F}$ belong to $\mathscr{Q}$ (resp. $\mathscr{M}$).

PART THREE

Chapter 9

The Axiom of Polarity and The Axiom of Domination

Throughout this chapter X will denote a harmonic space with a countable base.

The hypothesis of countable base is not necessary for the greater part of the results of this chapter. We have assumed it here only for the sake of simplicity and the reader who is interested in the uncountable case will find in Exercises 9.1.8 and 9.2.7 the counterparts of a number of statements proved in the text.

§ 9.1. Axiom of Polarity

We say that the *property of polarity* holds on an open set U of X if for every set $\mathscr{F}$ of positive, hyperharmonic functions on U, the set

$$\{x \in U \mid \bigwedge_{u \in \mathscr{F}} u(x) < \inf_{u \in \mathscr{F}} u(x)\}$$

is polar. The property of polarity holds on every open set of the harmonic space associated with the Laplace equation (H. Cartan 1942 [1]).

Theorem 9.1.1. *If X is a $\mathfrak{P}$-harmonic space then the following assertions are equivalent:*

a) the property of polarity holds on X;

b) (resp. b′)) Every semi-polar (resp. compact totally thin) set of X is polar;

c) (resp. c′)) for any subset (resp. compact set) A of X, the set $A \setminus b(A)$ is polar;

d) (resp. d′)) Every non-polar subset (resp. compact set) of X, possesses a point at which it is not thin;

e) (resp. e′)) for any subset (resp. compact set) A of X and any positive hyperharmonic function u on X, we have

$$\hat{R}^A_{\hat{R}^A_u} = \hat{R}^A_u;$$

f) (resp. f′)) for any subset (resp. compact set) A of X, and any $\mu\in\Lambda$, *we have*

$$(\mu^A)^A=\mu^A;$$

g) for any compact set K of X and any $p\in\mathcal{M}$, *we have* $\hat{R}_p^K\in\mathcal{M}$;

h) for any non-polar compact set K of X, there exists a finite, continuous potential p on X such that

$$\varnothing\neq S(p)\subset K;$$

h′) for compact set K of X with the property that for any open set U of X, $U\cap K$ *is either empty or non-polar, there exists a finite, continuous potential p on X such that*

$$S(p)=K;$$

i) for every open set U of X, the set of non-regular boundary points of U is polar;

j) for any relatively compact open set U of X and any $x\in U$, *the set of non-regular boundary points of U is of* μ_x^U*-measure zero;*

k) (resp. k′)), for any $\mu\in\Lambda$ *and any subset (resp. compact set) A of X, we have*

$$\mu^A(X\setminus b(A))=0$$

(R.-M. Hervé 1962 [4], H. Bauer 1965 [7], M. Brelot 1965 [19], N. Boboc-A. Cornea 1966 [1], 1968 [5]).

We prove the following system of implications:

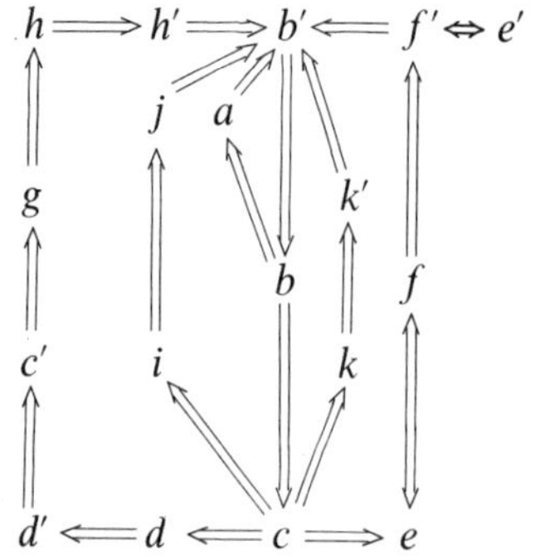

$a\Rightarrow b'$. Let K be a totally thin set of X and let p be a finite, strict potential on X (Proposition 7.2.1). Then

$$\hat{R}_p^K<R_p^K=p$$

on K (Proposition 7.2.2). From the property of polarity, it follows that K is polar.

$b' \Rightarrow b$. Let A be a totally thin G_δ set in X. Since every compact subset of A is polar and since A is K-analytic (Corollary 5.2.2), it follows that A is polar (Proposition 6.2.2). Now any totally thin set of X is contained in a totally thin G_δ set (Corollary 7.2.3) and since the countable union of polar sets is polar (Corollary 6.2.3), it follows that any semi-polar set is polar.

$b \Rightarrow a$ follows from Theorem 6.3.2.

$b \Rightarrow c$ follows from Corollary 7.2.1.

$c \Rightarrow d \Rightarrow d'$ are trivial.

$d' \Rightarrow c'$. Let K be a compact set. Since $b(K)$ is a G_δ set (Corollary 7.2.1), there exists a sequence $(K_n)_{n \in \mathbb{N}}$ of compact sets such that

$$K \setminus b(K) = \bigcup_{n \in \mathbb{N}} K_n.$$

Let $n \in \mathbb{N}$. Since

$$b(K_n) \subset K_n \cap b(K) = \varnothing$$

we deduce that K_n is polar and hence that $K \setminus b(K)$ is polar (Corollary 6.2.3).

$c' \Rightarrow g$. It is sufficient to prove the assertion when p is a finite, continuous potential. We set

$$q := \hat{R}_p^K.$$

Since q is locally bounded, we have

$$q_{K \setminus b(K)} = 0$$

(Proposition 8.2.3) and thus

$$q = q_{b(K)}.$$

Choose any $x \in b(K)$. We have

$$p(x) = q(x) = \liminf_{y \to x} q(y) \leq \lim_{y \to x} p(y) = p(x).$$

Hence q is continuous at all points of $b(K)$. For any compact set $L \subset b(K)$, the function q_L is continuous at every point of L since it is a specific minorant of q. Hence q_L is a finite continuous potential on X. Since $b(K)$ is a Borel set, we have

$$q = q_{b(K)} = \curlyvee_L q_L,$$

where L runs through the set of compact subsets of $b(K)$. Hence $q \in \mathcal{M}$.

$g \Rightarrow h$. Let K be a compact, non-polar set of X and let q be a finite, continuous potential on X which is strictly positive on K. Then $\hat{R}_q^K$ is not identically zero. Since $\hat{R}_q^K \in \mathcal{M}$, there exists a non-identically zero finite, continuous specific minorant of q and we obviously have

$$\emptyset \neq S(p) \subset K.$$

$h \Rightarrow h'$. Let $\mathfrak{B}$ be a countable base of X and let K be a compact set having the properties stated in h'. By h), for any $V \in \mathfrak{B}$ with $V \cap K \neq \emptyset$, there exists a finite, continuous potential, p_V, such that

$$\emptyset \neq S(p_V) \subset K \cap \overline{V}.$$

Let q be a finite, continuous potential on X which is strictly positive on K and let $(\alpha_V)_{V \in \mathfrak{B}}$ be a family of strictly positive, real numbers such that

$$\sum_{V \in \mathfrak{B}} \alpha_V < \infty.$$

The function

$$p := \sum_{\substack{V \in \mathfrak{B} \\ V \cap K \neq \emptyset}} \frac{\alpha_V}{\sup_K \frac{p_V}{q}} p_V$$

has the required properties.

$h' \Rightarrow b'$. Let K be a compact, totally thin set of X and let $\mathfrak{B}$ be a countable base of X. We denote by $\mathfrak{B}_0$, the set of $V \in \mathfrak{B}$ for which $K \cap V$ is polar, and set

$$K_0 := K \setminus \bigcup_{V \in \mathfrak{B}_0} V.$$

Let U be an open set of X for which $K_0 \cap U$ is not empty. Then there exists $V \in \mathfrak{B}$ such that

$$V \subset U, \qquad V \cap K_0 \neq \emptyset.$$

It follows that $V \notin \mathfrak{B}_0$. Hence $V \cap K$ is non-polar. Since

$$K \setminus K_0 = \bigcup_{V \in \mathfrak{B}_0} (K \cap V)$$

is polar (Corollary 6.2.3), we deduce that $K_0 \cap V$ and, a fortiori, $K_0 \cap U$ is non-polar. Hence there exists by h'), a finite continuous potential p on X for which

$$S(p) = K_0.$$

Since K_0 is totally thin,

$$p_{K_0} = 0$$

(Proposition 8.3.4). But $p = p_{K_0}$. We get therefore

$$K_0 = S(p) = \emptyset.$$

$c \Rightarrow e$. Obviously

$$\hat{R}^A_{\hat{R}^A_u} \leq \hat{R}^A_u.$$

Let v be any positive, hyperharmonic function on X which is equal to ∞ on $A \setminus b(A)$. Since A is not thin at any point of $b(A)$, we have

$$\hat{R}_u^A = u$$

on $b(A)$. We deduce

$$u \leq \hat{R}_u^A + v$$

on A. Hence

$$\hat{R}_u^A \leq \hat{R}^A_{\hat{R}_u^A + v} \leq \hat{R}^A_{\hat{R}_u^A} + \hat{R}_v^A \leq \hat{R}^A_{\hat{R}_u^A} + v$$

and since v is arbitrary, we obtain

$$\hat{R}_u^A \leq \hat{R}^A_{\hat{R}_u^A} + \hat{R}_\infty^{A \setminus b(A)}.$$

But $A \setminus b(A)$ is polar and thus

$$\hat{R}_u^A \leq \hat{R}^A_{\hat{R}_u^A}$$

(Corollary 6.2.2).

$e \Leftrightarrow f$ and $e' \Leftrightarrow f'$ are immediate consequences of Corollary 7.1.2.

$f \Rightarrow f'$ is trivial.

$f' \Rightarrow b'$. Let K be a totally thin, compact set of X and let p be a finite, strict potential on X (Proposition 7.2.1). Then

$$\hat{R}_p^K < p$$

on X (Proposition 7.2.2). We have for any $\mu \in \Lambda$,

$$\int p \, d\mu^K = \int p \, d(\mu^K)^K = \int \hat{R}_p^K \, d\mu^K$$

(Corollary 7.1.2). Hence

$$0 = \int (p - \hat{R}_p^K) \, d\mu^K, \qquad \mu^K = 0.$$

Since μ is arbitrary K is polar (Proposition 7.1.5).

$c \Rightarrow i$ follows from Theorem 6.3.3.

$i \Rightarrow j$ follows from Corollary 6.2.4.

$j \Rightarrow b'$. Let K be a totally thin compact set of X and let U be a relatively compact, open neighbourhood of K. Since the interior of K is empty, $K \subset \partial(U \setminus K)$. Let f be the function on $\partial(U \setminus K)$ which is equal to 1 on K and equal to 0 on ∂U. By Theorem 6.3.3 every point of K is a non-regular boundary point of $U \setminus K$. Hence for any $x \in U \setminus K$, K is of $\mu_x^{U \setminus K}$-measure zero and we have

$$H_f^{U \setminus K} = 0.$$

Let $x \in U \setminus K$. There exists a sequence $(u_n)_{n \in \mathbb{N}}$ in $\overline{\mathscr{U}}_f^{U \setminus K}$ such that

$$\sum_{n \in \mathbb{N}} u_n(x) < \infty.$$

The function on U, equal to $\sum_{n\in\mathbb{N}} u_n$ on $U\setminus K$ and equal to ∞ on K is hyperharmonic (Proposition 2.1.2) and thus we deduce

$${}^U\hat{R}^K_\infty = 0$$

But x is arbitrary and K nowhere dense. Hence

$${}^U\hat{R}^K_\infty = 0$$

and K is polar.

$c \Rightarrow k$. Obviously

$$\mu^A = \mu^{A\cup b(A)}.$$

Since $(A\cup b(A))\setminus b(A)$ is polar, we have

$$\mu^{A\cup b(A)} = \mu^{b(A)}$$

(Corollary 6.2.1) and we get

$$\mu^A(X\setminus b(A)) = \mu^{b(A)}(X\setminus b(A)) = 0$$

(Corollary 7.2.1, Theorem 7.2.1).

$k \Rightarrow k'$ is trivial.

$k' \Rightarrow b'$. Let K be a compact, totally thin set of X. For any $\mu\in\Lambda$ we have

$$\mu^K(X) = 0.$$

Hence K is polar (Proposition 7.1.5). □

Remark 1. O.D. Kellogg 1928 [1] and F. Vasilesco 1930 [1] proved that property *i)* holds for the harmonic space associated with the Laplace equation in the plane. For the harmonic space associated with the Laplace equation in $\mathbb{R}^3$ the properties *d')* and *i)* were proved by G.C. Evans 1933 [1] and the property *c)* by M. Brelot 1944 [6].

Remark 2. The relation $a \Rightarrow g$ is no longer true if in *g)*, we replace the compact set K by an open, fine closed set (see Exercise 9.1.3).

Axiom of Polarity. *The property of polarity holds for every open set.*

This axiom was first considered for harmonic spaces by R.-M. Hervé 1962 [4] which called it Axiom C. There exist Brelot spaces for which the axiom of polarity is not satisfied (see C. Constantinescu 1965 [1] or Exercise 6.3.10).

Corollary 9.1.1. *If there exists a covering* $\mathfrak{W}$ *of* X *by* $\mathfrak{P}$*-sets such that for every set of* $\mathfrak{W}$*, one of the above properties holds, then the axiom of polarity holds on* X. $\square$

Corollary 9.1.2. *If the axiom of polarity holds on* X*, then for any locally equally lower bounded set* $\mathscr{V}$ *of hyperharmonic functions on* X*, the set*

$$\{x\in X \mid \bigwedge_{u\in\mathscr{V}} u(x)<\inf_{u\in\mathscr{V}} u(x)\}$$

is polar.

Let U be an open set of X such that $\overline{U}$ is compact and contained in a $\mathfrak{P}$-set V. Then there exists a potential p on V such that

$$p+\bigwedge_{u\in\mathscr{V}} u$$

is positive on U. By hypothesis, the set

$$\{x\in X \mid \bigwedge_{u\in\mathscr{V}} (p+u)(x)<\inf_{u\in\mathscr{V}} (p(x)+u(x))\}$$

is polar and the assertion now follows from the relation

$$\bigwedge_{u\in\mathscr{V}} (p+u)=p+\bigwedge_{u\in\mathscr{V}} u. \quad \square$$

Exercises

9.1.1. If the axiom of polarity holds on X, then for any subset A of X,

$$b(b(A))=b(A).$$

9.1.2. Let $\mathfrak{F}$ be a lower directed set of fine closed subsets of X whose intersection is empty. If the axiom of polarity holds, then there exists a decreasing sequence $(F_n)_{n\in\mathbb{N}}$ in $\mathfrak{F}$ such that $\bigcap_{n\in\mathbb{N}} F_n$ is polar (use Exercise 7.2.6 and Theorem 9.1.1 $a \Rightarrow b$).

9.1.3. Show that the axiom of polarity holds on the harmonic space introduced in Exercise 3.1.7 and that the set $]-1, 0]$ is absorbent. Show that the function $x\mapsto 1-|x|$ on this space is a potential p for which $R_p^{]0,1[}$ does not belong to $\mathscr{M}$.

9.1.4. Show that the axiom of polarity does not hold on the harmonic space associated with the heat equation (use Exercise 6.3.8).

9.1.5. Let X be a $\mathfrak{P}$-harmonic space. The axiom of polarity holds on X if and only if the following assertion holds: for any decreasing sequence $(A_n)_{n\in\mathbb{N}}$ of subsets of X and for any decreasing sequence $(u_n)_{n\in\mathbb{N}}$ of positive, superharmonic functions on X, we have

$$\hat{R}_u^{A_n}=u$$

for any $n\in\mathbb{N}$, where

$$u:=\bigwedge_{n\in\mathbb{N}}\hat{R}_{u_n}^{A_n}$$

(N. Boboc-A. Cornea 1968 [5]).

9.1.6. Assume that X is a $\mathfrak{P}$-harmonic space for which the axiom of polarity holds and let A, B be subsets of X with $b(A)\subset b(B)$. If $\mu^B(X\setminus b(A))=0$, for a $\mu\in\mathcal{M}$, then $\mu^B=\mu^A$ (B. Fuglede 1971 [2]). (Let p be a potential on X. Then

$$\hat{R}_p^A=\hat{R}_p^{b(A)}=\hat{R}_{\hat{R}_p^{b(A)}}^{b(B)}=\hat{R}_{\hat{R}_p^A}^B$$

(Theorem 9.1.1 $a\Rightarrow c$, $a\Rightarrow e$, Corollary 6.2.1),

$$\int p\,d\mu^A=\int\hat{R}_p^A\,d\mu=\int\hat{R}_{\hat{R}_p^A}^B\,d\mu=\int\hat{R}_p^A\,d\mu^B=\int_{b(A)}\hat{R}_p^A\,d\mu^B$$

$$=\int_{b(A)}p\,d\mu^B=\int p\,d\mu^B.)$$

9.1.7. Let X be a $\mathfrak{P}$-harmonic space and let $(u_\iota)_{\iota\in I}$, $(v_\kappa)_{\kappa\in J}$ be two lower directed families of positive, superharmonic functions on X. If the axiom of polarity holds, then

$$\bigwedge_{(\iota,\kappa)\in I\times J}(u_\iota\vee v_\kappa)=\Big(\bigwedge_{\iota\in I}u_\iota\Big)\vee\Big(\bigwedge_{\kappa\in J}v_\kappa\Big).$$

(This relation is true for the harmonic space introduced in Theorem 2.1.2 although the axiom of polarity does not hold for it. We do not know if this assertion is true for any $\mathfrak{P}$-harmonic space with a countable base) (N. Boboc-A. Cornea 1968 [5]).

9.1.8. Let X be a $\mathfrak{P}$-harmonic space. The following assertions are equivalent without the hypothesis that X has a countable base:

a) every compact totally thin set of X is polar;

b) for any subset A of X, every compact subset of $A\setminus b(A)$ is polar;

c) for any finite continuous potential p on X and any compact set K of X, the set $\{x\in K\,|\,\hat{R}_p^K(x)<p(x)\}$ is polar;

d) for any compact set K of X and for any $p\in\mathcal{M}$ we have

$$\hat{R}_{\hat{R}_p^K}^K=\hat{R}_p^K;$$

e) for any compact set K of X and any $p\in\mathcal{M}$, we have $\hat{R}_p^K\in\mathcal{M}$;

e') for any compact set K of X and any $p \in \mathscr{P}$ such that $p|_K$ is continuous, we have $\hat{R}_p^K \in \mathscr{M}$;

f) for any compact non-polar set K, there exists a finite continuous potential p on X such that

$$\emptyset \neq S(p) \subset K;$$

g) for any compact set K of X and any $\mu \in \Lambda$, we have

$$(\mu^K)^K = \mu^K;$$

h) for any compact set K of X and any $\mu \in \Lambda$ whose carrier lies in K and for which every compact polar set is of μ-measure zero, we have

$$\mu^K = \mu;$$

i) for any open set U of X, every compact set of non-regular boundary points of U is a polar set.

j) for any relatively compact, open set U of X and any $x \in U$, the set of non-regular boundary points of U is of inner μ_x^U-measure zero;

k) for any relatively compact open set U of X, every lower bounded, hyperharmonic function on U is positive if its lower limit at any regular boundary point of U is positive.

(R.-M. Hervé 1962 [4], H. Bauer 1965 [7], N. Boboc-A. Cornea 1966 [1], 1968 [5].)

9.1.9. Assume that X is an elliptic $\mathfrak{P}$-harmonic space for which the equivalent properties of the preceding exercise hold, with no hypothesis about a countable base. Then, the set of non-regular boundary points of any open, connected set of X is an F_σ polar set (B. Collin 1964 [1]). (Let U be an open set of X and let p be a strictly positive potential on U which is harmonic outside of a compact set of U. For any $n \in \mathbb{N}$, set $F_n := \{y \in \partial U \mid \limsup_{x \to y} p(x) \geq n^{-1}\}$. F_n is closed and $\bigcup_{n \in \mathbb{N}} F_n$ is exactly the set of non-regular boundary points of U (Proposition 2.4.5 and Corollary 6.3.7. In order to show that F_n is polar, use the property *i)* of the preceding exercise and the Exercise 6.2.5).

§ 9.2. The Axiom of Domination

The Axiom of Domination. *For any open set U of X with $\overline{U}$ compact and contained in a $\mathfrak{P}$-set of X and for any locally bounded hyperharmonic function u defined on a neighbourhood of $\overline{U}$, H_u^U is the greatest harmonic minorant of the restriction of u to U.*

This axiom was introduced by M. Brelot 1958 [12] who called it axiom D.

It is obvious that if the axiom of domination holds on a harmonic space, then it holds on all of its harmonic subspaces.

Theorem 9.2.1. *If X is a $\mathfrak{P}$-harmonic space, then the following assertions are equivalent:*

a) the axiom of domination holds;

b) for any locally bounded potential p on X and any relatively compact, open set U of X, H_p^U is the greatest harmonic minorant of $p|_U$;

c) for any locally bounded potential p on X and for any compact set K of X, we have

$$\hat{R}_p^K = \bigwedge_U \hat{R}_p^U,$$

where U runs through the set of neighbourhoods of K;

d) for any locally bounded potential p on X and any positive hyperharmonic function u on X, we have

$$u \geq p \text{ on } S(p) \Rightarrow u \geq p \text{ on } X;$$

e) any locally bounded potential p on X is continuous if its restriction to $S(p)$ is continuous;

f) every locally bounded potential on X belongs to $\mathscr{M}$;

g) every hyperharmonic function on X is quasi-continuous;

h) for any subset A of X and any locally bounded potential p on X, we have

$$\bigwedge_U R_p^{A \cap U} = 0$$

where U runs through the set of neighbourhoods of $X \setminus A \cup b(A)$.

i) for any decreasing sequence $(F_n)_{n \in \mathbb{N}}$ of fine closed sets of X whose intersection is empty and any locally bounded potential p on X, we have

$$\bigwedge_{n \in \mathbb{N}} \hat{R}_p^{F_n} = 0;$$

j) for any decreasing sequence $(F_n)_{n \in \mathbb{N}}$ of fine closed sets of X and any locally bounded potential p on X, we have

$$\hat{R}_p^F = \bigwedge_{n \in \mathbb{N}} R_p^{F_n},$$

where

$$F := \bigcap_{n \in \mathbb{N}} F_n;$$

k) for any $p \in \mathscr{M}$ and any relatively compact, open set U, we have

$$R_p^U \in \mathscr{M};$$

l) for any relatively compact, open set U and any $x \in U$, the set of points of ∂U at which U is thin is of μ_x^U-measure zero.

(M. Brelot-R.-M. Hervé 1958 [1], M. Brelot 1958 [12], 1960 [15], 1965 [19], 1967 [22], R.-M. Hervé 1960 [2], 1962 [4], N. Boboc-C. Constantinescu-A. Cornea 1965 [3], N. Boboc-A. Cornea 1968 [5].)

We prove the following system of implications:

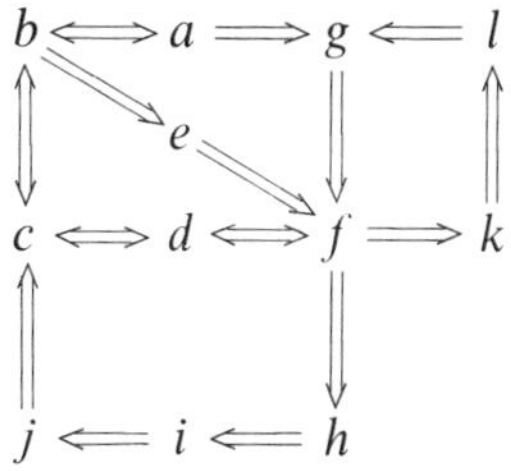

$a \Rightarrow b$ is trivial.

$b \Rightarrow a$. Let U be a relatively compact, open set of X and u be a locally bounded hyperharmonic function defined on an open neighbourhood of $\bar{U}$. Let V be relatively compact open neighbourhood of $\bar{U}$ such that u is defined on $\bar{V}$. By Theorem 2.3.2 there exist two potentials p, q on X such that p is finite and continuous on X, harmonic on V, and

$$q = u + p$$

on V. Since q is bounded on V, R_q^V is locally bounded and therefore, by hypothesis, $\bar{H}_{R_q^V}^U = \bar{H}_q^U$ is the greatest harmonic minorant of $R_q^V|_U = q|_U$. Since

$$\bar{H}_q^U = \bar{H}_u^U + p,$$

it follows that $\bar{H}_u^U$ is the greatest harmonic minorant of $u|_U$.

$b \Leftrightarrow c \Leftrightarrow d \Leftrightarrow f$ follows immediately from Theorem 8.3.2.

$b \Rightarrow e$ follows from Proposition 8.3.3.

$e \Rightarrow f$ follows from Theorem 8.2.2.

$a \Rightarrow g$. Let u be a hyperharmonic function on X and let U be a relatively compact, open set of X. Then there exists a finite, continuous potential, p, on X such that $u + p$ is positive on U. Let q be any finite, continuous potential on U. Then $(u+p)|_U \wedge q$ is a locally bounded potential on U. Since assertion *a)* is true on the harmonic subspace U, then, by $a \Rightarrow f$, it follows that $(u+p)|_U \wedge q$ belongs to $\mathcal{M}$ on U. Hence, this function is quasi-continuous (Theorem 8.4.1) and since q is arbitrary, we deduce (Proposition 8.4.3) that $(u+p)|_U$ is quasi-continuous. Therefore $u|_U$ is quasi-continuous and using Proposition 8.4.1, we obtain the quasi-continuity of u.

$g \Rightarrow f$ follows from Corollary 8.4.1.

$f \Rightarrow h$. Since every locally bounded potential belongs to $\mathcal{M}$, it follows from Theorem 9.1.1. $g \Rightarrow c$, that $A \setminus b(A)$ is polar. Let q be any finite continuous strict potential on X and set

$$B := A \cap b(A), \qquad q' := \hat{R}_q^B.$$

Since $A \setminus b(A)$ is polar,

$$b(B) \subset b(A) \subset b(B) \cup b(A \setminus b(A)) = b(B),$$
$$q' = q$$

on B. For any $x \in \bar{B} - B \cup b(B)$, we have

$$q'(x) < q(x) = \lim_{\substack{y \to x \\ y \in B}} q(y) = \lim_{\substack{y \to x \\ y \in B}} q'(y)$$

(Proposition 7.2.2). Since q' is quasi-continuous (Theorem 8.4.1) we get by Theorem 8.4.2 $a \Rightarrow b$

$$\bigwedge_U \hat{R}_p^{B \cap U} = 0,$$

where U runs through the set of neighbourhoods of $\bar{B} \setminus B \cup b(B)$. The above relation obviously remains true if U runs through the set of neighbourhoods of $X \setminus A \cup b(A)$. Again using the fact that $A \setminus b(A)$ is polar, we obtain

$$\bigwedge_U \hat{R}^{A \cap U} \leq \bigwedge_U \hat{R}_p^{B \cap U} + \hat{R}_p^{A \setminus b(A)} = 0.$$

$h \Rightarrow i$. Let K be a compact set of X. For any $n \in \mathbb{N}$, let $\mathfrak{V}_n$ be the set of neighbourhoods of $X \setminus F_n \cap K$. Then, by $h)$,

$$\bigwedge_{U \in \mathfrak{V}_n} \hat{R}_p^{F_n \cap K \cap U} = 0.$$

Let $\mathfrak{A}$ be the set of subsets of X of the form

$$\bigcup_{n \in \mathbb{N}} (F_n \cap K \cap U_n)$$

where, for all $n \in \mathbb{N}$, $U_n \in \mathfrak{V}_n$. By Proposition 4.1.3 and Theorem 5.1.1

$$\bigwedge_{C \in \mathfrak{A}} \hat{R}_p^C = 0.$$

If $(U_n)_{n \in \mathbb{N}}$ is any sequence of open sets of X such that $U_n \in \mathfrak{V}_n$ for every $n \in \mathbb{N}$, then $(F_n \cap K) \setminus U_n$ is a compact set for every $n \in \mathbb{N}$ and

$$\bigcap_{n \in \mathbb{N}} ((F_n \cap K) \setminus U_n) \subset \bigcap_{n \in \mathbb{N}} F_n = \emptyset.$$

Hence there exists an $m\in\mathbb{N}$ such that

$$\bigcap_{\substack{n\in\mathbb{N}\\ n\leq m}}((F_n\cap K)\setminus U_n)=\varnothing .$$

Obviously

$$F_m\cap K\subset\bigcup_{n\in\mathbb{N}}(F_n\cap K\cap U_n).$$

This means that for any $C\in\mathfrak{A}$, there exists an $m\in\mathbb{N}$ such that $F_m\cap K\subset C$. Thus

$$\bigwedge_{n\in\mathbb{N}}\hat{R}_p^{F_n}\leq\bigwedge_{n\in\mathbb{N}}\hat{R}_p^{F_n\cap K}+\hat{R}_p^{X\setminus K}\leq\bigwedge_{C\in\mathfrak{A}}\hat{R}_p^{C}+\hat{R}_p^{X\setminus K}=\hat{R}_p^{X\setminus K}.$$

The assertion follows now from the fact that K is arbitrary (Corollary 5.3.1).

$i\Rightarrow j$. Let U be a fine open neighbourhood of F. Then

$$\bigcap_{n\in\mathbb{N}}(F_n\setminus U)=\varnothing ,$$

and we get

$$\bigwedge_{n\in\mathbb{N}}\hat{R}_p^{F_n}\leq\bigwedge_{n\in\mathbb{N}}\hat{R}_p^{F_n\setminus U}+\hat{R}_p^{U}=\hat{R}_p^{U}.$$

Since U is arbitrary, we deduce further

$$\hat{R}_p^{F}\leq\bigwedge_{n\in\mathbb{N}}\hat{R}_p^{F_n}\leq\bigwedge_{U}\hat{R}_p^{U}=\hat{R}_p^{F}$$

(Proposition 4.2.1, Theorem 5.1.1).

$j\Rightarrow c$ is trivial.

$f\Rightarrow k$ is trivial.

$k\Rightarrow l$. Let p be a finite, strict potential on X. We set

$$q:=R_p^{U}.$$

At any point $y\in\partial U$ at which U is thin, we have

$$q(y)<p(y)\leq\liminf_{\substack{z\to y\\ z\in U}}p(z)=\liminf_{\substack{z\to y\\ z\in U}}q(z)$$

(Proposition 7.2.2). Since $q\in\mathcal{M}$, it is quasi-continuous (Corollary 8.4.1) and the assertion follows from Theorem 8.4.2 $a\Rightarrow c$.

$l\Rightarrow g$ follows immediately from Theorem 8.4.2 $c\Rightarrow a$ using Proposition 6.3.3 *b)*. □

Remark. In classical potential theory, *d)* was proved by A. J. Maria 1934 [1] and O. Frostman 1935 [1], *e)* by F. Vasilesco 1935 [2] and *h)* by G. Choquet 1959 [2].

Corollary 9.2.1. *If the axiom of domination holds on X, then assertions b)–k) of the theorem hold on X.*

Assertion *g)* follows from Proposition 8.4.1 and *f)* follows from *g)* and from Corollary 8.4.1. Assertions *b), c), d)*, and *k)* follow from *f)* and Theorem 8.3.2 while *e)* follows from *b)* by using Proposition 8.3.3. Assertions *h), i), j)* hold for a finite, continuous p by Corollary 2.2.2 and thus for a general p by using *f)*. □

Corollary 9.2.2. *Let $\mathfrak{W}$ be an open covering of X by $\mathfrak{P}$-sets such that for every $W \in \mathfrak{W}$, one of the assertions of the theorem holds. Then the axiom of domination holds on X.*

Let U be any $\mathfrak{P}$-set of X. By the theorem we have for any $W \in \mathfrak{W}$ that the axiom of domination holds on the harmonic subspace $U \cap W$. Hence, again by the theorem ($a \Rightarrow g$), every hyperharmonic function on $U \cap W$ is quasi-continuous and by Proposition 8.4.1, every hyperharmonic function on U is quasi-continuous. From $g \Rightarrow a$, we see that the axiom of domination holds on U. The corollary follows now from the fact that U is arbitrary. □

Corollary 9.2.3. *The axiom of domination implies the axiom of polarity* (M. Brelot 1958 [12], M. Brelot-R.-M. Hervé 1958 [1]).

The corollary follows from the implication $a \Rightarrow f$ of the theorem and from Theorem 9.1.1 $g \Rightarrow a$. □

Remark. There exist Brelot spaces on which the axiom of polarity holds but the axiom of domination does not (see N. Boboc-A. Cornea 1968 [5] or Exercise 9.2.6).

Proposition 9.2.1. *If the axiom of domination holds on X, then X is an elliptic harmonic space* (J. Köhn-M. Sieveking 1967 [1]).

Since every notion appearing in this statement is local we may assume that there exists a strictly positive, harmonic function h_0 on X. Let U be an open, connected set of X and let F be an absorbent set of U. By the definition of absorbent set, the function u on U which is equal to 0 on F and equal to ∞ on $U \setminus F$ is hyperharmonic. Let

$$h_0|_U \wedge u =: p + h$$

be the Riesz decomposition of the superharmonic function $h_0|_U \wedge u$. Obviously $S(p)$ is contained in F (here S is defined on the harmonic subspace U). Since p is 0 on $S(p)$, it vanishes identically (Theorem 9.2.1 $a \Rightarrow d$). Hence $h_0|_U \wedge u$ is continuous. It follows that F is either empty or equal to U and the proposition now follows from Proposition 6.1.3. □

Remark. The proposition no longer holds if the axiom of domination is replaced by the axiom of polarity (see Exercise 9.1.3).

Corollary 9.2.4. *If the axiom of domination holds on X, then X has a base of regular sets.*

This assertion follows immediately from the proposition with the aid of Proposition 3.1.4 $a \Rightarrow d$. □

Exercises

9.2.1. Let p, q be potentials on X with p finite, continuous and $q \leq p$. Let A be any subset of X such that

$$\{x \in X \mid q(x) < p(x)\} \setminus A$$

is a polar set. If the axiom of domination holds on X, then for any $x_0 \in X$ at which p and q are equal, we have

$$\hat{R}_p^A(x_0) = \hat{R}_q^A(x_0)$$

(R.-M. Hervé 1962 [4]). (Let $\alpha \in]0, 1[$. We set

$$U := \{x \in X \mid \alpha\, p(x) < q(x)\}, \qquad p' := (p - \hat{R}_p^{X \setminus U})|_U, \qquad q' := (q - \hat{R}_q^{X \setminus U})|_U.$$

By Exercise 5.3.1, we have

$$\begin{aligned}\hat{R}_q^A(x_0) &= \hat{R}_q^{X \setminus U}(x_0) + {}^U\hat{R}_{q'}^{A \cap U}(x_0)\\ &\geq \alpha\, \hat{R}_p^{X \setminus U}(x_0) + \alpha\, {}^U\hat{R}_{p'}^{A \cap U}(x_0) - (1-\alpha)\, p(x_0)\\ &= \alpha\, \hat{R}_p^A(x_0) - (1-\alpha)\, p(x_0).)\end{aligned}$$

9.2.2. Assume that X is a $\mathfrak{P}$-harmonic space. The axiom of domination holds if and only if one of the following assertions hold.

a) (resp. *a')*) for any fine open (resp. fine closed) set A and any locally bounded potential p on X, we have

$$\bigwedge_B \hat{R}_p^{B \setminus A} = 0 \qquad \left(\text{resp. } \bigwedge_B \hat{R}_p^{A \setminus B} = 0\right),$$

where B runs through the set of open (resp. closed) sets of X containing A (resp. contained in A) (M. Brelot 1967 [22]) (use Theorem 9.2.1 $a \Leftrightarrow h$);

b) for any lower directed set $\mathfrak{F}$ of fine closed sets of X and any locally bounded potential p on X, we have

$$\bigwedge_{F \in \mathfrak{F}} \hat{R}_p^F = \hat{R}_p^{F_0},$$

where

$$F_0 := \bigcap_{F \in \mathfrak{F}} F$$

(M. Brelot 1967 [21]) (use Exercise 9.1.2, Corollary 9.2.3 and Theorem 9.2.1 $a \Rightarrow j$).

c) for any lower directed set $\mathscr{F}$ of fine upper semi-continuous, positive real functions on X which is dominated by a locally bounded potential, we have

$$\bigwedge_{f\in\mathscr{F}} Rf = Rf_0,$$

where

$$f_0 := \inf_{f\in\mathscr{F}} f$$

(M. Brelot 1967 [21]) (use *b)*).

d) for any subset A of X and any $\mu\in\Lambda$, if a subset of the fine interior of A is of μ-measure zero, then it is also of μ^A-measure zero (R.-M. Hervé 1962 [4]). ($d \Rightarrow a$ follows from Theorem 9.2.1 $l \Rightarrow a$. For $a \Rightarrow d$, we remark first that by Proposition 7.1.4 it is sufficient to prove the assertion for $\mu := \varepsilon_x$, where $x\in X\setminus b(A)$. Let p be a finite, continuous, strict potential and $q := \hat{R}_p^{X\setminus B}$, where B is the fine interior of A. Then $q<p$ on B, and thus by the preceding exercise

$$\varepsilon_x^A(p-q) = \hat{R}_p^A(x) - \hat{R}_q^A(x) = 0.)$$

9.2.3. Assume that X is a $\mathfrak{P}$-harmonic space for which the axiom of domination holds.

a) Let U be a fine open set, A be a subset of $X\setminus b(U)$ and B be a polar set; then for any $x\in b(U)$ we have

$$\varepsilon_x^{X\setminus U} = \varepsilon_x^{X\setminus (U\cup A\cup B)}.$$

(It is sufficient to consider the case $B=\varnothing$ (Corollary 6.2.1) and $x\in U$ (Proposition 7.1.4). By Exercise 9.2.2 *d)* A is of $\varepsilon_x^{X\setminus U}$ measure zero. By Theorem 7.2.1 $a\Rightarrow k$, U is of $\varepsilon_x^{X\setminus U}$-measure zero. By Corollary 9.2.3 and Theorem 9.1.1 $a\Rightarrow c$ the set

$$(X\setminus(U\cup A))\setminus b(X\setminus(U\cup A))$$

is polar and therefore of $\varepsilon_x^{X\setminus U}$ measure zero (Corollary 7.1.3). Hence

$$\varepsilon_x^{X\setminus U}(X\setminus b(X\setminus(U\cup A))) = 0.$$

The assertion now follows from Exercise 9.1.6.)

b) Let A be a subbasic set of X (Exercise 6.3.2) and let $x\in X\setminus b(A)$. Denote by $\mathfrak{A}$, the set of subbasic sets, B of X containing A and such that $\varepsilon_x^A = \varepsilon_x^B$ and by A_x the union of $\mathfrak{A}$. Then $A_x\in\mathfrak{A}$, $b(A_x)=A_x$, $x\in X\setminus A_x$, and thus $X\setminus A_x$ is fine open and fine connected. If $X\setminus A$ is fine open and fine connected, then $A_x = A$. (By Theorem 4.2.2 and Theorem 5.1.1 the union of two elements of $\mathfrak{A}$ belongs to $\mathfrak{A}$. Now use

Exercise 6.3.2 to show that $A_x \in \mathfrak{A}$. Obviously $A_x = b(A_x)$ and therefore $X \setminus A_x$ is fine open and contains x. Let U, V be two disjoint, fine open sets such that $U \cup V = X \setminus A_x$, $x \in U$. By *a)*

$$\varepsilon_x^{X \setminus U} = \varepsilon_x^{X \setminus (U \cup V)} = \varepsilon_x^A.$$

Hence $A_x \cup V \in \mathfrak{A}$ and $V = \emptyset$. Assume now that $X \setminus A$ is fine open and fine connected and let p be a finite, strict potential on X. Since $\hat{R}^A_{\hat{R}_p^{A_x}} = \hat{R}_p^A$, it follows from Exercise 4.2.7 that the set

$$\{y \in X \mid \hat{R}_p^A(y) < p(y),\ \hat{R}_p^A(y) = \hat{R}_p^{A_x}(y)\}$$

is fine open. Since it contains x and since $X \setminus A$ is fine connected, it coincides with $X \setminus A$. Hence $A_x \setminus A$ must be empty (B. Fuglede 1971 [2]).)

9.2.4 (Theorem of Fuglede). Assume that the axiom of domination holds on X.

a) If X is connected, then every subset of X with polar fine boundary is either polar or has a polar complement. (Let A be such a set. Assume first that X is a $\mathfrak{P}$-harmonic space. We denote by U (resp. V) the fine interior of A (resp. $X \setminus A$) and by B the fine boundary of A. Let $x \in U$. By the preceding exercise *a)* $\varepsilon_x^{X \setminus U} = 0$. It follows that for any potential p on X, $\hat{R}_p^{X \setminus U}$ vanishes on U. If A is not polar, then U is non-empty and $\hat{R}_p^{X \setminus U}$ vanishes identically (Proposition 9.2.1). Hence $X \setminus U$ and, a fortiori, $X \setminus A$ is polar. If X is not a $\mathfrak{P}$-harmonic space, let $\mathfrak{W}$ (resp. $\mathfrak{W}'$) be the set of connected $\mathfrak{P}$-sets, W, of X such that $W \cap A$ (resp. $W \setminus A$) is polar. Show that for any $W \in \mathfrak{W}$, $W' \in \mathfrak{W}'$ we have $W \cap W' = \emptyset$. Since X is connected, it follows that either $\mathfrak{W}$ or $\mathfrak{W}'$ is empty. Hence either A on $X \setminus A$ is polar.)

b) If X is connected, then it is fine connected (use *a)*).

c) The fine topology is locally connected. (We may assume that X is a $\mathfrak{P}$-harmonic space. Let $x \in X$ and let U be a fine neighbourhood of x. Let K be a compact, fine neighbourhood of x with $K \subset U$ (Proposition 5.1.2). Then $A := X \setminus K$ is subbasic and $x \in X \setminus b(A)$. By the preceding exercise *b)*, $X \setminus A_x$ is a fine connected, fine neighbourhood of x and $X \setminus A_x \subset U$.) (We do not know whether *c)* holds on any elliptic harmonic space.)

d) If U is a fine open, fine connected set of X, then for every polar set A of X, $U \setminus A$ is fine open and fine connected. ($U \setminus A$ is obviously fine open. Assume first that X is a $\mathfrak{P}$-harmonic space and let U', U'' be two disjoint fine open sets of X such that $U \setminus A = U' \cup U''$. Let $x \in b(U') \cap b(U'') \cap U$. By the preceding exercise *c)*, we get the contradictory relation

$$\varepsilon_x = \varepsilon_x^{X \setminus U'} = \varepsilon_x^{X \setminus U} \neq \varepsilon_x.$$

Hence $b(U')\cap b(U'')\cap U$ is empty. Since

$$b(U')\cup b(U'')=b(U\setminus A)\supset U$$

and since U is fine connected, it follows that one of the sets U', U'' is empty. If X is not a $\mathfrak{P}$-harmonic space, let U', U'' be two disjoint fine open sets of X such that $U\setminus A=U'\cup U''$. Let x be a point of U which belongs to the fine closures of U' and U''. By *c)* there exists a fine open, fine connected set V of X which is contained in a $\mathfrak{P}$-set of X and such that $x\in V\subset U$. By the above proof, $V\setminus A$ is fine connected and it is therefore contained either in U' on in U''. Hence x cannot belong simultaneously to the fine closures of U' and U''.)

e) Assume that X is a $\mathfrak{P}$-harmonic space, let A be a subset of X and let B be its fine boundary.

α) $b(B)=b(A)\cap b(X\setminus A)$;

β) if $A=b(A)$, then $B=b(B)$;

γ) let $x\in X\setminus b(A)$, let U be the fine connected component of $X\setminus b(A)$ containing x and let C be the fine boundary of U. Then $X\setminus C$ is the greatest fine open set of X which is of ε_x^A-measure zero. (Show that $C_x=b(A)_x$. By the preceding exercise *b)* $b(A)_x=X\setminus U$).

(B. Fuglede 1971 [2].)

9.2.5. If X is an $\mathfrak{S}$-harmonic space on which the axiom of domination holds, then for any $\mathfrak{P}$-set U of X and any $x\in U$, the set of points of ∂U at which U is thin is of μ_x^U-measure zero (U is resolutive by Theorem 2.4.4). (Use Theorem 9.2.1 $a\Rightarrow l$ locally and Exercise 2.4.10.)

9.2.6. Let X be the Alexandroff compactification of the topological space $\{(x,y)\in\mathbb{R}^2\,|\,x^2+y^2<1\}$, let x_0 be the Alexandroff point and let μ be a measure on $\{(x,y)\in\mathbb{R}^2\,|\,x^2+y^2=1\}$ of total mass 1. For any open set U of X we denote by $\mathscr{H}(U)$ the set of continuous, real functions h on U such that $h|_{U\setminus\{x_0\}}$ is a solution of the Laplace equation and such that if $x_0\in U$, then

$$\lim_{r\to 1}\int\frac{\partial h}{\partial r}(r\,e^{i\theta})\,d\mu(\theta),$$

where we set

$$r\,e^{i\theta}:=x+iy.$$

a) $\mathscr{H}$ is a harmonic sheaf on X and X endowed with $\mathscr{H}$ is a Brelot space on which the axiom of polarity holds.

b) If the carrier of μ is not equal to $\{(x,y)\in\mathbb{R}^2\,|\,x^2+y^2=1\}$ then the axiom of domination does not hold.

c) Let g be a positive, continuous real function, different from 0 almost everywhere with respect to the measure $d\theta$; if $d\mu:=g\,d\theta$ then the axiom of domination holds.

(C. Constantinescu-A. Cornea 1968 [3].)

9.2.7. Let X be a harmonic space (not necessarily having a countable base). We say that the axiom of domination holds on X if, for any open set U of X with $\overline{U}$ compact and contained in a $\mathfrak{P}$-set of X and any locally bounded, hyperharmonic function, u, defined on a neighbourhood of $\overline{U}$, $\overline{H}_u^U$ is the greatest harmonic minorant of the restriction of u to U.

a) If X is a $\mathfrak{P}$-harmonic space, then the assertions *a)–g)* of Theorem 9.2.1 are equivalent provided one replaces H_p^U by $\overline{H}_p^U$ in *b)*. We also have $h \Rightarrow l \Rightarrow g$.

b) The axiom of domination implies all of the assertions of Exercise 9.1.8 and the fact that X is an elliptic harmonic space.

c) If the axiom of domination holds and $\mathscr{H}$ has the Doob convergence property, then $\mathscr{H}$ has the Brelot convergence property (use Exercise 1.1.17 and *b)*). Hence X endowed with $\mathscr{H}$ is a Brelot space.

9.2.8 (Keldych's lemma). Assume that X is a $\mathfrak{P}$-harmonic space on which there exists a strictly positive, harmonic function and let U be a relatively compact, open set of X. If the axiom of domination holds on X, then for any regular boundary point x of U, there exists a continuous, real function f on $\overline{U}$ which is equal to 0 at x, strictly positive on $\overline{U} \setminus \{x\}$ and harmonic on U. Hence the set of regular boundary points coincides with the set of extremal points in the sense of Exercise 2.4.7 (M.V. Keldych 1941 [2], M. Brelot 1961 [16], N. Boboc-A. Cornea 1967 [2]).

9.2.9. The axiom of domination holds on the harmonic space introduced in Exercise 3.2.7 (N. Boboc-P. Mustaţă 1968 [5]).

9.2.10. The axiom of domination holds on the harmonic space defined in Exercise 3.2.3 (M. Brelot 1945 [8]).

9.2.11. The following assertions are equivalent:

a) every potential p on X is continuous if its restriction to $S(p)$ is continuous;

b) every potential p on X is continuous if its restriction to $S(p)$ is finite and continuous;

c) every potential p on X belongs to $\mathscr{M}$ if $p_A = 0$, where

$$A := \{x \in X \mid p(x) = \infty\}.$$

d) every finite potential belongs to $\mathscr{M}$.

($b \Rightarrow c$ follows from Exercise 8.2.6. $c \Rightarrow d$ is trivial. For $d \Rightarrow a$, use Corollary 8.3.1.) These equivalent properties imply the axiom of domination if X is a $\mathfrak{P}$-harmonic space. The converse is not true. (Use the example of Exercise 9.2.6.) In classical potential theory all of these properties are true (A.J. Maria 1934 [1], O. Frostman 1935 [1]).

Chapter 10

Markov Processes on Harmonic Spaces

Throughout this chapter, X will denote a $\mathfrak{P}$-harmonic space with a countable base on which the function 1 *is superharmonic.*

§ 10.1. Sub-Markov Semi-Groups

In this section we define sub-Markov semi-group and some related notions. We then prove a few elementary results. The definitions as well as the theorems hold on a general, locally compact space with a countable base.

A (bounded Borel) diffusion on X is a family $\mathsf{V} := (\mathsf{V}_x)_{x \in X}$ of measures on X such that:

a) for any $f \in \mathscr{K}(X)$, the function $x \mapsto \mathsf{V}_x(f)$ is a Borel function;

b) the function $x \mapsto \mathsf{V}_x(X)$ is bounded.

For any numerical function f on X, we denote by $\mathsf{V} f$ the function on X

$$x \mapsto \int^{*} f \, d\mathsf{V}_x .$$

If f is a bounded Borel function, then so is $\mathsf{V} f$.

If p is a bounded potential on the harmonic space X, then $\mathsf{V}_p :=$ $(\mathsf{V}_{p,x})_{x \in X}$ (see page 195) is a diffusion on X, called *the diffusion generated by p*.

Let V', V'' be two diffusions on X and $\alpha \in \mathbb{R}_+$. For any $x \in X$ let V_x be the measure on X

$$f \mapsto \mathsf{V}' f(x) + \mathsf{V}'' f(x) \quad \big(\text{resp. } f \mapsto \alpha \mathsf{V}' f(x),\ f \mapsto (\mathsf{V}' \mathsf{V}'' f)(x)\big) \quad \big(f \in \mathscr{K}(X)\big).$$

The family $(\mathsf{V}_x)_{x \in X}$ is a diffusion on X which will be denoted by $\mathsf{V}' + \mathsf{V}''$ (resp. $\alpha \mathsf{V}'$, $\mathsf{V}' \mathsf{V}''$). For any bounded Borel function f on X we obviously have

$$(\mathsf{V}' + \mathsf{V}'') f = \mathsf{V}' f + \mathsf{V}'' f, \quad (\alpha \mathsf{V})' f = \alpha(\mathsf{V}' f), \quad (\mathsf{V}' \mathsf{V}'') f = \mathsf{V}'(\mathsf{V}'' f).$$

A (*weakly right continuous*) *sub-Markov semi-group on X* is a family $(\mathsf{P}_t)_{t\in\mathbb{R}_+}$ of diffusions on X such that:

a) for any $t\in\mathbb{R}_+$, $\mathsf{P}_t\,1\leq 1$;

b) for $f\in\mathcal{K}(X)$ and any $x\in X$, the function

$$t\mapsto \mathsf{P}_t\,f(x)$$

is continuous from the right;

c) for any $s,t\in\mathbb{R}_+$, $\mathsf{P}_s\,\mathsf{P}_t=\mathsf{P}_{s+t}$.

The semi-group is called *integrable* if

$$\sup_{x\in X}\int_0^\infty \mathsf{P}_t\,1(x)\,dt<\infty.$$

Proposition 10.1.1. *Let $\mathsf{P}:=(\mathsf{P}_t)_{t\in\mathbb{R}_+}$ be a sub-Markov semi-group on X and let f be a bounded Borel function on X. The function*

$$(x,t)\mapsto \mathsf{P}_t\,f(x)$$

is a Borel function on $X\times\mathbb{R}_+$.

It is sufficient to prove the proposition for $f\in\mathcal{K}(X)$. For $n\in\mathbb{N}$, we denote by τ_n the function on $\mathbb{R}_+$ such that for any $m\in\mathbb{N}$, τ_n is equal to $2^{-n}(m+1)$ on $[2^{-n}m,\,2^{-n}(m+1)[$. The proposition now follows from the fact that for all $n\in\mathbb{N}$, the function

$$(x,t)\mapsto \mathsf{P}_{\tau_n(t)}\,f(x)$$

is a Borel function on $X\times\mathbb{R}_+$ and from the relation

$$\mathsf{P}_t\,f(x)=\lim_{n\to\infty}\mathsf{P}_{\tau_n(t)}\,f(x)\qquad ((x,t)\in X\times\mathbb{R}_+).\quad \square$$

A resolvent on X is a family $(\mathsf{U}_\alpha)_{\alpha\in\mathbb{R}_+}$ of diffusions on X such that:

a) for any $\alpha\in\mathbb{R}_+$, $\alpha\,\mathsf{U}_\alpha\,1\leq 1$;

b) for any $\alpha,\beta\in\mathbb{R}_+$, $\alpha<\beta$,

$$\mathsf{U}_\alpha=\mathsf{U}_\beta+(\beta-\alpha)\,\mathsf{U}_\beta\,\mathsf{U}_\alpha.$$

For any positive Borel function f, $\alpha\mapsto\mathsf{U}_\alpha\,f$ is obviously decreasing.

A resolvent $(\mathsf{U}_\alpha)_{\alpha\in\mathbb{R}_+}$ is said to be *associated with a bounded potential p on X* if $\mathsf{U}_0=\mathsf{V}_p$.

Proposition 10.1.2. *Let $(\mathsf{U}_\alpha)_{\alpha\in\mathbb{R}_+}$ be a resolvent on X. Then for any $\alpha,\beta\in\mathbb{R}_+$*

$$\mathsf{U}_\alpha\,\mathsf{U}_\beta=\mathsf{U}_\beta\,\mathsf{U}_\alpha.$$

We may assume $\alpha < \beta$.

Assume first that $0 < \alpha$. Then for any $f \in \mathscr{K}_+(X)$ and any $n \in \mathbb{N}$, $n \neq 0$, we may prove inductively the relation

$$\mathsf{U}_\alpha f = \sum_{m=0}^{n-1} (\beta - \alpha)^m \, \mathsf{U}_\beta^{m+1} f + (\beta - \alpha)^n \, \mathsf{U}_\beta^n \, \mathsf{U}_\alpha f.$$

Since

$$\lim_{n \to \infty} (\beta - \alpha)^n \, \mathsf{U}_\beta^n \, \mathsf{U}_\alpha f \leq \lim_{n \to 0} \left(\frac{\beta - \alpha}{\beta} \right)^n \frac{1}{\alpha} \sup_X f = 0,$$

we get

$$\mathsf{U}_\alpha f = \sum_{n \in \mathbb{N}} (\beta - \alpha)^n \, \mathsf{U}_\beta^{n+1} f.$$

Hence

$$\mathsf{U}_\alpha \, \mathsf{U}_\beta = \mathsf{U}_\beta \, \mathsf{U}_\alpha .$$

Assume now that $\alpha = 0$. Then for any positive bounded Borel function f,

$$\lim_{\gamma \to 0} \gamma \, \mathsf{U}_\gamma \, \mathsf{U}_0 f \leq \lim_{\gamma \to 0} \gamma \, \mathsf{U}_0^2 f = 0.$$

Hence

$$\lim_{\gamma \to 0} \mathsf{U}_\gamma f = \mathsf{U}_0 f,$$

$$\mathsf{U}_0 \, \mathsf{U}_\beta f = \lim_{\substack{\gamma \to 0 \\ \gamma > 0}} \mathsf{U}_\gamma \, \mathsf{U}_\beta f = \lim_{\substack{\gamma \to 0 \\ \gamma > 0}} \mathsf{U}_\beta \, \mathsf{U}_\gamma f = \mathsf{U}_\beta \, \mathsf{U}_0 f. \quad \square$$

Proposition 10.1.3. *Let $(\mathsf{P}_t)_{t \in \mathbb{R}_+}$ be an integrable sub-Markov semigroup on X. For any $\alpha \in \mathbb{R}_+$ and $x \in X$, we denote by $\mathsf{U}_{\alpha, x}$ the measure on X,*

$$f \mapsto \int_0^\infty e^{-\alpha t} \, \mathsf{P}_t f(x) \, dt \qquad \big(f \in \mathscr{K}(X)\big).$$

Then for any $\alpha \in \mathbb{R}_+$, $\mathsf{U}_\alpha := (\mathsf{U}_{\alpha, x})_{x \in X}$ is a diffusion on X, $(\mathsf{U}_\alpha)_{\alpha \in \mathbb{R}_+}$ is a resolvent on X and for any $\alpha \in \mathbb{R}_+$, $t \in \mathbb{R}_+$, we have $\mathsf{P}_t \, \mathsf{U}_\alpha = \mathsf{U}_\alpha \, \mathsf{P}_t$.

The first assertion follows immediately from Proposition 10.1.1. The relation

$$\alpha \, \mathsf{U}_\alpha 1 \leq 1$$

is obvious for all $\alpha \in \mathbb{R}_+$. We remark also that for any $\alpha \in \mathbb{R}_+$, any $x \in X$ and any bounded Borel function f on X, we have

$$\mathsf{U}_\alpha f(x) = \int_0^\infty e^{-\alpha t} \, \mathsf{P}_t f(x) \, dt .$$

Let $f \in \mathscr{K}_+(X)$, $\alpha, \beta \in \mathbb{R}_+$, $\alpha < \beta$, and $x \in X$. We have by Fubini theorem

$$\mathsf{U}_\beta \mathsf{U}_\alpha f(x) = \int \mathsf{U}_\alpha f \, d\mathsf{U}_{\beta,x} = \int \left(\int_0^\infty e^{-\alpha t} \mathsf{P}_t f(y) \, dt \right) d\mathsf{U}_{\beta,x}(y)$$

$$= \int_0^\infty e^{-\alpha t} \left(\int \mathsf{P}_t f(y) \, d\mathsf{U}_{\beta,x}(y) \right) dt = \int_0^\infty e^{-\alpha t} \mathsf{U}_\beta \mathsf{P}_t f(x) \, dt$$

$$= \int_0^\infty e^{-\alpha t} \left(\int_0^\infty e^{-\beta s} \mathsf{P}_s \mathsf{P}_t f(x) \, ds \right) dt = \int_0^\infty e^{-\alpha t} \left(\int_0^\infty e^{-\beta s} \mathsf{P}_{s+t} f(x) \, ds \right) dt$$

$$= \int_0^\infty e^{-\alpha t} \left(\int_t^\infty e^{-\beta(s-t)} \mathsf{P}_s f(x) \, ds \right) dt = \int_0^\infty e^{(\beta-\alpha)t} \left(\int_t^\infty e^{-\beta s} \mathsf{P}_s f(x) \, ds \right) dt$$

$$= \int_0^\infty e^{-\beta s} \mathsf{P}_s f(x) \left(\int_0^s e^{(\beta-\alpha)t} \, dt \right) ds = \frac{1}{\beta-\alpha} \int_0^\infty e^{-\beta s} \mathsf{P}_s f(x) (e^{(\beta-\alpha)s} - 1) \, ds$$

$$= \frac{1}{\beta-\alpha} \left(\int_0^\infty e^{-\alpha s} \mathsf{P}_s f(x) \, ds - \int_0^\infty e^{-\beta s} \mathsf{P}_s f(x) \, ds \right) = \frac{1}{\beta-\alpha} (\mathsf{U}_\alpha f(x) - \mathsf{U}_\beta f(x)).$$

But f and x are arbitrary and thus

$$\mathsf{U}_\alpha = \mathsf{U}_\beta + (\beta-\alpha) \mathsf{U}_\beta \mathsf{U}_\alpha.$$

The last assertion follows from Fubini theorem. □

The resolvent defined in this proposition is called the resolvent generated by the (integrable) sub-Markov semi-group $(\mathsf{P}_t)_{t \in \mathbb{R}_+}$.

Proposition 10.1.4. *Two integrable sub-Markov semi-groups on X coincide if they generate the same resolvent.*

Let $(\mathsf{P}_t)_{t \in \mathbb{R}_+}$, $(\mathsf{Q}_t)_{t \in \mathbb{R}_+}$ be two integrable sub-Markov semi-groups on X generating the same resolvent. Let $f \in \mathscr{K}(X)$, $x \in X$ and denote by ν, the signed measure on the compact space $[0, \infty]$

$$g \mapsto \int_0^\infty g(t) e^{-t} (\mathsf{P}_t f(x) - \mathsf{Q}_t f(x)) \, dt : \mathscr{C}([0, \infty]) \to \mathbb{R}.$$

For any $\alpha \in]0, \infty[$ we denote by g_α the function of $\mathscr{C}([0, \infty])$ which is equal to 0 at ∞ and equal to $t \mapsto e^{-\alpha t}$ on $[0, \infty[$. By hypothesis, we have for any $\alpha \in]0, \infty[$

$$\nu(g_\alpha) = 0, \qquad \nu(1) = 0.$$

Since for any $\alpha, \beta \in]0, \infty[$ we have $g_\alpha g_\beta = g_{\alpha+\beta}$, it follows that

$$\nu(g) = 0$$

for any polynomial g (with real coefficients) in the functions $(g_\alpha)_{\alpha\in]0,\infty[}$. By Stone-Weierstrass theorem (N. Bourbaki [2]) $\nu=0$. Since the function

$$t\mapsto e^{-t}(\mathsf{P}_t f(x)-\mathsf{Q}_t f(x))$$

is continuous from the right, we immediately deduce that for any $t\in\mathbb{R}_+$

$$\mathsf{P}_t f(x)=\mathsf{Q}_t f(x).$$

Since f and x are arbitrary, we obtain for any $t\in\mathbb{R}_+$, $\mathsf{P}_t=\mathsf{Q}_t$. □

Proposition 10.1.5. *Let $(\mathsf{P}_t)_{t\in\mathbb{R}_+}$ (resp. $(\mathsf{U}_\alpha)_{\alpha\in\mathbb{R}_+}$) be a sub-Markov semi-group (resp. a resolvent) on X and let f be a positive Borel function on X such that for any $t\in\mathbb{R}_+$ (resp. $\alpha\in\mathbb{R}_+$)*

$$\mathsf{P}_t f\le f \quad (\text{resp. } \alpha\,\mathsf{U}_\alpha f\le f).$$

Then, for every $x\in X$, the function

$$t\mapsto \mathsf{P}_t f(x) \quad (\text{resp. } \alpha\mapsto\alpha\,\mathsf{U}_\alpha f(x))$$

is decreasing (resp. increasing).

Let $s,t\in\mathbb{R}_+$, $s<t$ (resp. $\alpha,\beta\in\mathbb{R}_+$, $\alpha<\beta$). We have

$$\mathsf{P}_t f=\mathsf{P}_s\,\mathsf{P}_{t-s} f\le\mathsf{P}_s f,$$

$$(\text{resp. } \alpha\,\mathsf{U}_\alpha f=\alpha\,\mathsf{U}_\beta f+\alpha(\beta-\alpha)\,\mathsf{U}_\beta\,\mathsf{U}_\alpha f\le\alpha\,\mathsf{U}_\beta f+(\beta-\alpha)\,\mathsf{U}_\beta f=\beta\,\mathsf{U}_\beta f). \ \square$$

Let $\mathsf{P}:=(\mathsf{P}_t)_{t\in\mathbb{R}_+}$ be a sub-Markov semi-group on X. A positive Borel numerical function f on X is called P-*excessive* if

a) for any $t\in\mathbb{R}_+$, $\mathsf{P}_t f\le f$;

b) for any $x\in X$, $\lim\limits_{t\to 0}\mathsf{P}_t f(x)=f(x)$.

Let $\mathsf{U}:=(\mathsf{U}_\alpha)_{\alpha\in\mathbb{R}_+}$ be a resolvent on X. A positive Borel numerical function f on X is called U-*excessive* if

a) for any $\alpha\in\mathbb{R}_+$, $\alpha\,\mathsf{U}_\alpha f\le f$;

b) for any $x\in X$, $\lim\limits_{\alpha\to\infty}\alpha\,\mathsf{U}_\alpha f(x)=f(x)$.

Proposition 10.1.6. *Let $\mathsf{P}:=(\mathsf{P}_t)_{t\in\mathbb{R}_+}$ be a sub-Markov semi-group on X and let f be a P-excessive function. Then for every $t\in\mathbb{R}_+$, $\mathsf{P}_t f$ is P-excessive.*

By the preceding proposition for any $s\in\mathbb{R}_+$

$$\mathsf{P}_s\,\mathsf{P}_t f=\mathsf{P}_t\,\mathsf{P}_s f\le\mathsf{P}_t f,$$

$$\lim_{s\to 0}\mathsf{P}_s\,\mathsf{P}_t f=\lim_{s\to 0}\mathsf{P}_t\,\mathsf{P}_s f=\mathsf{P}_t f. \quad \square$$

Proposition 10.1.7. *Let P be an integrable sub-Markov semi-group on X and let U be the resolvent generated by P. A function on X is U-excessive if and only if it is P-excessive.*

Let f be a P-excessive function and let $x \in X$ and $\alpha \in \mathbb{R}_+$. We have

$$\alpha \mathsf{U}_\alpha f(x) = \alpha \int_{[0,\infty[}^{*} e^{-\alpha t} \mathsf{P}_t f(x)\, dt \leq \left(\alpha \int_{[0,\infty[}^{*} e^{-\alpha t}\, dt\right) f(x) = f(x).$$

By the preceding proposition,

$$\lim_{\alpha\to\infty} \alpha \mathsf{U}_\alpha f(x) = \lim_{\alpha\to\infty} \alpha \int_{[0,\infty[}^{*} e^{-\alpha t} \mathsf{P}_t f(x)\, dt = \lim_{\alpha\to\infty} \int_{[0,\infty[}^{*} e^{-t} \mathsf{P}_{t/\alpha} f(x)\, dt$$
$$= \int_{[0,\infty[}^{*} e^{-t} f(x)\, dt = f(x).$$

In order to prove the converse relation, first let $\beta \in \mathbb{R}_+$ and let f be a positive Borel function. Then

$$e^{-\beta t} \mathsf{P}_t \mathsf{U}_\beta f(x) = e^{-\beta t} \mathsf{U}_\beta \mathsf{P}_t f(x) = e^{-\beta t} \int_{[0,\infty[}^{*} e^{-\beta s} \mathsf{P}_s \mathsf{P}_t f(x)\, ds$$
$$= \int_{[0,\infty[}^{*} e^{-\beta(s+t)} \mathsf{P}_{s+t} f(x)\, ds = \int_{[t,\infty[}^{*} e^{-\beta s} \mathsf{P}_s f(x)\, ds \leq \int_{[0,\infty[}^{*} e^{-\beta s} \mathsf{P}_s f(x)\, ds$$
$$= \mathsf{U}_\beta f(x).$$

Now let f be a U-excessive function. For any $\alpha, \beta \in \mathbb{R}_+$ we have

$$f + (\beta - \alpha) \mathsf{U}_\alpha f = (f - \alpha \mathsf{U}_\alpha f) + \beta \mathsf{U}_\alpha f \geq 0.$$

Hence, for any $t \in \mathbb{R}_+$ we get, by the above relation,

$$e^{-\beta t} \mathsf{P}_t \mathsf{U}_\alpha f = e^{-\beta t} \mathsf{P}_t \mathsf{U}_\beta (f + (\beta-\alpha) \mathsf{U}_\alpha f) \leq \mathsf{U}_\beta (f + (\beta - \alpha) \mathsf{U}_\alpha f) = \mathsf{U}_\alpha f.$$

Since β is arbitrary, it follows that

$$\mathsf{P}_t \mathsf{U}_\alpha f \leq \mathsf{U}_\alpha f.$$

Now for any $x \in X$,

$$\mathsf{P}_t f(x) = \int^{*} f\, d\mathsf{P}_{t,x} \leq \lim_{\alpha\to\infty} \int^{*} \alpha \mathsf{U}_\alpha f\, d\mathsf{P}_{t,x} \leq \lim_{\alpha\to\infty} \alpha \mathsf{U}_\alpha f(x) = f(x).$$

From the above relation we also get for any $\alpha \in \mathbb{R}_+$ and $x \in X$,

$$\lim_{t\to 0} \mathsf{P}_t f(x) \geq \liminf_{t\to 0} e^{-\alpha t} \mathsf{P}_t \alpha \mathsf{U}_\alpha f(x) = \liminf_{t\to 0} \alpha \int_{[t,\infty[}^{*} e^{-\alpha s} \mathsf{P}_s f(x)\, ds$$
$$= \alpha \mathsf{U}_\alpha f(x).$$

Since α is arbitrary, we deduce

$$\lim_{t\to 0} \mathsf{P}_t f = f. \quad \square$$

Exercises

10.1.1. Let N be a diffusion on X. For any positive Borel function f on X, we set

$$\mathsf{N}^0 f := f, \qquad \mathsf{G} f := \sum_{n \in \mathbb{N}} \mathsf{N}^n f.$$

Let u be a positive Borel function on X with

$$\mathsf{N} u \leq u.$$

Then, for any positive Borel function f on X such that

$$\mathsf{G} f \leq u$$

on $\{x \in X \mid f(x) > 0\}$, we have

$$\mathsf{G} f \leq u$$

on $\{x \in X \mid \mathsf{G} f(x) < \infty\}$ (P.-A. Meyer 1966 [2]). (Set

$$v := \inf(\mathsf{G} f, u)$$

and let w be the function which is equal to $\mathsf{G} f - v$ on $\{x \in X \mid v(x) < \infty\}$ and equal to 0 elsewhere. Show that

$$\mathsf{N}\mathsf{G} f \leq \mathsf{G} f, \qquad \mathsf{N} v \leq v$$

and that

$$\lim_{n \to \infty} \mathsf{N}^n \mathsf{G} f = 0$$

on $\{x \in X \mid \mathsf{G} f(x) < \infty\}$. From

$$v + w = \mathsf{G} f = f + \mathsf{N}\mathsf{G} f = f + \mathsf{N} v + \mathsf{N} w$$

it follows that

$$w \leq \mathsf{N} w.$$

Hence for any $n \in \mathbb{N}$

$$w \leq \mathsf{N}^n w \leq \mathsf{N}^n \mathsf{G} f, \qquad w \leq \lim_{n \to \infty} \mathsf{N}^n \mathsf{G} f.)$$

10.1.2. A family $\mathsf{U} := (\mathsf{U}_\alpha)_{\alpha \in]0, \infty[}$ of diffusions on X will be called an *open resolvent* provided:

a) for any $\alpha \in]0, \infty[$

$$\alpha \mathsf{U}_\alpha 1 \leq 1;$$

b) for any $\alpha, \beta \in]0, \infty[$, $\alpha < \beta$,

$$\mathsf{U}_\alpha = \mathsf{U}_\beta + (\beta - \alpha) \mathsf{U}_\beta \mathsf{U}_\alpha.$$

A positive Borel function f on x is called U-excessive if, for any $\alpha \in]0, \infty[$

$$\alpha \mathsf{U}_\alpha f \leq f, \qquad \lim_{\alpha \to \infty} \alpha \mathsf{U}_\alpha f = f.$$

Let f be a positive Borel function on X such that for any $\alpha \in]0, \infty[$

$$\alpha \mathsf{U}_\alpha f \leq f.$$

Then:

a) for every $x \in X$, the function

$$\alpha \mapsto \alpha \mathsf{U}_\alpha f(x)$$

is increasing;

b) if we set

$$f^* := \lim_{\alpha \to \infty} \alpha \mathsf{U}_\alpha f,$$

then f^* is U-excessive and for any $\alpha \in]0, \infty[$, $\mathsf{U}_\alpha f$ is U-excessive and

$$\mathsf{U}_\alpha f = \mathsf{U}_\alpha f^*.$$

10.1.3. We say that an open resolvent U (see the preceding exercise) is generated by the sub-Markov semi-group $\mathsf{P} := (\mathsf{P}_t)_{t \in \mathbb{R}_+}$ on X if, for any $f \in \mathscr{K}_+(X)$, any $x \in X$ and for any $\alpha \in]0, \infty[$

$$\mathsf{U}_\alpha f(x) = \int_0^\infty e^{-\alpha t} \mathsf{P}_t f(x)\, dt.$$

a) Prove all of the propositions of this section with open resolvents replacing the resolvents and arbitrary sub-Markov semi-groups replacing integrable sub-Markov semi-groups.

b) Let f be a positive Borel function such that for every $t \in \mathbb{R}_+$, $\mathsf{P}_t f \leq f$. Set

$$f^* := \lim_{\substack{t \to 0 \\ t > 0}} \mathsf{P}_t f.$$

Then f^* is P-excessive, for every $\alpha \in]0, \infty[$ we have $\alpha \mathsf{U}_\alpha f \leq f$, and

$$f^* = \lim_{\alpha \to \infty} \alpha \mathsf{U}_\alpha f.$$

c) Let f be a positive, lower semi-continuous numerical function on X such that for any $\alpha \in]0, \infty[$

$$\alpha \mathsf{U}_\alpha f \leq f.$$

Then for any $t \in \mathbb{R}_+$,

$$\mathsf{P}_t f \leq f, \qquad \mathsf{P}_t f^* = \mathsf{P}_t f$$

where

$$f^* = \lim_{t \to 0} \mathsf{P}_t f.$$

(Set

$$\bar{f} = \lim_{\alpha \to \infty} \alpha \mathsf{U}_\alpha f.$$

By the preceding exercise, for any $\alpha\in]0,\infty[$, $\mathsf{U}_\alpha f=\mathsf{U}_\alpha \bar{f}$. Hence $\bar{f}$ is U-excessive, hence P-excessive and therefore for any $x\in X$, the function

$$t\mapsto \mathsf{P}_t \bar{f}(x)$$

is continuous from the right. Since f is lower semi-continuous, the function

$$t\mapsto \mathsf{P}_t f(x)$$

is lower semi-continuous from the right. From

$$0=\int_0^\infty e^{-\alpha t}(\mathsf{P}_t f(x)-\mathsf{P}_t \bar{f}(x))\,dt$$

we obtain

$$\mathsf{P}_t f(x)=\mathsf{P}_t \bar{f}(x)\leq \bar{f}(x)\leq f(x).$$

Hence $\bar{f}=f^*$.)

10.1.4 (M. Motoo). Let $\mathsf{U}=(\mathsf{U}_\alpha)_{\alpha\in\mathbb{R}_+}$ be a resolvent on X for which there exists a measure μ on X such that for any $x\in X$, $\mathsf{U}_{0,x}$ is absolutely continuous with respect to μ. Let u, v be two U-excessive functions such that

$$\mathsf{U}_0 1=u+v.$$

Then there exists a Borel function f on X with $0\leq f\leq 1$ and such that

$$u=\mathsf{U}_0 f.$$

(For any $\alpha\in]0,\infty[$ we set

$$f_\alpha:=\alpha(u-\alpha\,\mathsf{U}_\alpha u).$$

From

$$u-\alpha\,\mathsf{U}_\alpha u\leq \mathsf{U}_0 1-\alpha\,\mathsf{U}_\alpha \mathsf{U}_0 1=\mathsf{U}_\alpha 1\leq\frac{1}{\alpha}$$

it follows that $0\leq f_\alpha\leq 1$. Let f be a limit function for α converging to ∞ in $L^\infty(\mu)$ with respect to the weak topology $\sigma(L^\infty(\mu), L^1(\mu))$. By the Radon-Nikodym theorem, there exists for any $x\in X$, $g_x\in L^1(\mu)$ such that $\mathsf{U}_{0,x}=g_x\cdot\mu$. Hence for any $x\in X$,

$$u(x)=\lim_{\alpha\to\infty}\alpha\,\mathsf{U}_\alpha u(x)=\lim_{\alpha\to\infty}\mathsf{U}_0 f_\alpha(x)=\lim_{\alpha\to\infty}\int f_\alpha g_x\,d\mu=\int f g_x\,d\mu=\mathsf{U}_0 f(x).)$$

§ 10.2. Sub-Markov Semi-Groups on Harmonic Spaces

Given a $\mathfrak{P}$-harmonic space with a countable base and for which the function 1 is superharmonic we construct, in this section, a sub-Markov semi-group P such that the set of P-excessive functions coincides with the set of positive hyperharmonic functions. This semi-group possesses certain properties which make possible the construction of a strong Markov process with continuous paths. This in turn links up the potential theory on harmonic spaces with Hunt's potential theory, by establishing a corespondence between the key notions of these two theories. This correspondence which is based on the construction of a semi-group associated in a certain way to a bounded, continuous strict potential, can in fact be carried out in far more general cases. An outline of the procedure will be sketched in some of the exercises.

The theory of Markov processes on harmonic spaces is not presented in this book since it forms by itself enough material for a whole monograph. The interested reader may find detailed information in N. Boboc-C. Constantinescu-A. Cornea 1967 [5], W. Hansen 1967 [1], C. Constantinescu 1968 [8], H. Bauer 1970 [10], J.C. Taylor 1972 [1], [2]. The research in this field in not finished important questions still being open.

Proposition 10.2.1. *Let p be a bounded potential of $\mathscr{M}$, let $\alpha \in \mathbb{R}_+$ and let f, g be bounded Borel functions on X. If*

$$\mathsf{V}_p f = g + \alpha \mathsf{V}_p g$$

then any positive hyperharmonic function on X which dominates f also dominates αg.

Let u be a positive hyperharmonic function on X dominating f. We set

$$f' := \sup(f - \alpha g, 0), \qquad f'' := -\inf(f - \alpha g, 0).$$

Let f_0 be a positive, upper semi-continuous real function on X with compact carrier and smaller than f', let ε be a strictly positive real number and let x be a point of X such that

$$f_0(x) \geq \varepsilon.$$

From

$$f(x) \leq u(x)$$

we obtain

$$0 \leq f(x) - \alpha g(x) \leq u(x) - \alpha g(x),$$

$$\alpha \mathsf{V}_p f'(x) = \alpha g(x) + \alpha \mathsf{V}_p f''(x) \leq u(x) + \alpha \mathsf{V}_p f''(x).$$

Since x is arbitrary,

$$\alpha \mathsf{V}_p f' \leq u + \alpha \mathsf{V}_p f''$$

on $\{x \in X \mid f_0(x) \geq \varepsilon\}$. Hence

$$\alpha \sup(f_0 - \varepsilon, 0) \cdot p \leq u + \alpha f'' \cdot p$$

(Theorem 8.2.1) on $\{x \in X \mid f_0(x) \geq \varepsilon\}$. Since

$$\sup(f_0 - \varepsilon, 0) \cdot p \in \mathcal{M}, \qquad S(\sup(f_0 - \varepsilon, 0) \cdot p) \subset \{x \in X \mid f_0(x) \geq \varepsilon\}$$

(Theorem 8.3.1, Corollary 8.2.1), it follows that

$$\alpha \mathsf{V}_p \sup(f_0 - \varepsilon, 0) = \alpha \sup(f_0 - \varepsilon, 0) \cdot p \leq u + \alpha f'' \cdot p$$

on X (Theorem 8.3.2 *c*)). Since ε and f_0 are arbitrary,

$$\alpha \mathsf{V}_p f' \leq u + \alpha \mathsf{V}_p f'',$$

$$\alpha g = \alpha \mathsf{V}_p f' - \alpha \mathsf{V}_p f'' \leq u. \quad \square$$

Corollary 10.2.1. *With the above notation, we have $g = 0$ if*

$$g + \alpha \mathsf{V}_p g = 0.$$

It is sufficient to take $u = 0$ in the proposition. $\square$

Proposition 10.2.2. *For any bounded potential p of $\mathcal{M}$, there exists a unique resolvent $\mathsf{U} := (\mathsf{U}_\alpha)_{\alpha \in \mathbb{R}_+}$ on X such that $\mathsf{U}_0 = \mathsf{V}_p$. If p is strict, then the set of positive hyperharmonic functions on X is equal to the set of U-excessive functions. If p is continuous, then for any $\alpha \in \mathbb{R}_+$ and any bounded Borel function f on X, $\mathsf{U}_\alpha f$ is a bounded continuous function.*

The unicity of U follows immediately from the preceding corollary. For the existence of U we need the following lemma.

Lemma. *Let E be a (real) Banach space with norm $\| \;\|$ and let V be a continuous linear map of E into itself such that for any $\alpha \in \mathbb{R}_+$ and any $x, y \in E$ with*

$$Vx = y + \alpha V y$$

we have

$$\alpha \|y\| \leq \|x\|.$$

Then there exists a family $(U_\alpha)_{\alpha \in \mathbb{R}_+}$ of continuous linear maps of E into itself such that $U_0 = V$ and for any $\alpha, \beta \in \mathbb{R}_+$

$$U_\alpha - U_\beta = (\beta - \alpha) U_\alpha U_\beta, \qquad \alpha \|U_\alpha\| \leq 1.$$

Let A be the set of $\alpha \in \mathbb{R}_+$ for which there exists a continuous linear map U_α of E into itself such that

$$V U_\alpha = U_\alpha V, \qquad V = U_\alpha + \alpha V U_\alpha.$$

Obviously $0 \in A$ and $U_0 = V$. From the hypothesis it follows immediately that for any $\alpha \in A$, U_α is uniquely determined and

$$\alpha \|U_\alpha\| \leq 1.$$

Let $\alpha \in A$ and let $\beta \in \mathbb{R}_+$, $|\beta-\alpha| < \|U_\alpha\|^{-1}$. We denote by U, the continuous linear map of E into itself

$$U := \sum_{n\in\mathbb{N}} (\alpha-\beta)^n U_\alpha^{n+1}.$$

We have

$$UV = VU, \qquad UU_\alpha = U_\alpha U,$$

$$U + (\beta-\alpha) UU_\alpha = \sum_{n\in\mathbb{N}} (\alpha-\beta)^n U_\alpha^{n+1} - \sum_{n\in\mathbb{N}} (\alpha-\beta)^{n+1} U_\alpha^{n+2} = U_\alpha.$$

If we denote by I the identity map on E, then we have

$$\begin{aligned} V &= (I+\alpha V) U_\alpha = \big(I+\beta V + (\alpha-\beta) V\big)\big(U + (\beta-\alpha) UU_\alpha\big) \\ &= (I+\beta V) U + (\alpha-\beta) VU + (\beta-\alpha)(I+\beta V) UU_\alpha - (\alpha-\beta)^2 VUU_\alpha \\ &= (I+\beta V) U + (\alpha-\beta) U\big(V - U_\alpha - \beta VU_\alpha - (\alpha-\beta) VU_\alpha\big) = U + \beta VU. \end{aligned}$$

Hence $\beta \in A$ and $U_\beta = U$. We have proved therefore that A is open and that if $\alpha \in A$, then $]0, 2\alpha[\subset A$. Since $0 \in A$ it follows that A contains a strictly positive real number and therefore coincides with $\mathbb{R}_+$.

Let $\alpha, \beta \in \mathbb{R}_+$, $0 < \beta < \alpha$. Then

$$|\beta-\alpha| < \|U_\alpha\|^{-1}$$

and by the above considerations

$$U_\beta = \sum_{n\in\mathbb{N}} (\alpha-\beta)^n U_\alpha^{n+1}.$$

Hence

$$U_\alpha U_\beta = U_\beta U_\alpha, \qquad U_\beta = U_\alpha + (\alpha-\beta) U_\beta U_\alpha. \quad \square$$

We prove now the proposition. Let $\mathscr{B}$ be the Banach space of bounded Borel functions f on X with

$$\|f\| = \sup_X |f|.$$

The map V_p

$$f \mapsto \mathsf{V}_p f \colon \mathscr{B} \to \mathscr{B}$$

is continuous and linear. By Proposition 10.2.1 and the above lemma there exists a family $(U_\alpha)_{\alpha\in\mathbb{R}_+}$ of continuous linear maps of $\mathscr{B}$ into itself such that $U_0 = V_p$ and for any $\alpha, \beta = \mathbb{R}_+$

$$U_\alpha - U_\beta = (\beta-\alpha) U_\alpha U_\beta, \qquad \alpha \|U_\alpha\| \le 1.$$

By Proposition 10.2.1, it follows immediately that for any $\alpha \in \mathbb{R}_+$ and any negative function $f \in \mathscr{B}$, $U_\alpha f$ is negative. Hence for any $f \in \mathscr{B}_+$ we have

$$0 \le U_\alpha f \le V_p f.$$

Let $\alpha \in \mathbb{R}_+$ and $x \in X$. The map

$$f \mapsto U_\alpha f(x): \mathcal{K}(X) \to \mathbb{R}$$

is a measure on X which will be denoted by $\mathsf{U}_{\alpha,x}$. Since for any decreasing sequence $(f_n)_{n\in\mathbb{N}}$ in $\mathcal{B}_+$ which converges to 0 at all points of X, we have

$$0 \leq \lim_{n\to\infty} U_\alpha f_n(x) \leq \lim_{n\to\infty} V_p f_n(x) = 0,$$

it follows that

$$U_\alpha f(x) = \int f \, d\mathsf{U}_{\alpha,x}$$

for every $f \in \mathcal{B}$. We deduce immediately that $\mathsf{U}_\alpha := (\mathsf{U}_{\alpha,x})_{x\in X}$ is a diffusion on X and that $(\mathsf{U}_\alpha)_{\alpha\in\mathbb{R}_+}$ is a resolvent on X associated with p.

Assume now that p is strict. Let u be a positive hyperharmonic function on X and let, for any $n \in \mathbb{N}$,

$$u_n := u \wedge n.$$

Then by Proposition 10.2.1, we get for any $\alpha \in \mathbb{R}_+$

$$\alpha \mathsf{U}_\alpha u_n \leq u.$$

Since n is arbitrary, we have

$$\alpha \mathsf{U}_\alpha u \leq u.$$

Let $x \in X$. By Proposition 10.1.5 the function

$$\alpha \mapsto \alpha \mathsf{U}_\alpha u(x)$$

is increasing. For any continuous finite potential q on X we set

$$\varphi(q) := \lim_{\alpha\to\infty} \alpha \mathsf{U}_\alpha q(x).$$

Now apply the lemma of Proposition 7.1.2, taking as $\mathcal{F}$ the convex cone of continuous, finite potentials on X. By Theorem 2.3.1, $\mathcal{F}$ satisfies the hypothesis of this lemma. The conditions *α), β)* of the lemma are obvious; *γ)* follows from the inequality

$$\varphi(q) \leq q(x)$$

and from Proposition 7.1.1. Hence there exists a measure μ on X such that

$$\varphi(q) = \int q \, d\mu$$

for every continuous finite potential q on X. Since every positive, hyperharmonic function u on X is the limit of an increasing sequence

$(q_n)_{n\in\mathbb{N}}$ of continuous finite potentials on X, we get

$$u(x)\geq \lim_{\alpha\to\infty} \alpha \mathsf{U}_\alpha u(x) = \lim_{\alpha\to\infty} \lim_{n\to\infty} \alpha \mathsf{U}_\alpha q_n(x) = \lim_{n\to\infty} \lim_{\alpha\to\infty} \alpha \mathsf{U}_\alpha q_n(x)$$

$$= \lim_{n\to\infty} \int q \, d\mu = \int^{*} u \, d\mu.$$

We have

$$\alpha \mathsf{U}_\alpha p(x) = \alpha \mathsf{U}_\alpha V_p 1(x) = \mathsf{V}_p 1(x) - \mathsf{U}_\alpha 1(x) \geq p(x) - \frac{1}{\alpha}.$$

Hence

$$\int p \, d\mu = \lim_{\alpha\to\infty} \alpha \mathsf{U}_\alpha p(x) = p(x).$$

p being strict we get $\mu = \varepsilon_x$. It follows immediately that every positive hyperharmonic function is U-excessive.

Let f be a U-excessive function. For any $n\in\mathbb{N}$, we set

$$f_n := \inf(f, n).$$

Then for any $\alpha\in\mathbb{R}_+$, we have

$$\alpha \mathsf{U}_\alpha f_n \leq f_n.$$

Hence

$$\mathsf{U}_\alpha f_n = \mathsf{U}_0(f_n - \alpha \mathsf{U}_\alpha f_n) = (f_n - \alpha \mathsf{U}_\alpha f_n)\cdot p.$$

It follows that $\mathsf{U}_\alpha f_n$ is a hyperharmonic function. From

$$\mathsf{U}_\alpha f = \lim_{n\to\infty} \mathsf{U}_\alpha f_n$$

it follows that $\mathsf{U}_\alpha f$ is also a hyperharmonic function. By Proposition 10.1.5, f is the limit of an increasing sequence of hyperharmonic functions and is therefore hyperharmonic.

Assume now that p is continuous and let g be a positive, bounded Borel function on X. From

$$p = \frac{g}{\sup_X g}\cdot p + \left(1 - \frac{g}{\sup_X g}\right)\cdot p$$

(Corollary 8.2.1) it follows immediately that $g\cdot p$ is continuous. Furthermore we see from

$$\mathsf{U}_\alpha f = \mathsf{U}_0(f - \alpha \mathsf{U}_\alpha f) = f\cdot p - \alpha \mathsf{U}_\alpha f \cdot p$$

that $\mathsf{U}_\alpha f$ is continuous for every positive bounded Borel function f on X. □

The following theorem is an adaptation of a result of G.A. Hunt 1957 [1] to harmonic spaces.

Theorem 10.2.1. *Let p be a bounded, continuous, strict potential on X. Then there exists a unique integrable sub-Markov semi-group $\mathsf{P}:=(\mathsf{P}_t)_{t\in\mathbb{R}_+}$ on X such that for any $f\in\mathscr{K}_+(X)$ and any $t\in\mathbb{R}_+$, $\mathsf{P}_t f$ is continuous and for any $x\in X$*

$$f\cdot p(x)=\int_0^\infty \mathsf{P}_t f(x)\,dt.$$

The set of P-excessive functions and the set of positive hyperharmonic functions on X coincide.

We prove first the following lemma.

Lemma (Hille-Yosida). *Let E be a (real) Banach space with norm $\|\ \|$ and C be a closed convex cone of E with vertex 0. Let $(U_\alpha)_{\alpha\in]0,\infty[}$ be a family of continuous linear maps of E into itself such that for any $\alpha,\beta\in]0,\infty[$:*

α) $\alpha\|U_\alpha\|\leq 1$;

β) $U_\alpha-U_\beta=(\beta-\alpha)\,U_\alpha U_\beta$;

γ) $\overline{U_\alpha(E)}=E$;

δ) $U_\alpha(C)\subset C$.

Then there exists a family $(P_t)_{t\in\mathbb{R}_+}$ of continuous linear maps of E into itself such that:

a) for any $t\in\mathbb{R}_+$, $\|P_t\|\leq 1$;

b) for any $x\in E$, $\lim\limits_{t\to 0}\|P_t x-x\|=0$;

c) for any $s,t\in\mathbb{R}_+$, $P_{s+t}=P_s P_t$;

d) for any $t\in\mathbb{R}_+$, $P_t(C)\subset C$;

e) for any $x\in E$, any element x' of the dual E' of E and any $\alpha\in]0,\infty[$

$$\langle U_\alpha x, x'\rangle=\int_0^\infty e^{-\alpha t}\langle P_t x, x'\rangle\,dt$$

(by a), b), c) the function $t\mapsto\langle P_t x, x'\rangle$ is continuous and bounded);

f) let x be an element of C such that for any $\alpha\in]0,\infty[$, $x-\alpha U_\alpha x\in C$; then for any $t\in\mathbb{R}_+$, $x-P_t x\in C$.

We remark first that from *β)* it follows immediately that

$$U_\alpha U_\beta=U_\beta U_\alpha,\qquad \alpha U_\alpha U_\beta-U_\beta=\beta U_\beta U_\alpha-U_\alpha$$

for all $\alpha,\beta\in]0,\infty[$. Hence by *α)*

$$\|\alpha U_\alpha U_\beta-U_\beta\|\leq\frac{2}{\alpha},\qquad \lim_{\alpha\to\infty}\|\alpha U_\alpha U_\beta-U_\beta\|=0.$$

With the aid of $\alpha)$ and $\gamma)$ we get further for any $x \in E$

$$\lim_{\alpha \to \infty} \alpha U_\alpha x = x.$$

The construction of the family $(P_t)_{t \in \mathbb{R}_+}$. For any $\alpha \in]0, \infty[$, $t \in \mathbb{R}_+$, we set

$$P_t^{(\alpha)} := \exp(-\alpha t) \exp(t \alpha^2 U_\alpha)$$

(N. Bourbaki [7], p. 50). We have for any $\alpha \in]0, \infty[$ and any $s, t \in \mathbb{R}_+$

$$\|P_t^{(\alpha)}\| \leq \exp(-\alpha t) \exp(t \alpha^2 \|U_\alpha\|) \leq 1,$$

$$P_{s+t}^{(\alpha)} = P_s^{(\alpha)} P_t^{(\alpha)}, \qquad P_t^{(\alpha)}(C) \subset C.$$

For any $x \in E$, we have

$$P_t^{(\alpha)} x - x = \sum_{\substack{n \in \mathbb{N} \\ n \neq 0}} \frac{\alpha^n t^n}{n!} (\alpha U_\alpha - I)^n x,$$

where I denotes the identity map of E. Hence

$$\begin{aligned} \|P_t^{(\alpha)} x - x\| &\leq \sum_{\substack{n \in \mathbb{N} \\ n \neq 0}} \frac{\alpha^n t^n}{n!} \|\alpha U_\alpha - I\|^n \|x\| \\ &= \|x\| \big(\exp(\alpha t \|\alpha U_\alpha - I\|) - 1\big), \\ \lim_{t \to 0} &\|P_t^{(\alpha)} x - x\| = 0. \end{aligned}$$

From this and the above relations, it follows that $t \mapsto P_t^{(\alpha)} x$ is a continuous map of $\mathbb{R}_+$ into E.

Let $x \in E$, $x' \in E'$. From $\alpha)$ and $\beta)$ it follows immediately that the function

$$\alpha \mapsto \langle U_\alpha x, x' \rangle$$

is continously differentiable and that

$$\frac{d}{d\alpha} \langle U_\alpha x, x' \rangle = -\langle U_\alpha^2 x, x' \rangle.$$

We deduce further that for any $t \in \mathbb{R}_+$, the function

$$\alpha \mapsto \langle P_t^{(\alpha)} x, x' \rangle$$

is also continously differentiable and

$$\begin{aligned} \frac{d}{d\alpha} \langle P_t^{(\alpha)} x, x' \rangle &= -t\, e^{-t\alpha} \langle e^{t\alpha^2 U_\alpha} x, x' \rangle \\ &\quad + e^{-t\alpha} \langle e^{t\alpha^2 U_\alpha} (2t\alpha U_\alpha - t\alpha^2 U_\alpha^2) x, x' \rangle \\ &= -t \langle P_t^{(\alpha)} (I - \alpha U_\alpha)^2 x, x' \rangle. \end{aligned}$$

For any $\gamma \in]0, \infty[$ we have

$$(I - \alpha U_\alpha) U_\gamma = U_\alpha (I - \gamma U_\gamma)$$

and therefore

$$\left| \frac{d}{d\alpha} \langle P_t^{(\alpha)} U_\gamma^2 x, x' \rangle \right| \leq \frac{t}{\alpha^2} \|x - \gamma U_\gamma x\|^2 \|x'\|.$$

Hence for any $\alpha, \beta \in]0, \infty[$

$$\langle P_t^{(\alpha)} U_\gamma^2 x - P_t^{(\beta)} U_\gamma^2 x, x' \rangle \leq t \left| \frac{1}{\alpha} - \frac{1}{\beta} \right| \|x - \gamma U_\gamma x\|^2 \|x'\|,$$

$$\|P_t^{(\alpha)} U_\gamma^2 x - P_t^{(\beta)} U_\gamma^2 x\| \leq t \left| \frac{1}{\alpha} - \frac{1}{\beta} \right| \|x - \gamma U_\gamma x\|^2.$$

Hence for any $x \in U_\gamma^2(E)$ the continuous functions

$$t \mapsto P_t^{(\alpha)} x \colon \mathbb{R}_+ \to E$$

converge, as α tends to ∞, uniformly on any bounded interval of $\mathbb{R}_+$. Hence, the limit function is a continuous function which will be denoted, for any $t \in \mathbb{R}_+$, by $P_t x$. From

$$\overline{U_\gamma^2(E)} \supset U_\gamma(\overline{U_\gamma(E)}) = \overline{U_\gamma(E)} = E$$

and from

$$\|P_t^{(\alpha)}\| \leq 1$$

for every $\alpha \in]0, \infty[$ and $t \in \mathbb{R}_+$, it follows that we may assume $x \in E$ in the above considerations. Obviously $x \mapsto P_t x$ is a continuous linear map, P_t, of E into itself. Now property *a)* is immediate and *b)* follows from the fact that $P_0 = I$ and $t \mapsto P_t x$ is continuous for every $x \in E$. Property *d)* is a consequence of the corresponding property of $P_t^{(\alpha)}$ since C is closed.

c) Let $s, t \in \mathbb{R}_+$ and $x \in E$. We have for any $\alpha \in]0, \infty[$

$$\begin{aligned} \|P_s^{(\alpha)} P_t^{(\alpha)} x - P_s P_t x\| &\leq \|P_s^{(\alpha)}(P_t^{(\alpha)} x - P_t x)\| + \|P_s^{(\alpha)} P_t x - P_s P_t x\| \\ &\leq \|P_t^{(\alpha)} x - P_t x\| + \|P_s^{(\alpha)} P_t x - P_s P_t x\|. \end{aligned}$$

Hence

$$\lim_{\alpha \to \infty} \|P_s^{(\alpha)} P_t^{(\alpha)} x - P_s P_t x\| = 0,$$

$$\begin{aligned} \|P_{s+t} x - P_s P_t x\| &\leq \|P_{s+t} x - P_{s+t}^{(\alpha)} x\| + \|P_s^{(\alpha)}(P_t^{(\alpha)} x - P_t x)\| \\ &\quad + \|P_s^{(\alpha)} P_t x - P_s P_t x\|, \end{aligned}$$

$$\begin{aligned} \|P_{s+t} x - P_s P_t x\| &\leq \lim_{\alpha \to \infty} \|P_{s+t} x - P_{s+t}^{(\alpha)} x\| + \lim_{\alpha \to \infty} \|P_t^{(\alpha)} x - P_t x\| \\ &\quad + \lim_{\alpha \to \infty} \|P_s^{(\alpha)} P_t x - P_s P_t x\| = 0. \end{aligned}$$

Since x is arbitrary, we get

$$P_{s+t} = P_s P_t.$$

e) Let $\alpha, \beta \in]0, \infty[$, $x \in E$, $x' \in E'$. We denote by f_α, the function on $\mathbb{R}_+$

$$t \mapsto \langle P_t^{(\alpha)} U_\beta x, x' \rangle.$$

From the definition of $P_t^{(\alpha)}$, it follows immediately that f is continuously differentiable and

$$\begin{aligned} \frac{df_\alpha}{dt}(t) &= \langle \alpha P_t^{(\alpha)} (\alpha U_\alpha U_\beta x - U_\beta x), x' \rangle \\ &= \langle P_t^{(\alpha)} \alpha U_\alpha (\beta U_\beta x - x), x' \rangle, \end{aligned}$$

$$\left| \frac{df_\alpha}{dt} \right| \leq \| \beta U_\beta x - x \| \; \| x' \|.$$

From

$$\begin{aligned} \| P_t^{(\alpha)} \alpha U_\alpha x - P_t x \| &\leq \| P_t^{(\alpha)} x - P_t x \| + \| P_t^{(\alpha)} (\alpha U_\alpha x - x) \| \\ &\leq \| P_t^{(\alpha)} x - P_t x \| + \| \alpha U_\alpha x - x \| \end{aligned}$$

we obtain

$$\lim_{\alpha \to \infty} \| P_t^{(\alpha)} \alpha U_\alpha x - P_t x \| = 0.$$

Hence $\left(\frac{df_\alpha}{dt} \right)_{\alpha \in]0, \infty[}$ is a uniformly bounded family of continuous real functions which converge, as α tends to ∞, to the function

$$t \mapsto \langle P_t (\beta U_\beta x - x), x' \rangle.$$

It follows that the function

$$t \mapsto \langle P_t U_\beta x, x' \rangle$$

is continously differentiable and

$$\frac{d}{dt} \langle P_t U_\beta x, x' \rangle = - \langle P_t (x - \beta U_\beta x), x' \rangle.$$

We get

$$\begin{aligned} \int_0^\infty e^{-\alpha t} \langle P_t (x - \beta U_\beta x), x' \rangle \, dt &= - \int_0^\infty e^{-\alpha t} \frac{d}{dt} \langle P_t U_\beta x, x' \rangle \, dt \\ &= \langle U_\beta x, x' \rangle - \alpha \int_0^\infty e^{-\alpha t} \langle P_t U_\beta x, x' \rangle \, dt, \end{aligned}$$

$$\langle U_\beta x, x' \rangle = \int_0^\infty e^{-\alpha t} \langle P_t (x + (\alpha - \beta) U_\beta x), x' \rangle \, dt,$$

$$\langle U_\beta U_\alpha x, x'\rangle = \int_0^\infty e^{-\alpha t} \langle P_t(U_\alpha x + (\alpha-\beta) U_\beta U_\alpha x), x'\rangle\, dt$$

$$= \int_0^\infty e^{-\alpha t} \langle P_t U_\beta x, x'\rangle\, dt.$$

Hence

$$\langle U_\alpha x, x'\rangle = \lim_{\beta\to\infty} \langle \beta U_\beta U_\alpha x, x'\rangle$$

$$= \lim_{\beta\to\infty} \int_0^\infty e^{-\alpha t} \langle P_t \beta U_\beta x, x'\rangle\, dt$$

$$= \int_0^\infty e^{-\alpha t} \langle P_t x, x'\rangle\, dt.$$

f) For any continuous linear map P of E into itself, we denote by P', the continuous linear map of E' into itself defined by

$$\langle x, P' x'\rangle = \langle P x, x'\rangle,$$

for all $x\in E$, $x'\in E'$. We denote by C', the set of elements of E' which are positive on C.

Let $x\in C$, let $\beta\in]0,\infty[$ and let $t\in\mathbb{R}_+$. For any $x'\in C'$ we have

$$\langle e^{-\beta t} P_t U_\beta x, x'\rangle = e^{-\beta t} \langle U_\beta x, P_t' x'\rangle = e^{-\beta t} \int_0^\infty e^{-\beta s} \langle P_s x, P_t' x'\rangle\, ds$$

$$= \int_0^\infty e^{-\beta(s+t)} \langle P_{s+t} x, x'\rangle\, ds = \int_t^\infty e^{-\beta s} \langle P_s x, x'\rangle\, ds.$$

Since $x\in C$, it follows that $P_s x\in C$ and therefore

$$\langle P_s x, x'\rangle \geq 0$$

for every $s\in\mathbb{R}_+$. Hence

$$\langle e^{-\beta t} P_t U_\beta x, x'\rangle \leq \int_0^\infty e^{-\beta s} \langle P_s x, x'\rangle\, ds = \langle U_\beta x, x'\rangle.$$

Since x' is arbitrary and C is convex and closed, we get by the bipolar theorem

$$U_\beta x - e^{-\beta t} P_t U_\beta x \in C$$

(N. Bourbaki [4], Ch. IV, §1, Proposition 3).

Let x be an element of C such that for every $\alpha\in]0,\infty[$, $x-\alpha U_\alpha x\in C$. For any $\alpha,\beta\in]0,\infty[$, we have

$$x + (\beta-\alpha) U_\alpha x = x - \alpha U_\alpha x + \beta U_\alpha x \in C$$

and therefore for any $t\in\mathbb{R}_+$

$$U_\alpha x - e^{-\beta t} P_t U_\alpha x = U_\beta(x + (\beta-\alpha) U_\alpha x) - e^{-\beta t} P_t U_\beta(x + (\beta-\alpha) U_\alpha x) \in C.$$

But β being arbitrary implies that

$$\alpha U_\alpha x - P_t \alpha U_\alpha x \in C.$$

Since α is arbitrary and C closed,

$$x - P_t x \in C. \quad \square$$

We now prove the theorem. By Proposition 10.2.2, there exists a resolvent $(\mathsf{U}_\alpha)_{\alpha\in\mathbb{R}_+}$ on X such that:

$\alpha_0)$ $\mathsf{U}_0 = \mathsf{V}_p$;

$\beta_0)$ the set of positive, hyperharmonic functions on X and the set of U-excessive functions coincide;

$\gamma_0)$ for any $\alpha\in\mathbb{R}_+$ and any bounded Borel function f on X, $\mathsf{U}_\alpha f$ is a bounded, continuous function.

Let u be a finite, continuous Evans function of p, greater than p (Proposition 2.2.4) and let $\mathscr{F}$ be the set of continuous real functions f on X such that

$$\sup_X \frac{|f|}{u} < \infty.$$

$\mathscr{F}$ is obviously a subspace of the real vector space $\mathscr{C}(X)$ containing $\mathscr{K}(X)$ and

$$f \mapsto \sup_X \frac{|f|}{u}$$

is a norm on $\mathscr{F}$. Endowed with this norm, $\mathscr{F}$ becomes a Banach space.

Let $\alpha\in\,]0,\infty[$. From

$$\alpha \mathsf{U}_\alpha u \le u$$

we deduce that

$$|\alpha \mathsf{U}_\alpha f| \le \alpha \mathsf{U}_\alpha(\|f\|\, u) \le \|f\|\, u,$$

for all $f\in\mathscr{F}$. From this inequality and from $\gamma_0)$, it follows that the map U_α

$$f \mapsto \mathsf{U}_\alpha f$$

is a continuous linear map of $\mathscr{F}$ into itself and

$$\alpha \|U_\alpha\| \le 1.$$

Since u is an Evans function of p, it follows immediately that $p\in\overline{\mathscr{K}(X)}$. Hence any function of $\mathscr{C}(X)$ dominated by a multiple of p belongs to $\overline{\mathscr{K}(X)}$. In particular, every continuous finite potential on X which is harmonic outside a compact set belongs to $\overline{\mathscr{K}(X)}$. We have also

$$U_\alpha(\mathscr{K}(X)) \subset \overline{\mathscr{K}(X)}, \qquad U_\alpha(\overline{\mathscr{K}(X)}) \subset \overline{\mathscr{K}(X)}.$$

We shall now apply the Hille-Yosida lemma to the Banach space $E := \overline{\mathscr{K}(X)}$. As the convex cone C, we take the set of positive functions of E. Properties α), β) and δ) of the lemma are obvious.

In order to prove the property γ), let us take $f \in \mathscr{K}(X)$ and let ε be a strictly positive, real number. By Theorem 2.3.1, there exist two continuous finite potentials p_1, p_2 on X such that $S(p_1)$, $S(p_2)$ are compact and

$$|f-(p_1-p_2)| < \frac{\varepsilon}{2} u.$$

Now $p_i \in E$ for $i=1,2$ and thus from β_0) we obtain

$$\lim_{\alpha \to \infty} \alpha \, \mathsf{U}_\alpha \, p_i = p_i.$$

By Dini's theorem and Proposition 10.1.5, there exists $\beta \in \mathbb{R}_+$ such that

$$p_i \leq \frac{\varepsilon}{4} u + \beta \, \mathsf{U}_\beta \, p_i$$

on $S(p_i)$. Since $p_i \in \mathscr{M}$, it follows that this inequality holds on X (Theorem 8.3.2 c)). We have

$$|f - \mathsf{U}_\beta \beta(p_1-p_2)| \leq |f-(p_1-p_2)| + |p_1 - \beta \, \mathsf{U}_\beta \, p_1| + |p_2 - \beta \, \mathsf{U}_\beta \, p_2| < \varepsilon u,$$

and therefore

$$\|f - U_\beta(\beta(p_1-p_2))\| < \varepsilon.$$

Let $\alpha \in]0, \infty[$. Since

$$U_\beta \, \beta(p_1-p_2) = U_\alpha(\beta(p_1-p_2) + (\alpha-\beta) \, U_\beta \, \beta(p_1-p_2))$$

we see that there exists $g \in E$ such that

$$\|f - U_\alpha g\| < \varepsilon.$$

Since ε and f are arbitrary, we deduce that

$$\mathscr{K}(X) \subset \overline{U_\alpha(E)}, \qquad E = \overline{\mathscr{K}(X)} \subset \overline{U_\alpha(E)} = E.$$

Let $(P_t)_{t \in \mathbb{R}_+}$ be the family whose existence was proved in the Hille-Yosida lemma. Let $x \in X$, let $f \in C$ and let C_f be the set of functions $g \in C$ such that $\{y \in X \mid g(y) < f(y)\}$ is relatively compact. Since

$$\overline{\mathscr{K}_+(X)} = C$$

we have

$$\inf_{g \in C_f} \|g\| = 0.$$

Hence by a) and d) of the lemma,

$$0 \leq \inf_{g \in C_f} P_t \, g(x) \leq \inf_{g \in C_f} \|g\| \, u(x) = 0.$$

From the lemma of Proposition 7.1.2, we deduce that there exists for any $x \in X$ and any $t \in \mathbb{R}_+$, a measure on X denoted by $\mathsf{P}_{t,x}$ such that

$$\int f \, d\mathsf{P}_{t,x} = P_t f(x)$$

for all $f \in C$.

Let $\mathscr{Q}$ be the family of continuous finite potentials on X which are harmonic outside a compact set and which are smaller than 1. Obviously $\mathscr{Q} \subset C$ and by $\beta_0)$

$$q - \alpha U_\alpha q \in C$$

for every $\alpha \in]0, \infty[$ and every $q \in \mathscr{Q}$. Hence by $f)$ of the lemma, for any $t \in \mathbb{R}_+$

$$q - P_t q \in C.$$

It follows that for any $x \in X$

$$\int q \, d\mathsf{P}_{t,x} \leq q(x) \leq 1,$$

$$\mathsf{P}_{t,x}(X) = \sup_{q \in \mathscr{Q}} \int q \, d\mathsf{P}_{t,x} \leq 1,$$

The family $\mathsf{P}_t := (\mathsf{P}_{t,x})_{x \in X}$ is therefore a diffusion on X for every $t \in \mathbb{R}_+$ and

$$\mathsf{P}_t 1 \leq 1.$$

For any $f \in \mathscr{K}_+(X)$ and any $s, t \in \mathbb{R}_+$, we have by $c)$ of the lemma

$$\mathsf{P}_{t+s} f = P_{t+s} f = P_t P_s f = \mathsf{P}_t \mathsf{P}_s f.$$

Moreover from $b)$ of the lemma it follows that for any $x \in X$ the function

$$t \mapsto \mathsf{P}_t f(x)$$

is continuous. Hence $(\mathsf{P}_t)_{t \in \mathbb{R}_+}$ is a sub-Markov semi-group on X.

Let $f \in \mathscr{K}_+(X)$ and let $x \in X$. We denote by x', the element of E'

$$g \mapsto g(x) \colon E \to \mathbb{R}.$$

By $e)$ of the lemma we have for any $\alpha \in]0, \infty[$

$$\mathsf{U}_\alpha f(x) = \langle U_\alpha f, x' \rangle = \int_0^\infty e^{-\alpha t} \langle P_t f, x' \rangle \, dt = \int_0^\infty e^{-\alpha t} \, \mathsf{P}_t f(x) \, dt.$$

Hence

$$f \cdot p(x) = \mathsf{U}_0 f(x) = \lim_{\alpha \to 0} \mathsf{U}_\alpha f(x) = \int_0^\infty \mathsf{P}_t f(x) \, dt.$$

Let P' be an integrable sub-Markov semi-group on X satisfying the property of the theorem. By Proposition 10.2.2, $(\mathsf{U}_\alpha)_{\alpha \in \mathbb{R}_+}$ is the resolvent generated by P'. By Proposition 10.1.4, P' and P coincide.

In order to prove the last assertion, we remark first that by the above relation $(\mathsf{U}_\alpha)_{\alpha\in\mathbb{R}_+}$ is the resolvent generated by P. Hence by Proposition 10.1.7 and by Proposition 10.2.2, the set of P-excessive functions, the set of U-excessive functions and the set of positive hyperharmonic functions coincide. □

Proposition 10.2.3. *Let $\mathsf{P}:=(\mathsf{P}_t)_{t\in\mathbb{R}_+}$ be a sub-Markov semi-group on X such that the set of P-excessive functions and the set of positive hyperharmonic functions on X coincide. Then for any $f\in\mathscr{K}(X)$*

$$\lim_{t\to 0}\sup_X|\mathsf{P}_t f - f| = 0.$$

Let p be a finite continuous potential on X for which $S(p)$ is compact. By Proposition 10.1.5, the family $(\mathsf{P}_t\, p)_{t\in\mathbb{R}_+}$ is decreasing. Hence by Proposition 10.1.6 and Dini's theorem,

$$\lim_{t\to 0}\sup_{S(p)}(p-\mathsf{P}_t\, p) = 0.$$

By Theorem 8.3.2, $a\Rightarrow c$, and by Proposition 10.1.6

$$p\le \mathsf{P}_t\, p + \sup_{S(p)}(p-\mathsf{P}_t\, p).$$

Hence

$$\lim_{t\to 0}\sup_X(p-\mathsf{P}_t\, p) = 0.$$

The assertion follows now using Theorem 2.3.1. □

Theorem 10.2.2. *There exists an integrable sub-Markov semi-group $\mathsf{P}:=(\mathsf{P}_t)_{t\in\mathbb{R}_+}$ on X such that:*

a) for any $f\in\mathscr{K}(X)$ and any $t\in\mathbb{R}_+$, $\mathsf{P}_t f$ is continuous;

b) for any $f\in\mathscr{K}(X)$

$$\lim_{t\to 0}\sup_X|\mathsf{P}_t f - f| = 0;$$

c) for any $f\in\mathscr{K}(X)$ and any closed set F disjoint from the carrier of f,

$$\lim_{\substack{t\to 0\\ t>0}}\frac{1}{t}\sup_F \mathsf{P}_t|f| = 0;$$

d) the set of P-excessive functions and the set of positive hyperharmonic functions on X coincide;

e) the function

$$x\mapsto\int_0^\infty \mathsf{P}_t 1(x)\,dt$$

is a bounded continuous hyperstrict potential on X.

(P.-A. Meyer 1963 [1], N. Boboc-C. Constantinescu-A. Cornea 1967 [5], G. Mokobodzki-D. Sibony 1967 [3].)

By Proposition 7.2.1, there exists a bounded continuous hyperstrict potential p on X. By Theorem 10.2.1 there exists an integrable sub-Markov semi-group $\mathsf{P} := (\mathsf{P}_t)_{t \in \mathbb{R}_+}$ on X which has properties *a)*, *d)* and *e)*. Property *b)* follows from *d)* and the preceding proposition.

c) Let α be any positive number, $q \in \mathscr{P}$ be a specific minorant of αp such that $S(q)$ is compact and let U be a neighbourhood of $S(q)$. We want to prove that

$$\lim_{\substack{t \to 0 \\ t > 0}} \frac{1}{t} \sup_{X \setminus U} (q - \mathsf{P}_t q) = 0.$$

Let g be a positive function of $\mathscr{K}(X)$ whose carrier lies in U, and which is equal to α on $S(q)$ and smaller than α everywhere else. Then

$$\left(1 - \frac{g}{\alpha}\right) \cdot q = 0$$

(Corollary 8.2.1). Hence

$$q = \left(1 - \frac{g}{\alpha}\right) \cdot q + \frac{g}{\alpha} \cdot q = \frac{g}{\alpha} \cdot q.$$

Since q is a specific minorant of αp, it follows that q is also a specific minorant of $g \cdot p$. We have therefore, using Fubini's theorem, for every strictly positive real number t,

$$\frac{1}{t} (q - \mathsf{P}_t q) \leq \frac{1}{t} (g \cdot p - \mathsf{P}_t g \cdot p) = \frac{1}{t} \int_0^t \mathsf{P}_s g \, ds.$$

Since by *b)*

$$\lim_{\substack{t \to 0 \\ t > 0}} \sup_X \left| \frac{1}{t} \int_0^t \mathsf{P}_s g \, ds - g \right| \leq \lim_{\substack{t \to 0 \\ t > 0}} \frac{1}{t} \int_0^t \sup_X |\mathsf{P}_s g - g| \, ds$$

$$\leq \lim_{t \to 0} \sup_{\substack{x \in X \\ s \in [0, t]}} |\mathsf{P}_s g(x) - g(x)| = 0$$

we get

$$\lim_{\substack{t \to 0 \\ t > 0}} \frac{1}{t} \sup_{X \setminus U} (q - \mathsf{P}_t q) = 0.$$

Let f and F be as in *c)* of the theorem. Since p is hyperstrict, there exists two potentials p', p'' on X such that:

a) there exists $\alpha \in \mathbb{R}_+$ with $p' + p'' \preccurlyeq \alpha p$;

b) $S(p') \cup S(p'') \subset X \setminus F$;

c) $|f| \leq p'' - p'$;

d) $p' = p''$ on F.

Then, for any strictly positive real number t, we have on F

$$\frac{1}{t}\mathsf{P}_t|f| \le \frac{1}{t}\mathsf{P}_t(p''-p') = \frac{1}{t}\big((p'-p'')-\mathsf{P}_t(p'-p'')\big) \le \frac{1}{t}(p'-\mathsf{P}_t p').$$

Hence

$$\lim_{\substack{t\to 0\\ t>0}} \frac{1}{t}\sup\nolimits_F \mathsf{P}_t|f| = 0. \quad \square$$

Remark 1. It would be interesting to characterise those sub-Markov semi-groups P on a locally compact space X with a countable base such that there exists a harmonic sheaf $\mathscr{H}$ on X with the property that X endowed with $\mathscr{H}$ is a $\mathfrak{P}$-harmonic space and the set of positive hyperharmonic functions coincides with the set of P-excessive functions.

Remark 2. J.C. Taylor, 1972 [unpublished] has extended Theorem 10.2.2 (with the corresponding modifications) to the case of a non-compact connected harmonic space for which the positive constant functions are superharmonic. The compact case (which contains for example the compact Riemann surfaces) is still open.

Exercises

10.2.1. Let p be a potential of $\mathscr{M}$. Then for any potential $q \preccurlyeq p$, there exists a Borel function f on X, $0 \le f \le 1$, such that

$$q = f \cdot p$$

(G. Mokobodzki 1967 [unpublished]). (Let $(p_n)_{n\in\mathbb{N}}$ be the sequence considered in Exercise 8.3.8 for p. Then there exists a sequence $(q_n)_{n\in\mathbb{N}}$ of potentials such that

$$q = \sum_{n\in\mathbb{N}} q_n$$

and such that for any $n\in\mathbb{N}$, $q_n \preccurlyeq p_n$. By Proposition 10.2.2, Exercise 10.1.4 and by using the measure constructed in Exercise 8.3.9, there exists for any $n\in\mathbb{N}$, a Borel function f_n on X, $0\le f_n \le 1$, such that

$$q_n = f_n \cdot p_n.$$

We may assume $f_n = 0$ on $X \setminus S(p_n)$. We set

$$f := \sum_{n\in\mathbb{N}} f_n.$$

Then

$$q = f \cdot p.)$$

10.2.2. Let V be a kernel on X and $\mathsf{U} := (\mathsf{U}_\alpha)_{\alpha \in]0, \infty[}$ be an open resolvent (see Exercise 10.1.2) such that for any positive Borel function f on X and any $x \in X$,

$$\lim_{\alpha \to 0} \mathsf{U}_\alpha f(x) = \mathsf{V} f(x).$$

Then:

a) $\mathsf{V} \in \mathscr{V}_{\mathscr{M}}$.

b) For any positive hyperharmonic function u on X and any $\alpha \in]0, \infty[$,

$$\alpha\, \mathsf{U}_\alpha u \leq u.$$

c) For any locally bounded potential p on X and any $x \in X$,

$$\lim_{\alpha \to 0} \alpha\, \mathsf{U}_\alpha p(x) = 0.$$

d) Let f be any positive Borel function on X which is dominated by a locally bounded potential and such that for all $\alpha \in]0, \infty[$

$$\alpha\, \mathsf{U}_\alpha f \leq f;$$

then for all $\alpha \in]0, \infty[$,

$$\mathsf{U}_\alpha f = \mathsf{V}(f - \alpha\, \mathsf{U}_\alpha f).$$

e) Let F be the fine carrier of V (Exercise 8.3.7); for any positive hyperharmonic function u on X

$$\lim_{\alpha \to \infty} \alpha\, \mathsf{U}_\alpha u = \hat{R}_u^F.$$

f) For any $x \in X$, the measures $\alpha\, \mathsf{U}_{\alpha, x}$ ($\alpha \in]0, \infty[$) converge vaguely to ε_x^F as α tends to ∞. Moreover $\varepsilon^F\, \mathsf{U}_\alpha = \mathsf{U}_\alpha\, \varepsilon^F = \mathsf{U}_\alpha$.

g) $\mathscr{U}_{\mathsf{V}}$ (Exercise 8.3.7), the set of U-excessive functions and the set of positive hyperharmonic functions u on X for which $u = \hat{R}_u^F$, coincide. In particular, the following relations are equivalent: *α)* V is strict, (Exercise 8.2.3); *β)* $F = X$; *γ)* every positive hyperharmonic function on X belongs to $\mathscr{U}_{\mathsf{V}}$.

h) The open resolvent U is uniquely determined by V.

i) Let $x \in X$ and let $(f_n)_{n \in \mathbb{N}}$ be a uniformly bounded sequence of positive Borel functions converging to 0 for which the sequence $(\mathsf{V} f_n)_{n \in \mathbb{N}}$ is uniformly bounded and converges (resp. does not converge) to 0 on $X \setminus \{x\}$ (resp. at x); then for any $f \in \mathscr{K}_+(X)$, $\mathsf{V} f(x) < \infty$.

(For any positive Borel function f on X, the relation

$$\mathsf{V} f = \mathsf{U}_\alpha f + \alpha\, \mathsf{U}_\alpha \mathsf{V} f$$

holds. Hence

$$\alpha\, \mathsf{U}_\alpha \mathsf{V} f \leq \mathsf{V} f,$$

and

$$\lim_{\alpha\to 0} \alpha\,\mathsf{U}_\alpha \mathsf{V} f(x)=0$$

for all $x\in X$ for which $\mathsf{V} f(x)<\infty$.

Let A_{V} be the set of points $x\in X$ for which there exists $f\in\mathscr{K}_+(X)$ such that $\mathsf{V} f(x)=\infty$. Let K be a compact subset of A_{V}, let $x\in X\setminus K$ and let $f\in\mathscr{K}_+(X)$ such that $x\notin \operatorname{Supp} f$ and $f\geq 1$ on K. Then $\mathsf{V} f=\infty$ on K and for all $\alpha\in]0,\infty[$

$$\alpha\,\mathsf{U}_\alpha \mathsf{V} f\leq \mathsf{V} f(x)<\infty.$$

Hence

$$\mathsf{U}_{\alpha,x}(K)=0, \qquad \mathsf{V} K(x)=0.$$

Since K is polar and x arbitrary, we have

$$\mathsf{V} K=0.$$

Let $\alpha\in]0,\infty[$ and $\mathsf{N}:=\alpha\,\mathsf{U}_\alpha$. Then for any positive Borel function f on X,

$$\mathsf{V} f+\frac{f}{\alpha}=\sum_{n\in\mathbb{N}} \mathsf{N}^n \frac{f}{\alpha}.$$

Let K be a compact set of X, let f be its characteristic function and assume there exists $\beta\in\mathbb{R}_+$ such that

$$\mathsf{V} f\leq\beta$$

on K. Since

$$\mathsf{N}\left(\beta+\frac{1}{\alpha}\right)\leq\beta+\frac{1}{\alpha}$$

one obtains

$$\mathsf{V} f+\frac{f}{\alpha}\leq\beta+\frac{1}{\alpha}$$

on $\{x\in X\mid \mathsf{V} f(x)<\infty\}$ (Exercise 10.1.1). Since α is arbitrary and $\{x\in X\mid \mathsf{V} f(x)=\infty\}$ is polar, we have

$$\mathsf{V} f\leq\beta.$$

a) follows now from Exercise 8.3.6.

b) Let u be a positive hyperharmonic function on X. Then for any $f\in\mathscr{K}_+(X)$, $f\leq u$, and any $\alpha\in]0,\infty[$,

$$\mathsf{V} f=\mathsf{U}_\alpha f+\alpha\,\mathsf{V}\mathsf{U}_\alpha f.$$

By *a)* and Proposition 10.2.1

$$\alpha\,\mathsf{U}_\alpha u\leq u.$$

c) Let $f \in \mathscr{K}_+(X)$ and $x \in X \setminus A_{\mathsf{V}}$. Then $\mathsf{V} f(x) < \infty$ and

$$\lim_{\alpha \to 0} \alpha \, \mathsf{U}_\alpha f(x) \leq \lim_{\alpha \to 0} \alpha \, \mathsf{V} f(x) = 0.$$

Set $\varphi := \lim_{\alpha \to 0} \alpha \, \mathsf{U}_\alpha p$. Let q be a potential on X which is greater than p outside of a compact set of X and $f \in \mathscr{K}_+(X)$, $f \geq p - q$. Then for any $x \in X \setminus A_{\mathsf{V}}$,

$$\lim_{\alpha \to 0} \alpha \, \mathsf{U}_\alpha p(x) \leq \lim_{\alpha \to 0} \alpha \, \mathsf{U}_\alpha f(x) + \lim_{\alpha \to 0} \alpha \, \mathsf{U}_\alpha q(x) \leq q(x).$$

Since q is aribitrary, $\varphi = 0$ on $X \setminus A_{\mathsf{V}}$. By *a)* and Exercises 8.2.3 and 8.3.7 *a)*, $\mathsf{V} \varphi = 0$. Show that for any $\alpha \in]0, \infty[$

$$\alpha \, \mathsf{U}_\alpha \varphi = \varphi.$$

Deduce that $\varphi = 0$.

d) By *c)*

$$\lim_{\alpha \to 0} \alpha \, \mathsf{U}_\alpha f = 0.$$

For any $\alpha \in]0, \infty[$, we have

$$\mathsf{U}_\alpha f = \lim_{\beta \to 0} \mathsf{U}_\beta (f - \alpha \, \mathsf{U}_\alpha f) + \lim_{\beta \to 0} \beta \, \mathsf{U}_\beta \, \mathsf{U}_\alpha f = \mathsf{V}(f - \alpha \, \mathsf{U}_\alpha f).$$

e) & f) Let p be a finite continuous potential on X. By *d)* and by Exercise 8.3.7 *d)* for any $\alpha \in]0, \infty[$,

$$\mathsf{U}_\alpha p = \hat{R}^F_{\mathsf{U}_\alpha p}.$$

Hence

$$p^* := \lim_{\alpha \to \infty} \alpha \, \mathsf{U}_\alpha p \leq \hat{R}^F_p.$$

By the Lemma of Proposition 7.1.2 there exists for any $x \in X$, a measure ν_x on X such that for any finite continuous potential p on X

$$\int p \, d\nu_x = p^*(x).$$

Let q be a finite continuous strict potential. Set

$$F' := \{x \in X \mid q^*(x) = q(x)\}.$$

Obviously $F' \subset F$. For any $\alpha \in]0, \infty[$

$$\mathsf{U}_\alpha (q - q^*) = 0, \qquad \mathsf{V}(q - q^*) = 0$$

(Exercise 10.1.2). Hence

$$\mathsf{V}(X \setminus F') = 0, \qquad F \subset F', \qquad q^* = \hat{R}^F_q.$$

Since q is strict, $\nu_x = \varepsilon^F_x$ for all $x \in X$.

g) By Exercise 8.3.7 *e)* for any function u of $\mathscr{U}_{\mathsf{V}}$, we have

$$u = \hat{R}_u^F.$$

By *e)* every positive hyperharmonic function u on X for which this relation holds is U-excessive. Let f be a U-excessive function and let p be a locally bounded potential on X. For any $n \in \mathbb{N}$, set

$$f_n := \inf(f, n\,p).$$

By *d)* for all $\alpha \in]0, \infty[$

$$\alpha\, \mathsf{U}_\alpha f_n = \mathsf{V}\, \alpha (f_n - \alpha\, \mathsf{U}_\alpha f_n).$$

Hence $f \in \mathscr{U}_{\mathsf{V}}$.

h) follows from Corollary 10.2.1.

i) We have for any $\alpha \in]0, \infty[$,

$$\mathsf{V} f_n(x) = \mathsf{U}_\alpha f_n(x) + \alpha\, \mathsf{U}_\alpha\, \mathsf{V} f_n(x).$$

If $x \in A_{\mathsf{V}}$, then $\mathsf{U}_\alpha \{x\} = 0$ and we get the contradictory relation

$$\lim_{n \to \infty} \mathsf{U}_\alpha f_n(x) = \lim_{n \to \infty} \alpha\, \mathsf{U}_\alpha\, \mathsf{V} f_n(x) = 0.$$

10.2.3. Let V be a kernel on X and let $\mathsf{P} := (\mathsf{P}_t)_{t \in \mathbb{R}_+}$ be a sub-Markov semi-group on X associated with V, i.e. such that for any positive Borel function f on X and any $x \in X$,

$$\mathsf{V} f(x) = \int_0^\infty \mathsf{P}_t f(x)\, dt.$$

Let $\mathsf{U} := (\mathsf{U}_\alpha)_{\alpha \in]0, \infty[}$ be the open resolvent generated by P. (See Exercise 10.1.3.)

a) For any positive Borel function f on X and any $x \in X$

$$\lim_{\alpha \to 0} \mathsf{U}_\alpha f(x) = \mathsf{V} f(x).$$

b) $\mathsf{V} \in \mathscr{V}_{\mathscr{M}}$.

c) Let f be a lower semi-continuous, positive numerical function on X such that for any $\alpha \in]0, \infty[$, $\alpha\, \mathsf{U}_\alpha f \leq f$. Set

$$f^* := \lim_{\alpha \to \infty} \alpha\, \mathsf{U}_\alpha f.$$

Then f^* is P-excessive, and for any $t \in \mathbb{R}_+$

$$\mathsf{P}_t f = \mathsf{P}_t f^*.$$

d) If F denotes the fine carrier of V (Exercise 8.3.7) then for any $x \in X$ and any $t \in \mathbb{R}_+$, $\mathsf{P}_{t,x}(X \setminus F) = 0$ and $\mathsf{P}_{0,x} = \varepsilon_x^F$.

e) For any positive hyperharmonic function u on X and any $t \in \mathbb{R}_+$, $\mathsf{P}_t u \in \mathscr{U}_{\mathsf{V}}$ (Exercise 8.3.7) and

$$\mathsf{P}_t u \leq u .$$

f) The set of P-excessive functions and $\mathscr{U}_{\mathsf{V}}$ coincide.

g) P is uniquely determined by V.

h) If A_{V} denotes the set of points $x \in X$ for which there exists $f \in \mathscr{K}_+(X)$ such that $\mathsf{V} f(x) = \infty$, then the interior of $\bar{A}_{\mathsf{V}}$ is empty.

(*b)* follows from Exercise 10.2.2. For *c)* use *c)* of Exercise 10.1.3. *h)* Let $f \in \mathscr{K}_+(X)$ and let $x \in X$. If there exists $\alpha \in]0, \infty[$ such that $\mathsf{U}_\alpha f(x) = 0$, then $\mathsf{P}_t f(x) = 0$ for all $t \in \mathbb{R}_+$ and therefore $\mathsf{V} f(x) = 0$. Let x be an interior point of $\bar{A}_{\mathsf{V}}$. There exists $f \in \mathscr{K}_+(X)$ such that $x \notin \operatorname{Supp} f$ and $\mathsf{V} f(x) \neq 0$. Let K be a compact neighbourhood of x, contained in the interior of $\bar{A}_{\mathsf{V}}$, on which $\mathsf{V} f$ does not vanish. Let $\alpha \in]0, \infty[$. Then $\mathsf{U}_\alpha f$ does not vanish on K and from

$$\mathsf{U}_\alpha f = \mathsf{V} f - \alpha \mathsf{V} \mathsf{U}_\alpha f$$

it follows that f is subharmonic and therefore upper semi-continuous on $X \setminus \operatorname{Supp} f$. By the Baire property, there exists a strictly positive real number ε such that the set $L := \{y \in K | \mathsf{U}_\alpha f(y) \geq \varepsilon\}$ has a non-empty interior. We therefore get the contradictory relation

$$\infty > \mathsf{V} f(y) \geq \alpha \mathsf{V} \mathsf{U}_\alpha f(y) \geq \alpha \varepsilon \mathsf{V} L(y) = \infty$$

for any $y \in A_{\mathsf{V}}$ which is contained in the interior of L.)

10.2.4. Let $(\mathsf{P}_t)_{t \in \mathbb{R}_+}$ be a sub-Markov semi-group on X such that for any potential p on X and any $t \in \mathbb{R}_+$, we have $\mathsf{P}_t p \leq p$. Let $\mathsf{V} \in \mathscr{V}_{\mathscr{M}}$ such that for any $x \in X$ and any positive Borel function f on X,

$$\int_0^\infty \mathsf{P}_t f(x)\, dt \leq \mathsf{V} f(x).$$

Then for any $x \in X$ and any locally bounded potential p on X

$$\lim_{t \to \infty} \mathsf{P}_t p(x) = 0.$$

and

$$\mathsf{P}_{t,x}(A) = 0$$

for every strictly positive real number t and every polar set A. (See Proposition 2.5 of N. Boboc-C. Constantinescu-A. Cornea 1967 [5].)

10.2.5. Let p, q be two bounded potentials on X, $p \preccurlyeq q$, such that there exists two sub-Markov semi-groups $(\mathsf{P}_t)_{t \in \mathbb{R}_+}$, $(\mathsf{Q}_t)_{t \in \mathbb{R}_+}$ on X associated, in the sense of Exercise 10.2.3, with V_p, V_q respectively. Then, for any positive hyperharmonic function u on X and any $t \in \mathbb{R}_+$,

$$\mathsf{P}_t u \leq \mathsf{Q}_t u.$$

(Use Exercise 10.2.3 *e)* and Proposition 1.3 of N. Boboc-C. Constantinescu-A. Cornea 1967 [5].)

10.2.6. Let p be a finite continuous potential on X such that $S(p)$ is compact. Then there exists a sub-Markov semi-group on X associated with V_p, in the sense of Exercise 10.2.3 (N. Boboc-C. Constantinescu-A. Cornea 1967 [5]).

10.2.7. Let V be a kernel on X such that $\mathsf{V} \in \mathscr{V}_{\mathscr{M}}$ and let F be its fine carrier (see Exercise 8.3.7). If $A_{\mathsf{V}} \cap F = \emptyset$ (Exercise 8.2.3) then:

a) There exists a sub-Markov semi-group on X, $(\mathsf{P}_t)_{t \in \mathbb{R}_+}$, associated, in the sense of the Exercise 10.2.3, with V.

b) If the axiom of domination holds on X, then for any locally bounded potential p on X and any $x \in X$, the function $t \mapsto \mathsf{P}_t p(x)$ is continuous.

c) If for any locally bounded potential p on X and any $x \in X$, the function $t \mapsto \mathsf{P}_t p(x)$ is continuous, then for any $x \in X$, and any strictly positive real number t, the semi-polar sets are of $\mathsf{P}_{t,x}$-measure zero (N. Boboc-C. Constantinescu-A. Cornea 1967 [5], C. Constantinescu 1968 [8]).

(*a)* Let $(p_n)_{n \in \mathbb{N}}$ be a specifically increasing sequence of finite continuous potentials on X such that for any $n \in \mathbb{N}$, $S(p_n)$ is compact and such that $\mathsf{V} = \bigvee_{n \in \mathbb{N}} \mathsf{V}_{p_n}$ (Exercise 8.2.4). By Exercise 10.2.6, there exists a sub-Markov semi-group $(\mathsf{P}_t^{(n)})_{t \in \mathbb{R}_+}$ on X associated with V_{p_n}, in the sense of Exercise 10.2.3. By Exercise 10.2.5, for any finite continuous potential p on X,

$$\lim_{n \to \infty} \mathsf{P}_t^{(n)} p$$

exists. Using the Lemma of Proposition 7.1.2, construct a semi-group $(\mathsf{P}_t)_{t \in \mathbb{R}_+}$ on X such that for any finite continuous potential p on X,

$$\mathsf{P}_t p = \lim_{n \to \infty} \mathsf{P}_t^{(n)} p.$$

Show that for any $t \in \mathbb{R}_+$

$$\varepsilon^F \mathsf{P}_t = \mathsf{P}_t \varepsilon^F = \mathsf{P}_t$$

(Exercise 10.2.3 *d)*). Let $f \in \mathscr{K}_+(X)$. Use Exercise 10.2.4 and the proof of Theorem 2.3 from N. Boboc-C. Constantinescu-A. Cornea 1967 [5] in order to show that for any $x \in X \setminus A_{\mathsf{V}}$,

$$\mathsf{V} f(x) = \int_0^\infty \mathsf{P}_t f(x)\, dt.$$

Extend this relation to the case where f is a positive Borel function on X. Now let $x \in X$. Then

$$\mathsf{V} f(x) = \int \mathsf{V} f \, d\varepsilon_x^F = \int \left(\int_0^\infty \mathsf{P}_t f(y) \, dt \right) d\varepsilon_x^F(y)$$

$$= \int_0^\infty \varepsilon^F \, \mathsf{P}_t f(x) \, dt = \int_0^\infty \mathsf{P}_t f(x) \, dt.)$$

10.2.8. Let X be the locally compact space $]0, \infty]$. For any open set U of X, let $\mathscr{H}(U)$ be the set of finite continuous functions which are locally linear on $U \cap]0, \infty[$. Then $\mathscr{H}$ is a harmonic sheaf and X endowed with $\mathscr{H}$ is a $\mathfrak{P}$-Brelot space. For any $n \in \mathbb{N}$, $n \neq 0$, we denote by p_n the function on X which is equal to 1 on $[n^2, \infty]$ and equal to $x \mapsto \dfrac{x}{n^2}$ on $]0, n[$.

$$p := \sum_{\substack{n \in \mathbb{N} \\ n \neq 0}} p_n$$

is a continuous potential of $\mathscr{M}$ which is equal to ∞ at ∞. There exists no open resolvent $(\mathsf{U}_\alpha)_{\alpha \in]0, \infty[}$ on X such that for any positive Borel function f on X and any $x \in X$

$$\lim_{\alpha \to 0} \mathsf{U}_\alpha f(x) = \mathsf{V} f(x).$$

(Use *i)* of Exercise 10.2.2.) (This exercise shows that we may not drop the hypothesis $A_{\mathsf{V}} \cap F = \emptyset$ of the preceding exercise. However we do not know whether or not it is a consequence of *a)* of the same exercise.)

10.2.9. There exists a $\mathfrak{P}$-Brelot space with countable base and a potential $p \in \mathscr{M}$ which is finite on its fine carrier and ∞ at a point of X. (Take the harmonic space associated with the Laplace equation on $\{(x, y) \in \mathbb{R}^2 \mid x^2 + y^2 < 1\}$. Let $(K_n)_{n \in \mathbb{N}}$ be a sequence of pairwise disjoint discs such that:

a) $\overline{\bigcup_{n \in \mathbb{N}} K_n} \setminus \bigcup_{n \in \mathbb{N}} K_n = \{(0, 0)\}$;

b) $\bigcup_{n \in \mathbb{N}} K_n$ is thin at $(0, 0)$;

c) there exists a sequence $(\alpha_n)_{n \in \mathbb{N}}$ in $\mathbb{R}_+$ such that

$$\sum_{n \in \mathbb{N}} \hat{R}_{\alpha_n}^{K_n}$$

is a potential on X which is equal to ∞ at $(0, 0)$.)

10.2.10. Let P be a sub-Markov semi-group on X such that for any positive hyperharmonic function, u, on X and any $t \in \mathbb{R}_+$, $\mathsf{P}_t u \leq u$.

A P-*process* is a system

$$(E, \mathfrak{F}, \mu;\ (E_t)_{t\in\mathbb{R}_+},\ (\mathfrak{F}_t)_{t\in\mathbb{R}_+},\ (x_t)_{t\in\mathbb{R}_+})$$

where:

a) E is a set and $\mathfrak{F}$ is a σ-algebra of subsets of E;

b) μ is an abstract measure on the measurable space $(E, \mathfrak{F})$ with $\mu(E)<\infty$;

c) $(E_t)_{t\in\mathbb{R}_+}$ is a decreasing family in $\mathfrak{F}$ and $E_0=E$;

d) $(\mathfrak{F}_t)_{t\in\mathbb{R}_+}$ is an increasing family of σ-algebras such that for any $t\in\mathbb{R}_+$, $E_t\in\mathfrak{F}_t\subset\mathfrak{F}$.

e) for any $t\in\mathbb{R}_+$, x_t is an $\mathfrak{F}_t$-measurable map of E_t into X;

f) for any $s, t\in\mathbb{R}_+$, any $f\in\mathscr{K}_+(X)$ and any $A\in\mathfrak{F}_t$

$$\int_{A\cap E_{s+t}} f(x_{s+t}(\xi))\,d\mu(\xi)=\int_{A\cap E_t} \mathsf{P}_s f(x_t(\xi))\,d\mu(\xi).$$

We say that this process is *separable* if there exists a set $A\in\mathfrak{F}$ of μ-measure zero and a countable set $S\subset\mathbb{R}_+$ such that for any $t\in\mathbb{R}_+$ and any $\xi\in E_t\setminus A$, there exists a sequence $(t_n)_{n\in\mathbb{N}}$ in S, converging to t and such that for any $n\in\mathbb{N}$, $\xi\in E_{t_n}$ and

$$x_t(\xi)=\lim_{n\to\infty} x_{t_n}(\xi).$$

We set, for any $\xi\in E$,

$$\zeta(\xi)=\sup\{t\in\mathbb{R}_+ \mid \xi\in E_t\}.$$

For any bounded measure ν on X such that for all $f\in\mathscr{K}_+(X)$

$$\int \mathsf{P}_0 f\,d\nu=\int f\,d\nu,$$

there exists a separable P-process

$$(E, \mathfrak{F}, \mu;\ (E_t)_{t\in\mathbb{R}_+},\ (\mathfrak{F}_t)_{t\in\mathbb{R}_+},\ (x_t)_{t\in\mathbb{R}_+})$$

such that for all $f\in\mathscr{K}_+(X)$

$$\int f(x_0(\xi))\,d\mu(\xi)=\int f\,d\nu.$$

Let $(E, \mathfrak{F}, \mu; (E_t)_{t\in\mathbb{R}_+}, (\mathfrak{F}_t)_{t\in\mathbb{R}_+}, (x_t)_{t\in\mathbb{R}_+})$ be a separable P-process. Then:

a) There exists a set $A\in\mathfrak{F}$ of μ-measure zero such that for any $t\in\mathbb{R}_+$ and any $\xi\in E_t\setminus A$, the set

$$\{x_s(\xi) \mid s\in[0, t]\}$$

is a relatively compact set of X.

b) For any positive hyperharmonic function u on X and any $t\in\mathbb{R}_+$, there exists a set $A_{u,t}\in\mathfrak{F}$ of μ-measure zero such that for any $\xi\in E\setminus A_{u,t}$, the function

$$s \mapsto u(x_s(\xi))$$

is bounded on $[t, \zeta(\xi)[$ if $u(x_t(\xi))$ is finite.

c) If P has property *b)* of Theorem 10.2.2, then for any $t \in \mathbb{R}_+$, we have

$$\mu(E_t \setminus \bigcup_{s>t} E_s) = 0.$$

d) If P has properties *b)* and *c)* of Theorem 10.2.2, then there exists a set $A \in \mathfrak{F}$ of μ-measure zero such that for any $\xi \in E \setminus A$, the map

$$t \mapsto x_t(\xi)$$

is continuous on $[0, \zeta(\xi)[$.

(N. Boboc-C. Constantinescu-A. Cornea 1967 [5], W. Hansen 1967 [1].)

Chapter 11

Integral Representation of Positive Superharmonic Functions

In classical potential theory an integral representation for a positive harmonic function h on a ball (resp. for a potential p (in the sense of this book) on an open set of $\mathbb{R}^3$) is a proof that there exists a unique measure μ such that

$$h(x)=\int g(x,y)\,d\mu(y) \qquad (\text{resp. } p(x)=\int g(x,y)\,d\mu(y)),$$

where g is the Poisson kernel (resp. the Green function). The first such representations were obtained by G. Herglotz in 1911 [1] (resp. F. Riesz in 1925 [1]). A similar integral representation was given by R.S. Martin 1941 [1] for any positive superharmonic function on an open set of $\mathbb{R}^3$, where the measure μ and the function g are no longer defined on the closure of the basic space but on an abstract one, now called the Martin space. These results were extended to $\mathfrak{P}$-Brelot spaces with countable base by M. Brelot 1959 [14] and R.-M. Hervé 1960 [3]. In the last two sections of this chapter, we extend these results to more general $\mathfrak{P}$-harmonic spaces with countable base. In the first three sections of the chapter, we develop some theories which are needed for the integral representation and which may be of some interest by themselves. A number of results from the theory of locally convex vector spaces and measure theory will be used.

§ 11.1. Locally Convex Vector Spaces of Harmonic Functions

This section is a continuation of § 1.1. Its aim is to prove two theorems which throw some light on the convergence properties of the harmonic sheaves.

Theorem 11.1.1. *Let $\mathscr{H}$ be a harmonic sheaf on a locally compact space X. If there exists an $\mathscr{H}$-sweeping system, then the following assertions are equivalent:*

a) $\mathscr{H}$ has the Bauer convergence property;

b) every set of uniformly bounded $\mathscr{H}$-functions on an open set of X is equicontinuous;

c) for any $\mathscr{H}$-sweeping $(\omega_x)_{x\in V}$ and any bounded real function f on ∂V, ωf is an $\mathscr{H}$-function;

d) there exists an $\mathscr{H}$-sweeping system $((\omega_{\iota x})_{x\in V_\iota})_{\iota\in I}$ such that for any $\iota\in I$, $\omega_\iota f$ is continuous where f is the characteristic function of an arbitrary compact G_δ-set of ∂V_ι.

(A. Grothendieck 1953 [2], G. Mokobodzki 1963 [unpublished], 1968 [1], P. Loeb-B. Walsh 1965 [1], C. Constantinescu 1966 [4], A. Ionescu-Tulcea 1967 [1].)

We first prove the following lemma.

Lemma. *Let Y be a compact space and $\mathscr{F}$ be a subset of $\mathscr{C}(Y)$ such that every sequence in $\mathscr{F}$ possesses an adherent point in $\mathscr{C}(Y)$ in the topology of pointwise convergence. Then for any sequence in $\mathscr{F}$, there exists a subsequence converging to an element of $\mathscr{C}(Y)$ in the topology of pointwise convergence* (A. Grothendieck 1952 [1]).

Let $(f_n)_{n\in\mathbb{N}}$ be a sequence in $\mathscr{F}$ and φ: $Y\to\mathbb{R}^{\mathbb{N}}$ be the map $y\mapsto(f_n(y))_{n\in\mathbb{N}}$. Since φ is continuous and Y compact, $\varphi(Y)$ is compact and a function g: $\varphi(Y)\to\mathbb{R}$ is continuous if and only if $g\circ\varphi$ is continuous. Since $\varphi(Y)$ is metrisable, it contains a countable dense subset A. Let B be a countable subset of Y such that $\varphi(B)=A$. By hypothesis, $(f_n(y))_{n\in\mathbb{N}}$ is a bounded subset of $\mathbb{R}$ for all $y\in Y$. Hence, there exists a subsequence $(f_{n_k})_{k\in\mathbb{N}}$ of the sequence $(f_n)_{n\in\mathbb{N}}$ converging at every point of B. Let f be a limit function of the sequence $(f_{n_k})_{k\in\mathbb{N}}$ with respect to the topology of pointwise convergence. It is obvious that for any y, $y'\in Y$ such that $\varphi(y)=\varphi(y')$, we have $f(y)=f(y')$. Hence, there exists a continuous function g: $\varphi(Y)\to\mathbb{R}$ such that $f=g\circ\varphi$. Since A is dense in $\varphi(Y)$, it follows that f is the unique limit point of the sequence $(f_{n_k})_{k\in\mathbb{N}}$ in $\mathscr{C}(Y)$ with respect to the topology of pointwise convergence. Using the fact that every subsequence of the sequence $(f_{n_k})_{k\in\mathbb{N}}$ has a limit point in $\mathscr{C}(Y)$, we deduce that $(f_{n_k})_{k\in\mathbb{N}}$ converges to f in the topology of pointwise convergence. □

$a\Rightarrow b$. We show first that for any $\mathscr{H}$-sweeping $(\omega_x)_{x\in V}$ and any compact set $K\subset V$, there exists a countable subset A of K such that for every bounded positive real function f on ∂V, we have $\omega f=0$ on K if this equality holds on A. Let $\mathfrak{A}$ be the set of finite subsets of K. For any $B\in\mathfrak{A}$, we denote by $\mathscr{F}_B$, the set of real functions f on ∂V such that

$$\omega f=0 \quad \text{on } B,$$

$$0\leq f\leq 1.$$

By Proposition 1.1.4, ωf is an $\mathscr{H}$-function on V for every $f \in \mathscr{F}_B$. $\mathscr{F}_B$ is upper directed. Indeed if $f, g \in \mathscr{F}_B$, then

$$0 \leq \sup(f, g) \leq 1, \qquad \omega \sup(f, g) \leq \omega(f+g) \leq \omega f + \omega g = 0 \qquad \text{on } B.$$

Hence, the family $(\omega f)_{f \in \mathscr{F}_B}$ is an upper directed family of $\mathscr{H}$-functions on V. Let h_B be its supremum. By Proposition 1.1.2, h_B is an $\mathscr{H}$-function since it is dominated by $\omega 1$. The family $(h_B)_{B \in \mathfrak{A}}$ is obviously a lower directed family of positive $\mathscr{H}$-functions on V. Since its infimum is obviously 0 on K, there exists by Dini's convergence theorem, an increasing sequence $(B_n)_{n \in \mathbb{N}}$ in $\mathfrak{A}$ such that $(h_{B_n})_{n \in \mathbb{N}}$ converges uniformly to 0 on K. We set

$$A := \bigcup_{n \in \mathbb{N}} B_n.$$

Let f be a bounded positive real function on ∂V such that $\omega f = 0$ on A. We may suppose $0 \leq f \leq 1$. Then $f \in \mathscr{F}_{B_n}$ for any $n \in \mathbb{N}$ and therefore $0 \leq \omega f = \lim_{n \to \infty} h_{B_n} = 0$ on K.

Let U be an open set of X, $\mathscr{F}$ be a uniformly bounded set of $\mathscr{H}$-functions on U and x be a point of U. We shall prove that there exists a compact neighbourhood K of x which is contained in U and such that the set $\{h|_K \mid h \in \mathscr{F}\}$ is a relatively compact set of the Banach space $\mathscr{C}(K)$. Let $(\omega_x)_{x \in V}$ (resp. $(\omega'_x)_{x \in V'}$) be an $\mathscr{H}$-sweeping such that

$$x \in V', \qquad \overline{V'} \subset V, \qquad \overline{V} \subset U.$$

Let K be a compact neighbourhood of x which is contained in V'. By the preceding considerations, there exists a sequence $(x_n)_{n \in \mathbb{N}}$ in $\overline{V'}$ such that for any bounded positive real function f on ∂V, we have $\omega f = 0$ on $\overline{V'}$ if $\omega f(x_n) = 0$ for all $n \in \mathbb{N}$. We set

$$\mu := \sum_{n \in \mathbb{N}} \frac{1}{2^n} \omega_{x_n}.$$

It follows immediately that for any $y \in \overline{V'}$, the measure ω_y is absolutely continuous with respect to μ. By the Radon Nikodym theorem, there exists for any $y \in \overline{V'}$, a μ-integrable function f_y such that

$$\omega_y = f_y \cdot \mu.$$

(For any positive μ-integrable function g, $g \cdot \mu$ denotes the measure $f \mapsto \int f g \, d\mu$.) Let us denote by φ, the map of $\mathscr{F}$ into $L^\infty(\mu)$ which associates to any function h of $\mathscr{F}$, the equivalence class (with respect to μ) of $h|_{\partial V}$. Since $\mathscr{F}$ is a uniformly bounded set of functions, $\varphi(\mathscr{F})$ is a bounded set in $L^\infty(\mu)$ and therefore relatively compact with respect to the weak topology $\sigma(L^\infty(\mu), L^1(\mu))$ associated with the duality between $L^1(\mu)$

and L^∞ (i.e. the coarsest topology on $L^\infty(\mu)$ for which the maps of the form $f \mapsto \int f g \, d\mu$: $L^\infty(\mu) \to \mathbb{R}$, where $g \in L^1(\mu)$, are continuous). Let $(h_n)_{n\in\mathbb{N}}$ be a sequence in $\mathscr{F}$. There exists then, a bounded Borel function f on ∂V such that the equivalence class of f (with respect to μ) is an adherent point of the sequence $(\varphi(h_n))_{n\in\mathbb{N}}$ with respect to the $\sigma(L^\infty(\mu), L^1(\mu))$ topology. Since by Proposition 1.1.4, the function $\omega f|_{\overline{V}'}$ is continuous, it is a limit point in $\mathscr{C}(\overline{V}')$ with respect to the topology of pointwise convergence of the sequence $(h_n|_{\overline{V}'})_{n\in\mathbb{N}}$. Indeed, let B be a finite subset of $\overline{V}'$ and let ε be a strictly positive real number. By the definition of f, there exists for any $m\in\mathbb{N}$, an $n\in\mathbb{N}$, $n\geq m$, such for any $y\in B$,

$$\left|\int (h_n - f) f_y \, d\mu\right| < \varepsilon.$$

Hence, for any $y\in B$

$$|h_n(y) - \omega f(y)| = |\omega(h_n - f)(y)| = \left|\int (h_n - f) f_y \, d\mu\right| < \varepsilon.$$

By the above Lemma, there exists a subsequence $(h_{n_k})_{k\in\mathbb{N}}$ of $(h_n)_{n\in\mathbb{N}}$ such that for any $y\in\overline{V}'$,

$$\lim_{k\to\infty} h_{n_k}(y) = \omega f(y).$$

For any $m\in\mathbb{N}$, we set

$$\bar{u}_m := \sup_{k\geq m} h_{n_k}, \qquad \underline{u}_m := \inf_{k\geq m} h_{n_k}.$$

The sequence $(\bar{u}_m)_{m\in\mathbb{N}}$ (resp. $(\underline{u}_m)_{m\in\mathbb{N}}$) is decreasing (resp. increasing) and converges to ωf on $\overline{V}'$. Hence $(\omega'(\bar{u}_m - \underline{u}_m))_{m\in\mathbb{N}}$ is a decreasing sequence of continuous functions (Proposition 1.1.4) on V converging to 0. By Dini's convergence theorem, it converges to 0 uniformly on K. From the inequalities

$$\omega' \underline{u}_m \leq h_{n_k} = \omega' h_{n_k} \leq \omega' \bar{u}_m$$

for any $k\geq m$, we deduce that the sequence $(h_{n_k}|_K)_{k\in\mathbb{N}}$ is a Cauchy sequence in $\mathscr{C}(K)$. This fact proves the above assertion that $\{h|_K \mid h\in\mathscr{F}\}$ is a relatively compact set of $\mathscr{C}(K)$.

The equicontinuity of $\mathscr{F}$ now follows from Ascoli's theorem.

$b \Rightarrow a$. Let $(h_n)_{n\in\mathbb{N}}$ be an increasing sequence of $\mathscr{H}$-functions whose limit function is locally bounded. Since this family is uniformly bounded on any compact set of U, it is equicontinuous. Hence the limit is continuous and therefore harmonic (Proposition 1.1.5).

$a \Rightarrow c$ follows from Proposition 1.1.4.

$c \Rightarrow d$ follows from the fact that there exists an $\mathscr{H}$-sweeping system on X.

$d \Rightarrow a$. Let $\iota\in I$ and let f be a lower semi-continuous, bounded Baire function on ∂V_ι. For any strictly positive real number ε, there exists a real function f_ε on ∂V_ι which is a linear combination of characteristic functions of compact G_δ-sets and such that $|f - f_\varepsilon| < \varepsilon$ on ∂V_ι. Since

$\omega_\iota f_\varepsilon$ is a continuous function on V_ι with

$$|\omega_\iota f - \omega_\iota f_\varepsilon| \leq \varepsilon\, \omega_\iota 1$$

and since ε is arbitrary, it follows that $\omega_\iota f$ is continuous.

Let $(h_n)_{n\in\mathbb{N}}$ be an increasing sequence of $\mathscr{H}$-functions on open set U of X such that its supremum h is locally bounded. Let $\iota \in I$, $\overline{V}_\iota \subset U$. Since

$$h = \lim_{n\to\infty} h_n = \lim_{n\to\infty} \omega_\iota h_n = \omega_\iota h$$

on V_ι and since h is a lower semi-continuous, bounded Baire function, it follows by the above considerations that h is continuous on V_ι. Since ι is arbitrary, h is continuous on U. Hence, h is an $\mathscr{H}$-function on U (Proposition 1.1.5). □

Remark. The assertion *b)* was proved in classical potential theory by W. F. Osgood, 1900 [1].

We say that a harmonic sheaf $\mathscr{H}$ on a locally compact space X has *the property of nuclearity* if, for any open set U of X, the vector space $\mathscr{H}(U)$ endowed with the topology of compact convergence is nuclear.

Theorem 11.1.2. *Let $\mathscr{H}$ be a harmonic sheaf on a locally compact space X. If there exists an $\mathscr{H}$-sweeping system on X, then the following assertions are equivalent:*

a) $\mathscr{H}$ possesses the property of nuclearity;

a′) there exists a base $\mathfrak{B}$ of X such that for any $U \in \mathfrak{B}$, the vector space $\mathscr{H}(U)$ endowed with the topology of compact convergence is nuclear;

b) (resp. b′)) for any open set U of X and any compact set K of U, there exists a measure μ on U with compact carrier such that for every $\mathscr{H}$-function (resp. positive $\mathscr{H}$-function) h on U, we have

$$\sup_K |h| \leq \mu(|h|);$$

c) for any open set U of X and any compact set K of U, there exists a measure μ on U with compact carrier such that if the limit function of an increasing sequence of harmonic functions on U is μ-integrable then it is harmonic in the interior of K;

d) for any open set U of X, any compact set K of U and any sequence $(h_n)_{n\in\mathbb{N}}$ of positive $\mathscr{H}$-functions on U such that $\sum_{n\in\mathbb{N}} h_n$ is locally bounded, we have

$$\sum_{n\in\mathbb{N}} \sup_K h_n < \infty;$$

e) for any open set U of X, any compact set K of U and any positive $\mathscr{H}$-function h on U, we have

$$\sup \sum_{\iota \in I} h_\iota(x_\iota) < \infty,$$

where $(x_\iota)_{\iota \in I}$ is an arbitrary finite family in K and $(h_\iota)_{\iota \in I}$ is an arbitrary finite family of positive $\mathscr{H}$-functions on U with $h = \sum_{\iota \in I} h_\iota$;

f) for any $\mathscr{H}$-sweeping $(\omega_x)_{x \in V}$ and any compact set K of V, there exists a measure μ on ∂V such that $\omega_x \leq \mu$ for all $x \in K$;

g) there exists an $\mathscr{H}$-sweeping system $((\omega_{\iota x})_{x \in V_\iota})_{\iota \in I}$ such that for any $\iota \in I$ and any compact set K of V_ι, there exists a measure μ on ∂V_ι such that $\omega_{\iota x} \leq \mu$ for all $x \in K$.

(D. Hinrichsen 1967 [1], C. Constantinescu-A. Cornea 1969 [4].)

$a \Rightarrow a'$ is trivial.

$a' \Rightarrow b$. Assume first that $U \in \mathfrak{B}$. The set $\{h \in \mathscr{H}(U) | \sup_K |h| \leq 1\}$ is a circled closed convex neighbourhood of the origin of $\mathscr{H}(U)$. Since $\mathscr{H}(U)$ is nuclear, there exists a compact set L of U, $K \subset L$, such that the map $u|_L \mapsto u|_K$: $\mathscr{H}(U)|_L \to \mathscr{H}(U)|_K$ is nuclear where $\mathscr{H}(U)|_L$ (resp. $\mathscr{H}(U)|_K$) is the normed space of the restrictions of the functions of $\mathscr{H}(U)$ to L (resp. K) with the norm $\sup_L |h|$ (resp. $\sup_K |h|$) (A. Pietsch [1] p. 63 or H.H. Schaefer [1] p. 100). Hence, there exist a sequence $(l_n)_{n \in \mathbb{N}}$ of continuous linear functionals, of norm at most 1, on the normed space $\mathscr{H}(U)|_L$ and a sequence $(h_n)_{n \in \mathbb{N}}$ in $\mathscr{H}(U)$ such that

$$\sum_{n \in \mathbb{N}} \sup_K |h_n| < \infty, \qquad h|_K = \sum_{n \in \mathbb{N}} l_n(h|_L)\, h_n|_K$$

for every $h \in \mathscr{H}(U)$. By the Hahn-Banach theorem, there exists for any $n \in \mathbb{N}$, a signed measure μ_n on L such that $\|\mu_n\| \leq 1$ and $\mu_n(h|_L) = l_n(h|_L)$ for every $h \in \mathscr{H}(U)$. We set

$$\mu := \sum_{n \in \mathbb{N}} (\sup_K |h_n|)\, |\mu_n|.$$

Obviously

$$\sup_K |h| \leq \sum_{n \in \mathbb{N}} |l_n(h|_L)| \sup_K |h_n| \leq \mu(|h|).$$

Now let U be an arbitrary open set of X. There exists a finite family $(U_\iota)_{\iota \in I}$ in $\mathfrak{B}$ and a finite family of compact sets $(K_\iota)_{\iota \in I}$ such that $K_\iota \subset U_\iota \subset U$ for all $\iota \in I$ and

$$K = \bigcup_{\iota \in I} K_\iota.$$

By the above considerations, there exists for any $\iota \in I$, a measure μ_ι on U_ι with compact carrier such that

$$\sup_{K_\iota} |h| \leq \mu_\iota(|h|)$$

for every $h \in \mathscr{H}(U_\iota)$. The measure

$$\mu := \sum_{\iota \in I} \mu_\iota$$

has the required properties.

$b \Rightarrow a$. Let U be an open set of X. For any $x \in U$, we denote by l_x, the element of the dual, $\mathscr{H}(U)'$, of $\mathscr{H}(U)$ defined by $h \mapsto h(x)$. It is obvious that the map

$$x \mapsto l_x \colon U \to \mathscr{H}(U)'$$

is continuous for the $\sigma(\mathscr{H}(U)', \mathscr{H}(U))$ topology. Let L be a compact set of U and W be the circled closed convex neighbourhood of the origin of $\mathscr{H}(V)$

$$W := \{h \in \mathscr{H}(U) \mid \sup_L |h| \leq 1\}.$$

For any signed measure μ on L, the map

$$f \mapsto \int f(l_x)\, d\mu(x) \colon \mathscr{C}(W^0) \to \mathbb{R}$$

defines a signed measure on W^0, the polar of W.

Let K be a compact set of U. From *b)*, there exists a measure μ on U with compact carrier such that

$$\sup_K |h| \leq \mu(|h|)$$

for every $h \in \mathscr{H}(U)$. The required implication now follows from the above considerations by using A. Pietsch [1] Satz 4.1.5 or H. H. Schaefer [1] p. 178.

$b \Rightarrow b'$ is trivial.

$b' \Rightarrow c$. Let μ be one of the measures of *b')* for U and K and let $(h_n)_{n \in \mathbb{N}}$ be an increasing sequence in $\mathscr{H}(U)$ such that its supremum is μ-integrable. Then for all natural numbers m, n with $m < n$, we have

$$\sup_K (h_n - h_m) \leq \mu(h_n - h_m).$$

Hence the sequence $(h_n|_K)_{n \in \mathbb{N}}$ is uniformly convergent. We deduce that the restriction of the limit function to the interior of K is continuous and therefore is an $\mathscr{H}$-function (Proposition 1.1.5).

$c \Rightarrow b'$. Let L be a compact neighbourhood of K with $L \subset U$, and let μ be a measure satisfying property *c)* for U and L. We want to show that there exists a positive real number α such that the measure $\alpha\mu$ satisfies the conditions of *b')*. Assume the contrary. Then for any $m \in \mathbb{N}$, there exists a positive $\mathscr{H}$-function h_m on U such that

$$\sup_K h_m \geq m, \qquad \mu(h_m) \leq \frac{1}{m^2}.$$

Then $(\sum_{m\leq n} h_m)_{n\in\mathbb{N}}$ is an increasing sequence of $\mathscr{H}$-functions on U whose limit is μ-integrable but not bounded on K. This is a contradiction since by *c)* this limit function is an $\mathscr{H}$-function on the interior of L.

$b' \Rightarrow d$. Let K be a compact set of U, μ be a measure satisfying property *b')* for U and K and $(h_n)_{n\in\mathbb{N}}$ be a sequence of positive $\mathscr{H}$-functions on U such that $\sum_{n\in\mathbb{N}} h_n$ is locally bounded. Then

$$\sum_{n\in\mathbb{N}} \sup_K h_n \leq \sum_{n\in\mathbb{N}} \int h_n \, d\mu = \int \Big(\sum_{n\in\mathbb{N}} h_n\Big) d\mu < \infty .$$

$d \Rightarrow e$. Suppose that *e)* is not true. Then, there exists an open set U of X, a compact set K of U and a positive $\mathscr{H}$-function h on U such that for any $n\in\mathbb{N}$, there exists a finite family $(h_{n,\iota})_{\iota\in I_n}$ of positive $\mathscr{H}$-functions on U with

$$\sum_{\iota\in I_n} \sup_K h_{n,\iota} > 2^n, \qquad \sum_{\iota\in I_n} h_{n,\iota} = h .$$

This contradicts *d)* since

$$\sum_{n\in\mathbb{N}} \sum_{\iota\in I_n} \frac{1}{2^n} h_{n,\iota}$$

is locally bounded and

$$\sum_{n\in\mathbb{N}} \sum_{\iota\in I_n} \sup_K \frac{1}{2^n} h_{n,\iota} = \infty .$$

$e \Rightarrow f$. Let f be a positive, real, continuous function on ∂V. By *e)*

$$\sup \sum_{\iota\in I} \omega \, f_\iota(x_\iota) < \infty ,$$

where $(x_\iota)_{\iota\in I}$ is an arbitrary finite family in K and $(f_\iota)_{\iota\in I}$ is an arbitrary family of positive, real, continuous functions on ∂V such that

$$f = \sum_{\iota\in I} f_\iota .$$

The map

$$f \mapsto \sup \sum_{\iota\in I} \omega \, f_\iota(x_\iota)$$

yields the required measure μ.

$f \Rightarrow g$ follows from the fact that there exists an $\mathscr{H}$-sweeping system on X.

$g \Rightarrow b$. Let U be an open set of X and K be a compact set of U. By *g)*, there exists a finite set $J \subset I$ and a family $(K_\iota, V_\iota)_{\iota\in J}$ such that for any $\iota \in I$, $\overline{V}_\iota \subset U$ and K_ι are compact sets of V_ι with

$$K = \bigcup_{\iota\in J} K_\iota .$$

Further there exists for any $\iota \in J$, a positive measure μ_ι on ∂V_ι such that $\omega_{\iota,x} \leq \mu_\iota$ for all $x \in K_\iota$. We set

$$\mu := \sum_{\iota \in J} \mu_\iota .$$

Then for any $\mathscr{H}$-function h on U, we have

$$\sup_K |h| \leq \sum_{\iota \in J} \sup_{K_\iota} |h| = \sum_{\iota \in J} \sup_{x \in K_\iota} |\omega_\iota h(x)|$$
$$\leq \sum_{\iota \in J} \sup_{x \in K_\iota} \omega_\iota |h| (x) \leq \sum_{\iota \in J} \mu_\iota(|h|) = \mu(|h|). \quad \square$$

Corollary 11.1.1. *Let $\mathscr{H}$ be a harmonic sheaf on a locally compact space such that there exists an $\mathscr{H}$-sweeping system. If $\mathscr{H}$ has the property of nuclearity then $\mathscr{H}$ has the Bauer convergence property.*

The assertion follows from $a \Rightarrow c$. $\square$

Corollary 11.1.2. *Let $\mathscr{H}$ be a harmonic sheaf on a locally compact space X such that there exists an $\mathscr{H}$-sweeping system. If $\mathscr{H}$ has Brelot convergence property or if X has a countable base and $\mathscr{H}$ has the Doob convergence property, then $\mathscr{H}$ has the property of nuclearity* (P. Loeb-B. Walsh 1966 [2], H. Bauer 1966 [9]).

Let U be an open set of X and let K be a compact set of U.

Assume first that $\mathscr{H}$ has the Brelot convergence property. Then there exists a finite set $\mathfrak{V}$ of nonempty open connected subsets of U such that

$$K \subset \bigcup_{V \in \mathfrak{V}} V.$$

For any $V \in \mathfrak{V}$, let x_V be a point of V. We set

$$\mu := \sum_{V \in \mathfrak{V}} \varepsilon_{x_V} .$$

Assume now that X has a countable base and that $\mathscr{H}$ has the Doob convergence property. Let A be a countable dense subset and let $(\alpha_x)_{x \in A}$ be a family of strictly positive, real numbers such that

$$\sum_{x \in A} \alpha_x < \infty .$$

We set

$$\mu := \sum_{x \in A} \alpha_x \varepsilon_x .$$

Let $(h_n)_{n \in \mathbb{N}}$ be an increasing sequence of $\mathscr{H}$-functions on U such that its limit, h, is μ-integrable. Then in both cases, h is an $\mathscr{H}$-function on the interior of K. The corollary follows now from $c \Rightarrow a$. $\square$

Remark. We do not know whether or not we may drop the hypothesis that X has a countable base in the above corollary.

Proposition 11.1.1. *Assume that X is a $\mathfrak{P}$-harmonic space such that the sheaf of harmonic functions on X has the Doob convergence property. Let x be a point for which there exists a measure μ, with compact carrier K on $X \setminus \{x\}$, such that the smallest absorbent set containing K is a neighbourhood of x. Then there exists a potential p on X such that $S(p)=\{x\}$.*

Let q be a locally bounded potential on X which is strictly positive at x and let $\mathfrak{B}_x$ be the set of neighbourhoods of x, ordered by the converse inclusion relation. For any $U \in \mathfrak{B}_x$, we set

$$q_U := \left(\int \hat{R}_q^U \, d\mu\right)^{-1} \hat{R}_q^U.$$

Let U_1 be the interior of the smallest absorbent set of X containing K and let U_2, U_3 be open, relatively compact sets of X such that

$$x \in U_3 \subset \overline{U}_3 \subset U_2 \subset \overline{U}_2 \subset U_1.$$

By Proposition 6.1.5, the set

$$\{q_U|_{U_1 \setminus \overline{U}_3} \mid U \in \mathfrak{B}_x,\ U \subset U_3\}$$

is locally uniformly bounded. For every $U \in \mathfrak{B}_x$, with $U \subset U_3$, we have

$$R_{q_U}^{U_2} = q_U$$

(Proposition 8.2.2). From this fact, it follows that the set

$$\{q_U|_{X \setminus \overline{U}_2} \mid U \in \mathfrak{B}_x,\ U \subset U_3\}$$

is uniformly bounded by a potential and therefore locally uniformly bounded and equicontinuous (Theorem 11.1.1).

Let $\mathfrak{U}$ be an ultrafilter on $\mathfrak{B}_x$, finer then the section filter of $\mathfrak{B}_x$. By the above considerations, the function p on $X \setminus \{x\}$

$$y \mapsto \lim_{U, \mathfrak{U}} q_U(y)$$

is harmonic. We set

$$p(x) := \liminf_{\substack{y \to x \\ y \neq x}} p(y).$$

We want to prove that p is a superharmonic function. Let V be an open relatively compact, neighbourhood of x. We have, for any $U \in \mathfrak{B}_x$, $\overline{U} \subset V$,

$$\mu^V q_U \leq q_U$$

on V. Since q_U converges to p uniformly on ∂V along $\mathfrak{U}$, we get

$$\mu^V p \leq p$$

on $V \setminus \{x\}$. Since $\mu^V p$ is continuous at x, we deduce

$$\mu^V p(x) \leq p(x).$$

Hence p is superharmonic; being dominated by a potential, it must be a potential. Obviously

$$S(p) \subset \{x\}.$$

Since

$$\int p \, d\mu = \lim_{U, \mathfrak{U}} \int q_U \, d\mu = 1,$$

p does not vanish identically and

$$S(p) \neq \varnothing, \quad S(p) = \{x\}. \quad \square$$

Corollary 11.1.3. *If X is a $\mathfrak{P}$-Brelot space, then for any $x \in X$, there exists a potential p on X such that $S(p) = \{x\}$* (R.-M. Hervé 1962 [4]).

The assertion follows immediately from the proposition using Corollary 6.1.1. $\square$

Remark 1. This corollary is no longer true if X is a $\mathfrak{P}$-Bauer space (Exercise 11.1.9).

Remark 2. There is no relation between the polarity (resp. the non-polarity) of $\{x\}$ and the value of p at x (resp. the boundedness of p in a neighbourhood of x); i.e. $\{x\}$ may be polar and we may have

$$p(x) < \infty$$

(Exercise 11.4.9) or $\{x\}$ may be non-polar and

$$\limsup_{y \to x} p(y) = \infty$$

(Exercise 9.2.6).

Exercises

11.1.1. Let $\mathscr{H}$ be a harmonic sheaf on a locally compact space X such that there exists an $\mathscr{H}$-sweeping system on X. Assume that $\mathscr{H}$ has the Bauer convergence property and let us endow $\mathscr{H}(X)$ with the topology of compact convergence. Then every bounded set of $\mathscr{H}(X)$ is relatively compact (use Theorem 11.1.1). Hence, if $\mathscr{H}(X)$ is a Frechet space, it is a Montel space.

11.1.2. If X is a connected Brelot space, then for every $x \in X$, the set of positive harmonic functions on X, smaller than 1 at x, is equicontinuous.

11.1.3. Let U be an open set of a harmonic space X and let $\mathscr{F}$ be a locally uniformly bounded set of continuous real functions on $\bar{U}$ which are harmonic on U. If $\mathscr{F}|_{\partial U}$ is equicontinuous at a point $x_0 \in \partial U$, then $\mathscr{F}$ is equicontinuous at x_0 (N. Boboc-P. Mustaţă 1968 [4]).

11.1.4. Let U be an open set of a harmonic space X such that every continuous real function on $\bar{U}$ which vanishes on ∂U and is harmonic on U, vanishes identically. Then there exists an $\alpha \in \mathbb{R}_+$ such that

$$\sup_{\bar{U}} |u| \leq \alpha \sup_{\partial U} |u|$$

for every continuous, real function h on $\bar{U}$ which is harmonic on U (N. Boboc-P. Mustaţă 1968 [4]) (use the preceding exercise).

11.1.5. Let U be an open, relatively compact set of a harmonic space X such that every continuous real function on $\bar{U}$ which vanishes on ∂U and is harmonic on U, vanishes identically. If the set of functions $f \in \mathscr{C}(\partial U)$ which have continuous extensions to $\bar{U}$ which are harmonic on U is dense in $\mathscr{C}(\partial U)$ (with respect to the topology of uniform convergence), then every function of $\mathscr{C}(\partial U)$ has a unique continuous extension to $\bar{U}$ which is harmonic on U (N. Boboc-P. Mustaţă 1967 [3]) (use the preceding exercise).

11.1.6. If every open, relatively compact set of a Brelot space X is an $\mathfrak{S}$-set, then X is an $\mathfrak{S}$-Brelot space (C. Constantinescu-A. Cornea 1963 [1]).

11.1.7. On every connected, non-compact $\mathfrak{P}$-Brelot space, there exists a strictly positive harmonic function (C. Constantinescu-A. Cornea 1963 [1]).

11.1.8. With the notation of Exercise 6.1.8, if the sheaf of harmonic functions on X has the property of nuclearity, then so does the sheaf $\mathscr{H}_{\mathscr{U}_F}$.

11.1.9. Let X be the set $\{(x,t) \in \mathbb{R} \times \mathbb{R} \mid t \leq 0\}$. For any open set U of X, we denote by $\mathscr{H}(U)$, the set of continuous real functions h on U such that the restriction of h to $\{(x,t) \in U \mid t < 0\}$ is a solution of the heat equation. Then $\mathscr{H}$ is a harmonic sheaf which has the property of nuclearity, but not the Doob convergence property and X endowed with $\mathscr{H}$ is a Bauer space (C. Constantinescu-A. Cornea 1968 [3]). (Consider the open set $U := \{(x,t) \in X \mid -1 < t\}$ and the sequence of harmonic functions

$$\left((x,t) \mapsto \int_{-n}^{n} \frac{1}{\sqrt{t+1}} e^{-\frac{(x-\xi)^2}{4(t+1)}} e^{\frac{\xi^2}{4}} d\xi\right)_{n \in \mathbb{N}}$$

on U in order to show that $\mathscr{H}$ does not have the Doob convergence property. For the property of nuclearity, use Theorem 11.1.2, Theorem 3.3.1 and the fact that the Doob convergence property implies the property of nuclearity).

11.1.10. There exist elliptic Bauer spaces whose sheaf of harmonic functions does not possess the property of nuclearity (C. Constantinescu-A. Cornea 1969 [4]).

11.1.11. There exist elliptic Bauer spaces whose sheaf of harmonic functions possesses the property of nuclearity but not the Doob convergence property (C. Constantinescu-A. Cornea 1969 [4]).

11.1.12. Let X be a locally compact space, let $\mathscr{C}(X)$ be the locally convex space of continuous real functions on X (endowed with the compact convergence topology) and let $\mathscr{C}(X)'$ (resp. $\mathscr{C}(X)''$) be the strong dual of $\mathscr{C}(X)$ (resp. $\mathscr{C}(X)'$). Let $\mathscr{H}$ be a harmonic sheaf on X having the Bauer convergence property and such that there exists an $\mathscr{H}$-sweeping system on X. If we identify $\mathscr{H}(X)$ with its image under the evaluation map $\mathscr{C}(X) \to \mathscr{C}(X)''$, then $\mathscr{H}(X)$ is closed with respect to the $\sigma(\mathscr{C}(X)'', \mathscr{C}(X)')$ topology. (Let $(\omega_x)_{x \in V}$ be an $\mathscr{H}$-sweeping. For any measure λ on X whose carrier is contained in V, we denote by $\lambda \cdot \omega$, the measure on ∂V

$$\int \omega_x \, d\lambda(x).$$

Choose $\varphi \in \mathscr{C}(X)''$ so that for any $\mathscr{H}$-sweeping $(\omega_x)_{x \in V}$ and any measure λ on X whose carrier is contained in V, we have

$$(*) \qquad \varphi(\lambda) = \varphi(\lambda \cdot \omega).$$

Let $(\omega_x)_{x \in V}$ be an $\mathscr{H}$-sweeping and let λ be a measure on X whose carrier is contained in V. Show that the function

$$x \mapsto \omega_x \colon V \to \mathscr{C}(X)'$$

is continuous (use Theorem 11.1.1 $a \Rightarrow b$). It follows that the function

$$x \mapsto \varphi(\omega_x) \colon V \to \mathbb{R}$$

is continuous and that

$$\varphi(\lambda \cdot \omega) = \int \varphi(\omega_x) \, d\lambda(x).$$

Hence the function $\hat{\varphi}$ on X

$$x \mapsto \varphi(\varepsilon_x)$$

belongs to $\mathscr{H}(X)$. Show that for any $\mu \in \mathscr{C}(X)'$, we have

$$\int \hat{\varphi} \, d\mu = \varphi(\mu).$$

Hence $\mathscr{H}(X)$ coincides with the set of $\varphi \in \mathscr{C}(X)''$ for which the above relations (*) hold and is therefore closed.)

11.1.13. Let U be a relatively compact open set of a harmonic space X such that $X \setminus U$ is fine closed and let $\mathscr{L}$ be a linear space of continuous functions on $\bar{U}$, whose restrictions to U are harmonic. Then we have:

a) If $\mathscr{L}|_{\partial U}$ is finite dimensional, then $\mathscr{L}$ is finite dimensional.

b) There exists a finite set $A \subset U$ such that any continuous function h on $\bar{U}$ which is harmonic on U and is zero on $A \cup \partial U$ is zero.

c) If the points of X are polar, then there exists a finite set $A \subset U$ such that $U \setminus A$ is a set of unicity (i.e. any continuous function on $\overline{U \setminus A}$ which is harmonic on $U \setminus A$ and vanishes on $\partial(U \setminus A)$ is identically zero).

(G. Albinus-N. Boboc-P. Mustaţă 1973 [1].)

§ 11.2. Locally Convex Topologies on the Convex Cone of Positive Superharmonic Functions

Throughout this section X will denote a $\mathfrak{P}$-harmonic space.

Proposition 11.2.1. *For any prevector lattice C, there exists a vector lattice $[C]$ and an injection, i, of C into $[C]$ such that $i(C)$ is the set of positive elements of $[C]$ and such that*

$$i(u+v)=i(u)+i(v), \qquad i(\alpha u)=\alpha\, i(u),$$
$$u \preccurlyeq v \Leftrightarrow i(u) \preccurlyeq i(v)$$

for all $u, v \in C$, $\alpha \in \mathbb{R}_+$. Moreover, if C is lower complete with respect to the order relation $\preccurlyeq$, then $[C]$ is a conditionally complete vector lattice.

Let $\sim$ be the equivalence relation on $C \times C$ defined by

$$(u, v) \sim (u', v') :\Leftrightarrow u+v' = u'+v.$$

If $u_i, v_i, u'_i, v'_i \in C$ $(i=1, 2)$ and

$$(u_i, v_i) \sim (u'_i, v'_i)$$

for any i, then for any $\alpha \in \mathbb{R}_+$, we have

$$(u_1+u_2, v_1+v_2) \sim (u'_1+u'_2, v'_1+v'_2),$$
$$(\alpha u_1, \alpha v_1) \sim (\alpha u'_1, \alpha v'_1).$$

Let $[C]$ be the quotient space of $C \times C$ with respect to the equivalence relation $\sim$ and let $[u, v]$ be the equivalence class of the element (u, v) of $C \times C$. We define on $[C]$, an addition and a multiplication by real numbers α as follows:

$$[u, v]+[u', v']:=[u+u', v+v']$$

$$\alpha[u, v]:=\begin{cases}[\alpha u, \alpha v] & \text{if } \alpha \geq 0\\ [-\alpha v, -\alpha u] & \text{if } \alpha<0.\end{cases}$$

By the above considerations, these laws of composition are well defined and it is easy to verify that $[C]$ endowed with them becomes a real vector space.

The map $i: u \mapsto [u, 0]$ of C into $[C]$ is an injection such that

$$i(u+v)=i(u)+i(v)$$

$$i(\alpha u)=\alpha i(u)$$

for all $u, v \in C$ and $\alpha \in \mathbb{R}_+$. Clearly $i(C)$ is a convex cone in $[C]$ with the property that if the sum of two of its elements is 0, then both elements are 0. Hence, there exists an order relation on $[C]$, denoted by $\preccurlyeq$, for which $i(C)$ is the set of positive elements. Since for any $[u, v] \in [C]$, we have

$$[u, v]=i(u)-i(v)$$

and since

$$u \preccurlyeq v \Leftrightarrow i(u) \preccurlyeq i(v)$$

it follows immediately that $[C]$ is a vector lattice which is conditionally complete if C is lower complete. □

According to the preceding proposition, we shall denote by $[\mathcal{S}]$ the vector lattice generated by the prevector lattice $\mathcal{S}$ of positive superharmonic functions on X and by $\preccurlyeq$, $\curlywedge$, $\curlyvee$, the order relation and the lattice operations on $[\mathcal{S}]$. We remark that $[\mathcal{S}]$ is conditionally complete (Theorem 4.1.1 and Theorem 5.1.1). We shall identify $\mathcal{S}$ with the set of positive elements of $[\mathcal{S}]$. According to this identification, for any $u \in [\mathcal{S}]$ and any $x \in X$, we shall denote by $u(x)$, the number

$$(u \curlyvee 0)(x)-((-u) \curlyvee 0)(x)$$

whenever this difference makes sense. In this way, every element u of $[\mathcal{S}]$ will be considered to be identified with a numerical function defined on X (outside a polar set) which will also be denoted by u. For any $f \in \mathscr{C}_+(X_0)$ (where X_0 denotes the Alexandroff compactification of X) and any $u \in [\mathcal{S}]$, we set

$$f \cdot u := f \cdot (u \curlyvee 0) - f \cdot ((-u) \curlyvee 0).$$

Let $x \in X$. We denote by I_x, the set of pairs (V, f) such that

a) V is an open relatively compact neighbourhood of x;

b) $f \in \mathscr{K}_+(X)$;

c) $f \leq 1$;
d) $\operatorname{Supp} f \subset V$;
e) $f = 1$ on a neighbourhood of x.

The set I_x will be considered ordered by the order relation:

$$(V, f) \leq (V', f') :\Leftrightarrow V' \subset V,\ f' \leq f.$$

It is obvious that I_x is upper directed with respect to this order relation. We denote by $\mathfrak{J}_x$, the section filter of I_x and by $A_{(V,f)}u$, the superharmonic function on V

$$\mu^V(f \cdot u) + (1-f) \cdot u|_V,$$

for any $(V, f) \in I_x$ and $u \in \mathscr{S}$. $A_{(V,f)}$ is obviously additive.

Proposition 11.2.2. *Let $x \in X$ and $u \in \mathscr{S}$. Then:*

a) $(V, f) \in I_x \Rightarrow \mu^V u \leq A_{(V,f)} u \leq u|_V$.
b) $(V, f), (V', f') \in I_x$ *and* $(V, f) \leq (V', f') \Rightarrow A_{(V,f)} u|_{V'} \leq A_{(V',f')} u$.
c) $\lim_{(V,f),\mathfrak{J}_x} A_{(V,f)} u(x) = \sup_{(V,f) \in I_x} A_{(V,f)} u(x) = u(x)$.
d) $A_{(V,f)} u$ *is harmonic on* $V \setminus \operatorname{Supp}(1-f)$.

a) We have on V

$$\begin{aligned}\mu^V u = \mu^V(f \cdot u) + \mu^V((1-f) \cdot u) &\leq \mu^V(f \cdot u) + (1-f) \cdot u \\ &\leq (f \cdot u) + (1-f) \cdot u = u.\end{aligned}$$

b) We have on V'

$$\begin{aligned}A_{(V,f)} u = \mu^V(f \cdot u) + (1-f) \cdot u &= \mu^V(f' \cdot u) + \mu^V((f-f') \cdot u) + (1-f) \cdot u \\ &\leq \mu^{V'}(f' \cdot u) + (f-f') \cdot u + (1-f) \cdot u = A_{(V',f')} u.\end{aligned}$$

c) follows from *a)*.
d) is trivial. □

Proposition 11.2.3. *Let $\mathscr{F}$ be a subset of $\mathscr{C}_+(X_0)$ such that every function of $\mathscr{C}_+(X_0)$ is the supremum of an upper directed subset of $\mathscr{F}$ and let $u \in [\mathscr{S}]$. If the function $f \cdot u$ vanishes on $X \setminus \operatorname{Supp} f$ for every $f \in \mathscr{F}$, then $u = 0$.*

We set

$$v := u \vee 0, \quad w := (-u) \vee 0.$$

By the hypothesis about u,

$$f \cdot v = f \cdot w$$

on $X \setminus \operatorname{Supp} f$ for every $f \in \mathscr{F}$. By the hypothesis about $\mathscr{F}$ and by Theorem 8.1.3 and Proposition 8.2.1, the same relation holds for every $f \in \mathscr{C}_+(X_0)$.

Let $x \in X$ and $(V, f) \in I_x$. Then

$$A_{(V, f)} v(x) = A_{(V, f)} w(x).$$

By the preceding proposition,

$$v(x) = w(x). \quad \square$$

Let $f \in \mathscr{C}_+(X_0)$ and let K be a compact subset of $X \setminus \operatorname{Supp} f$. We set

$$\mathscr{V}(f, K) := \{u \in [\mathscr{S}] \mid \sup_K |f \cdot u| \leq 1\}.$$

Now, finite intersections of sets of the form $\mathscr{V}(f, K)$ form a fundamental system of neighbourhoods of the origin of $[\mathscr{S}]$ for a locally convex topology on $[\mathscr{S}]$ which will by denoted by $\mathscr{T}$. By the above proposition, it is a Hausdorff topology. In the sequel, $[\mathscr{S}]$ will denote the corresponding locally convex space. The topology $\mathscr{T}$ is a modified version of a topology introduced by R.-M. Hervé 1960 [3] for $\mathfrak{P}$-Brelot spaces.

Proposition 11.2.4. *Let $(f_\iota, K_\iota, \mu_\iota)_{\iota \in I}$ be a finite family such that for any $\iota \in I$ and $f_\iota \in \mathscr{C}_+(X_0)$, K_ι is a compact subset of $X \setminus \operatorname{Supp} f_\iota$ and μ_ι is a signed measure on K_ι. Then the map*

$$u \mapsto \sum_{\iota \in I} \mu_\iota(f_\iota \cdot u)$$

is a continuous linear form on $[\mathscr{S}]$. Conversely every continuous linear form on $[\mathscr{S}]$ may be written in this way.

The first assertion is trivial. Let l be a continuous linear map of $[\mathscr{S}]$ into $\mathbb{R}$. There exists a finite family $(f_\iota, K_\iota)_{\iota \in I}$ such that for any $\iota \in I$, $f_\iota \in \mathscr{C}_+(X_0)$ and K_ι is a compact subset of $X \setminus \operatorname{Supp} f_\iota$ and such that for any

$$u \in \bigcap_{\iota \in I} \mathscr{V}(f_\iota, K_\iota),$$

we have

$$|l(u)| \leq 1.$$

Let K be the topological sum of the family $(K_\iota)_{\iota \in I}$ of topological spaces. For any $u \in [\mathscr{S}]$, we denote by $\psi(u)$, the function on K which is equal to $f_\iota \cdot u_\iota$ on K_ι. ψ is obviously a linear map of $[\mathscr{S}]$ into $\mathscr{C}(K)$. Choose $u, v \in [\mathscr{S}]$ such that

$$\psi(u) = \psi(v).$$

Then for any $\alpha \in \mathbb{R}$, we have

$$\alpha(u-v) \in \bigcap_{\iota \in I} \mathscr{V}(f_\iota, K_\iota).$$

Hence

$$l(u) = l(v)$$

and we may define a linear map $\rho: \psi([\mathscr{S}]) \to \mathbb{R}$ such that

$$l = \rho \circ \psi.$$

For any $f \in \mathscr{C}(K)$ for which ρ is defined and for which $\|f\| \leq 1$, we have $|\rho(f)| \leq 1$. Hence by the Hahn-Banach theorem, there exists a signed measure μ on K which is an extention of ρ. If, for any $\iota \in I$, we denote by μ_ι the restriction of μ to K_ι, then we have for any $u \in [\mathscr{S}]$

$$l(u) = \sum_{\iota \in I} \mu_\iota(f_\iota \cdot u). \quad \square$$

Proposition 11.2.5. *Let $x \in X$, let $(V, f) \in I_x$ and let U be an open set of X whose closure is compact and contained in $X \setminus \operatorname{Supp}(1-f)$.*

a) For any $\alpha \in \mathbb{R}_+$,

$$\{A_{(V,f)} u|_U \mid u \in \mathscr{S} \cap \alpha \mathscr{V}(f, \partial V) \cap \alpha \mathscr{V}(1-f, \bar{U})\}$$

is a bounded, equicontinuous set of harmonic functions on U.

b) The map

$$(u, y) \mapsto A_{(V,f)} u(y): \mathscr{S} \times U \to \mathbb{R}$$

is continuous.

a) For any $u \in \alpha \mathscr{V}(f, \partial V) \cap \alpha \mathscr{V}(1-f, \bar{U})$, we have on U

$$0 \leq A_{(V,f)} u \leq \alpha(\mu^V 1 + 1).$$

The equicontinuity follows from Theorem 11.1.1.

b) By Proposition 11.2.4, for any $y \in U$, the map

$$u \mapsto A_{(V,f)} u(y): \mathscr{S} \to \mathbb{R}$$

is continuous. Let $(u_0, y_0) \in \mathscr{S} \times U$ and let β be a positive real number such that

$$u_0 \in \beta \mathscr{V}(f, \partial V) \cap \beta \mathscr{V}(1-f, \bar{U}).$$

By *a)*, the set

$$\{A_{(V,f)} u|_U \mid u \in \mathscr{S} \cap (\beta+1) \mathscr{V}(f, \partial V) \cap (\beta+1) \mathscr{V}(1-f, \bar{U})\}$$

is equicontinuous. Hence

$$(u, y) \mapsto A_{(V,f)} u(y): \mathscr{S} \cap (\beta+1) \mathscr{V}(f, \partial V) \cap (\beta+1) \mathscr{V}(1-f, \bar{U}) \to \mathbb{R}$$

is continuous (N. Bourbaki [2], § 2, Corollaire 3 de la Proposition 1). $\square$

Corollary 11.2.1. *The map*

$$(u, x) \mapsto u(x) \colon \mathscr{S} \times X \to [0, \infty]$$

is lower semi-continuous (R.-M. Hervé 1962 [4], C. Constantinescu 1965 [2]).

The corollary follows from *b)* of the proposition and *c)* of Proposition 11.2.2. □

For any Cauchy filter $\mathfrak{F}$ on $\mathscr{S}$, we denote by $L(\mathfrak{F})$, the positive hyperharmonic function on X

$$\bigvee_{\mathscr{F} \in \mathfrak{F}} \wedge \mathscr{F}.$$

For any $g \in \mathscr{C}_+(X_0)$, we denote by $g \cdot \mathfrak{F}$, the Cauchy filter on $\mathscr{S}$ which is the image of $\mathfrak{F}$ with respect to the map

$$u \mapsto g \cdot u \colon \mathscr{S} \to \mathscr{S}.$$

Proposition 11.2.6. *Let $\mathfrak{F}$ be a Cauchy filter on $\mathscr{S}$ and let $x \in X$. Then for any $(V, f) \in I_x$,*

$$\lim_{u, \mathfrak{F}} A_{(V,f)} u(x)$$

exists and

$$\big(L(\mathfrak{F})\big)(x) = \lim_{(V,f), \mathfrak{I}_x} \lim_{u, \mathfrak{F}} A_{(V,f)} u(x).$$

By Proposition 11.2.4, the map

$$u \mapsto A_{(V,f)} u(x) \colon \mathscr{S} \to \mathbb{R}$$

is the restriction to $\mathscr{S}$ of a continuous linear form on $[\mathscr{S}]$. $\mathfrak{F}$ being a Cauchy filter, we have that

$$\lim_{u, \mathfrak{F}} A_{(V,f)} u(x)$$

exists.

Let U be an open set of X whose closure is compact and contained in $X \setminus \operatorname{Supp}(1-f)$. Since $\mathfrak{F}$ is a Cauchy filter, there exists $\mathscr{F} \in \mathfrak{F}$ such that

$$\mathscr{F} - \mathscr{F} \subset \mathscr{V}(f, \partial V) \cap \mathscr{V}(1-f, \bar{U}).$$

Hence there exists $\alpha \in \mathbb{R}_+$ for which

$$\mathscr{F} \subset \alpha \mathscr{V}(f, \partial V) \cap \alpha \mathscr{V}(1-f, \bar{U}).$$

By Proposition 11.2.5, *a)*

$$\{A_{(V,f)} u|_U \mid u \in \mathscr{F}\}$$

is an equicontinuous set of harmonic functions on U. Then

$$\mu_x^V(\wedge \mathscr{F}) \le \inf_{u \in \mathscr{F}} \mu_x^V u \le \inf_{u,\, \mathscr{F}} A_{(V,f)} u(x) \le \wedge \mathscr{F}(x)$$

(Proposition 11.2.2 *a)*). We obtain

$$\mu_x^V(L(\mathfrak{F})) \le \lim_{u,\, \mathfrak{F}} A_{(V,f)} u(x) \le (L(\mathfrak{F}))(x),$$

$$(L(\mathfrak{F}))(x) \le \lim_{(V,f),\, \mathfrak{J}_x} \lim_{u,\, \mathfrak{F}} A_{(V,f)} u(x) \le (L(\mathfrak{F}))(x). \quad \square$$

Proposition 11.2.7. *Let $\mathfrak{F}$ be a Cauchy filter on $\mathscr{S}$ and let $g \in \mathscr{C}_+(X_0)$. Then $L(g \cdot \mathfrak{F})$ is harmonic on $X \setminus \operatorname{Supp} g$ and the filter $g \cdot \mathfrak{F}$ converges uniformly to $L(g \cdot \mathfrak{F})$ on every compact subset of $X \setminus \operatorname{Supp} g$.*

Let K be a compact subset of $X \setminus \operatorname{Supp} g$ and U be an open neighbourhood of K such that $\bar{U}$ is compact and contained in $X \setminus \operatorname{Supp} g$. Since $\mathfrak{F}$ is a Cauchy filter, there exists $\mathscr{F} \in \mathfrak{F}$ such that

$$\mathscr{F} - \mathscr{F} \subset \mathscr{V}(g, \bar{U}).$$

Then $((g \cdot u)|_U)_{u \in \mathscr{F}}$ is a uniformly bounded family of harmonic functions on U. By Theorem 11.1.1, it is equicontinuous. Since $\mathfrak{F}$ is a Cauchy filter,

$$\lim_{u,\, g \cdot \mathfrak{F}} u(x) = \lim_{u,\, \mathfrak{F}} (g \cdot u)(x)$$

exists for any $x \in U$ and from equicontinuity, it follows that the convergence is uniform on K (N. Bourbaki [2], § 2, Théorème 1). Again using equicontinuity, we deduce that for any $\mathscr{F}' \in \mathfrak{F}$, $\mathscr{F}' \subset \mathscr{F}$, we have

$$\bigwedge_{u \in \mathscr{F}'} (g \cdot u) = \inf_{u \in \mathscr{F}'} (g \cdot u)$$

on U. Hence

$$L(g \cdot \mathfrak{F}) = \bigvee_{\mathscr{F}' \in \mathfrak{F}} \bigwedge_{u \in \mathscr{F}'} (g \cdot u) = \lim_{u,\, \mathfrak{F}} (g \cdot u) = \lim_{u,\, g \cdot \mathfrak{F}} u$$

on K. Thus $L(g \cdot \mathfrak{F})$ is harmonic since it is a uniform limit of harmonic functions. $\square$

Proposition 11.2.8. *Let $\mathfrak{F}$ be a Cauchy filter on $\mathscr{S}$. Then $L(\mathfrak{F})$ is superharmonic and $\mathfrak{F}$ converges to $L(\mathfrak{F})$.*

Let $g, g' \in \mathscr{C}_+(X_0)$ and $x \in X$. Then by Proposition 11.2.6,

$$\begin{aligned}(L(g+g') \cdot \mathfrak{F})(x) &= \lim_{(V,f),\, \mathfrak{J}_x} \lim_{u,\, \mathfrak{F}} A_{(V,f)} (g+g') \cdot u(x) \\ &= \lim_{(V,f),\, \mathfrak{J}_x} \lim_{u,\, \mathfrak{F}} A_{(V,f)} g \cdot u(x) + \lim_{(V,f),\, \mathfrak{J}_x} \lim_{u,\, \mathfrak{F}} A_{(V,f)} g' \cdot u(x) \\ &= (L(g \cdot \mathfrak{F}))(x) + (L(g' \cdot \mathfrak{F}))(x).\end{aligned}$$

Since x is arbitrary, we get

$$L((g+g')\cdot\mathfrak{F})=L(g\cdot\mathfrak{F})+L(g'\cdot\mathfrak{F}).$$

Let $x\in X$ and $(V,f)\in I_x$. We have

$$\mu^V L(\mathfrak{F})=\mu^V L(f\cdot\mathfrak{F})+\mu^V L((1-f)\cdot\mathfrak{F})\leq\mu^V(L(f\cdot\mathfrak{F}))+L((1-f)\cdot\mathfrak{F}).$$

By Proposition 11.2.7, $L(f\cdot\mathfrak{F})$ is bounded on ∂V. Hence $\mu^V L(f\cdot\mathfrak{F})$ is harmonic. By the same proposition, $L((1-f)\cdot\mathfrak{F})$ is harmonic on $X\setminus\operatorname{Supp}(1-f)$. Hence $\mu^V L(\mathfrak{F})$ is harmonic on $V\setminus\operatorname{Supp}(1-f)$. Since f is arbitrary, $\mu^V L(\mathfrak{F})$ is harmonic and $L(\mathfrak{F})$ is superharmonic.

Let $g\in\mathscr{C}_+(X_0)$. By the above considerations, $L(g\cdot\mathfrak{F})$ is a superharmonic function. We want to show that

$$S_0(L(g\cdot\mathfrak{F}))\subset\operatorname{Supp} g.$$

This is obvious if $\operatorname{Supp} g\not\subset X$ from the preceding proposition. Assume now that $\operatorname{Supp} g\subset X$ and let K be a compact neighbourhood of $\operatorname{Supp} g$. Then the filter $g\cdot\mathfrak{F}$ converges uniformly to $L(g\cdot\mathfrak{F})$ on ∂K (Proposition 11.2.7). Hence for any $x\in X\setminus K$, we have

$$\begin{aligned}\hat{R}^K_{L(g\cdot\mathfrak{F})}(x)&=H^{X\setminus K}_{L(g\cdot\mathfrak{F})}(x)=\int L(g\cdot\mathfrak{F})\,d\mu_x^{X\setminus K}=\lim_{u,\mathfrak{F}}\int g\cdot u\,d\mu_x^{X\setminus K}\\&=\lim_{u,\mathfrak{F}}H^{X\setminus K}_{g\cdot u}(x)=\lim_{u,\mathfrak{F}}\hat{R}^K_{g\cdot u}(x)=\lim_{u,\mathfrak{F}}g\cdot u(x)\\&=(L(g\cdot\mathfrak{F}))(x)\end{aligned}$$

(Proposition 5.3.3, Proposition 8.2.2). But x is arbitrary, and therefore $L(g\cdot\mathfrak{F})$ and $\hat{R}^K_{L(g\cdot\mathfrak{F})}$ coincide outside of a compact set. It follows that $L(g\cdot\mathfrak{F})$ is a potential (Proposition 5.3.5). Hence

$$S_0(L(g\cdot\mathfrak{F}))\subset\operatorname{Supp} g.$$

From these considerations, we get, using Theorem 8.1.1 and Proposition 8.2.1, that for any $g\in\mathscr{C}_+(X_0)$

$$g\cdot L(\mathfrak{F})=L(g\cdot\mathfrak{F}).$$

We want to show now that $\mathfrak{F}$ converges to $L(\mathfrak{F})$. Let $g\in\mathscr{C}_+(X_0)$ and let K be a compact subset of $X\setminus\operatorname{Supp} g$. By the preceding proposition, there exists $\mathscr{F}\in\mathfrak{F}$ such that for all $u\in\mathscr{F}$

$$|g\cdot u-g\cdot L(\mathfrak{F})|=|g\cdot u-L(g\cdot\mathfrak{F})|\leq 1$$

on K. □

Corollary 11.2.2. *$\mathscr{S}$ is complete in $[\mathscr{S}]$.* □

Proposition 11.2.9. *On any bounded set of $[\mathscr{S}]$ the uniform structure induced by the topology* T *coincides with the uniform structure induced by the weak topology.*

Let $\mathscr{B}$ be a bounded set of $[\mathscr{S}]$ and let $\mathscr{V}(f, K)$ be a neighbourhood of 0. Since $\mathscr{B}$ is bounded, $\{f \cdot u|_{X \setminus \operatorname{Supp} f} \mid u \in \mathscr{B} - \mathscr{B}\}$ is a locally bounded set of harmonic functions and hence, it is equicontinuous (Theorem 11.1.1). There exists then, a finite subset A of K such that for any $x \in K$, there exists $y \in A$ with the property that

$$|(f \cdot u)(x) - (f \cdot u)(y)| < \tfrac{1}{3}$$

for all $u \in \mathscr{B} - \mathscr{B}$. The set

$$\bigcap_{y \in A} \{u \in [\mathscr{S}] \mid |(f \cdot u)(y)| < \tfrac{1}{3}\}$$

is a weak neighbourhood $\mathscr{W}$ of the origin. By the above considerations,

$$(\mathscr{B} - \mathscr{B}) \cap \mathscr{W} \subset \mathscr{V}(f, K). \quad \square$$

Corollary 11.2.3. *Every bounded subset of $\mathscr{S}$ is relatively compact. Hence the closed convex hull of a compact set of $\mathscr{S}$ is a compact set of $\mathscr{S}$.*

Let $\mathscr{B}$ be a bounded subset of $\mathscr{S}$. By the preceding corollary and Proposition 11.2.9, $\bar{\mathscr{B}}$ is weakly complete. Being bounded, $\bar{\mathscr{B}}$ is weakly precompact (H. Schaefer [1], Ch. IV.5.5, Corollary 2) and therefore weakly compact. Again by Proposition 11.2.9, it is a compact set. $\square$

Remark. There exist bounded sets of $[\mathscr{S}]$ which are not relatively compact (see Exercise 11.2.10).

Corollary 11.2.4. *Let $\mathscr{F}$ be a naturaly upper directed set in $\mathscr{S}$ such that for every positive continuous linear form l on $[\mathscr{S}]$,*

$$\sup_{u \in \mathscr{F}} l(u) < \infty.$$

Then $\bigvee \mathscr{F}$ is superharmonic.

By Proposition 11.2.4, the set $\mathscr{F}$ is weakly bounded and therefore bounded (N. Bourbaki [4], Ch. IV, § 2, Corollaire du Théorème 3). By the preceding corollary, it is relatively compact. Let $\mathfrak{F}$ be an ultrafilter on $\mathscr{F}$ which is finer than the section filter of $\mathscr{F}$. Then $\mathfrak{F}$ is a Cauchy filter. By Proposition 11.2.8, $L(\mathfrak{F})$ is superharmonic. Let $u \in \mathscr{F}$. Since $\{v \in \mathscr{F} \mid v \geq u\} \in \mathfrak{F}$, we get that

$$u = \bigwedge \{v \in \mathscr{F} \mid v \geq u\} \leq L(\mathfrak{F}).$$

Hence $\bigvee \mathscr{F}$ is superharmonic since it is dominated by $L(\mathfrak{F})$. $\square$

Remark. In fact, the section filter of $\mathscr{F}$ is convergent and its limit is $\bigvee \mathscr{F}$.

Proposition 11.2.10. *If X has a countable base then $\mathscr{S}$ is metrisable and has a countable base.*

Let $\mathscr{F}$ be a countable set in $\mathscr{C}_+(X_0)$ such that every function of $\mathscr{C}_+(X_0)$ is the supremum of an upper directed subset of $\mathscr{F}$ (Lemma of Proposition 7.2.1) and let $\mathfrak{V}$ be a countable base of X of open relatively compact sets. Finite intersections of sets of the form $\mathscr{V}(f, \overline{V})$, where $f \in \mathscr{F}$, $V \in \mathfrak{V}$ and $\operatorname{Supp} f \cap \overline{V} = \varnothing$, form a fundamental system of neighbourhoods of the origin of $[\mathscr{S}]$ for a locally convex topology $\mathscr{T}_0$. By Proposition 11.2.3, $\mathscr{T}_0$ is Hausdorff and therefore, metrisable.

We want to show that $\mathscr{T}$ and $\mathscr{T}_0$ coincide on $\mathscr{S}$. This will show in particular, the metrisability of $\mathscr{S}$. Obviously $\mathscr{T}_0$ is coarser than $\mathscr{T}$. Let $u_0 \in \mathscr{S}$, let $f \in \mathscr{C}_+(X_0)$ and let K be a compact subset of $X \setminus \operatorname{Supp} f$. There exists a finite family $(V_\iota)_{\iota \in I}$ in $\mathfrak{V}$ such that

$$K \subset \bigcup_{\iota \in I} \overline{V}_\iota \subset X \setminus \operatorname{Supp} f.$$

Choose $g \in \mathscr{F}$ greater than 1 on $\operatorname{Supp} f$ and such that

$$\operatorname{Supp} g \cap \Big(\bigcup_{\iota \in I} \overline{V}_\iota\Big) = \varnothing.$$

Let

$$u \in \mathscr{S} \cap \Big(u_0 + \bigcap_{\iota \in I} \mathscr{V}(g, \overline{V}_\iota)\Big).$$

Then

$$\sup_K g \cdot u \leq 1 + \sup_K g \cdot u_0.$$

We set

$$\varepsilon := \frac{1}{1 + \sup_K g \cdot u_0}.$$

Choose $g' \in \mathscr{F}$ such that $|g' - 3f| \leq \varepsilon g$. Let

$$u \in \mathscr{S} \cap \Big(u_0 + \bigcap_{\iota \in I} \big(\mathscr{V}(g, \overline{V}_\iota) \cap \mathscr{V}(g', \overline{V}_\iota)\big)\Big).$$

Then

$$|f \cdot (u - u_0)| \leq |f - \tfrac{1}{3} g'| \cdot u + |f - \tfrac{1}{3} g'| \cdot u_0 + |\tfrac{1}{3} g' \cdot (u - u_0)|$$

$$\leq \frac{\varepsilon}{3} g \cdot u + \frac{\varepsilon}{3} g \cdot u_0 + \tfrac{1}{3} \leq 1$$

on K. Since u is arbitrary, we have

$$\mathscr{S} \cap \Big(u_0 + \bigcap_{\iota \in I} \big(\mathscr{V}(g, \overline{V}_\iota) \cap \mathscr{V}(g', \overline{V}_\iota)\big)\Big) \subset u_0 + \mathscr{V}(f, K).$$

In order to prove the existence of a countable base for $\mathscr{S}$, it is sufficient to show that there exists a countable subset of $\mathscr{S}$ which is dense for $\mathscr{T}_0$. Let $(A_n)_{n \in \mathbb{N}}$ be an increasing sequence of finite subsets of X such that

$\bigcup_{n\in\mathbb{N}} A_n$ is dense in X. For any $f\in\mathscr{F}$, let $(U_{f,n})_{n\in\mathbb{N}}$ be an increasing sequence of open sets of X such that for any $n\in\mathbb{N}$, $\overline{U}_{f,n}$ is compact and contained in $X\setminus \operatorname{Supp} f$ and such that

$$\bigcup_{n\in\mathbb{N}} U_{f,n} = X\setminus \operatorname{Supp} f.$$

For any finite subset $\mathscr{F}_0$ of $\mathscr{F}$ and for any $(m,n)\in\mathbb{N}^2$, we denote by $\mathscr{S}(\mathscr{F}_0,m,n)$, a countable subset of $\mathscr{S}\cap(\bigcap_{f\in\mathscr{F}_0}\mathscr{V}(f,\overline{U}_{f,n}))$ such that for any $u\in\mathscr{S}\cap(\bigcap_{f\in\mathscr{F}_0}\mathscr{V}(f,\overline{U}_{f,n}))$ and for any strictly positive real number ε, there exists $v\in\mathscr{S}(\mathscr{F}_0,m,n)$ such that for any $f\in\mathscr{F}_0$ and any $x\in A_m\cap U_{f,n}$, we have

$$|(f\cdot u)(x)-(f\cdot v)(x)|<\varepsilon.$$

The union $\mathscr{S}_0$ of the sets $\mathscr{S}(\mathscr{F}_0,m,n)$ is obviously countable.

We want to show that the union of the sets, $\alpha\mathscr{S}_0$, where α runs through the set of positive rational numbers is dense in $\mathscr{S}$ for $\mathscr{T}_0$. Let $u\in\mathscr{S}$ and let $(f_\iota)_{\iota\in I}$, $(V_\iota)_{\iota\in I}$ be finite families in $\mathscr{F}$ and $\mathfrak{V}$ respectively and such that for any $\iota\in I$,

$$\overline{V}_\iota\cap \operatorname{Supp} f_\iota=\emptyset.$$

There exists $n\in\mathbb{N}$ such that for any $\iota\in I$,

$$\overline{V}_\iota\subset U_{f_\iota,n}.$$

We set

$$\mathscr{F}_0:=\{f_\iota\mid\iota\in I\}.$$

Let α be a strictly positive rational number such that

$$\alpha u\in\bigcap_{f\in\mathscr{F}_0}\mathscr{V}(f,\overline{U}_{f,n}).$$

By Theorem 11.1.1 $a\Rightarrow b$, for any $g\in\mathscr{F}_0$, the set

$$\{(g\cdot v)|_{U_{g,n}}\mid v\in\bigcap_{f\in\mathscr{F}_0}\mathscr{V}(f,\overline{U}_{f,n})\}$$

is equicontinuous. Hence, there exists $m\in\mathbb{N}$ such that for any $\iota\in I$ and any $y\in\overline{V}_\iota$, there exists $x\in V_\iota\cap A_m$ with the property that for all $v\in\bigcap_{f\in\mathscr{F}_0}\mathscr{V}(f,\overline{U}_{f,n})$, we have

$$|(f_\iota\cdot v)(x)-(f_\iota\cdot v)(y)|<\frac{1}{3\alpha}.$$

Choose $v\in\mathscr{S}(\mathscr{F}_0,m,n)$ such that for any $f\in\mathscr{F}_0$ and any $x\in A_m\cap U_{f,n}$, we have

$$|(f\cdot\alpha u)(x)-(f\cdot v)(x)|<\frac{1}{3\alpha}.$$

Then, for any $\iota \in I$ and any $y \in \overline{V}_\iota$, we have

$$|(f_\iota \cdot u)(y) - (f_\iota \cdot \alpha v)(y)| < 1, \qquad \alpha v \in u + \bigcap_{\iota \in I} \mathscr{V}(f_\iota, \overline{V}_\iota).$$

The assertion follows now from the fact that u, $(f_\iota)_{\iota \in I}$ and $(V_\iota)_{\iota \in I}$ are arbitrary. □

Proposition 11.2.11. *If X is a connected Brelot space, then there exists a continuous linear form l on $[\mathscr{S}]$ such that l is strictly positive on $\mathscr{S} \setminus \{0\}$ and*

$$\{u \in \mathscr{S} \mid l(u) = 1\}$$

is compact (R.-M. Hervé 1960 [3]).

Let x, y be two different points of X and let $f \in \mathscr{C}_+(X_0)$ such that

$$f \leq 1, \qquad x \notin \operatorname{Supp} f, \qquad y \notin \operatorname{Supp}(1-f).$$

We denote by l, the continuous linear form on $[\mathscr{S}]$

$$u \mapsto (f \cdot u)(x) + ((1-f) \cdot u)(y).$$

Let $u \in \mathscr{S}$. If

$$l(u) = 0,$$

then

$$(f \cdot u)(x) = ((1-f) \cdot u)(y) = 0.$$

Since every non-empty absorbent set of X is equal to X, we get

$$f \cdot u = (1-f) \cdot u = 0$$

and therefore

$$u = 0.$$

In order to show that $\{u \in \mathscr{S} \mid l(u) = 1\}$ is compact, it is sufficient by Corollary 11.2.3 to show that it is bounded. Assume the contrary. Then there exists $g \in \mathscr{C}_+(X_0)$ and a compact subset K of $X \setminus \operatorname{Supp} g$ such that for any $n \in \mathbb{N}$, $n \neq 0$, there exists $u_n \in \mathscr{S}$ for which

$$l(u_n) = 1, \qquad \frac{1}{n^3} u_n \notin \mathscr{V}(g, K).$$

We set

$$v' := \sum_{\substack{n \in \mathbb{N} \\ n \neq 0}} \frac{1}{n^2} f \cdot u_n, \qquad v'' := \sum_{\substack{n \in \mathbb{N} \\ n \neq 0}} \frac{1}{n^2} (1-f) \cdot u_n.$$

Since

$$v'(x) \leq \sum_{\substack{n \in \mathbb{N} \\ n \neq 0}} \frac{1}{n^2} l(u_n) < \infty, \qquad v''(y) \leq \sum_{\substack{n \in \mathbb{N} \\ n \neq 0}} \frac{1}{n^2} l(u_n) < \infty$$

it follows that the sets

$$\overline{\{z \in X \mid v'(z) < \infty\}}, \qquad \overline{\{z \in X \mid v''(z) < \infty\}}$$

are not empty. Being absorbent sets (Proposition 6.1.4) and X being a connected elliptic harmonic space, (Theorem 3.1.2), these sets must be equal to X (Proposition 6.1.3). We deduce that v' and v'' are superharmonic functions and this leads to the contradictory relation

$$n \in \mathbb{N} \Rightarrow n \leq \frac{1}{n^2} \sup_K g \cdot u_n \leq \sup_K g \cdot (v' + v'') < \infty . \quad \square$$

Exercises

11.2.1. Let $\mathscr{F}$ be a naturally lower directed (resp. naturally upper directed and naturally upper bounded) subset of $\mathscr{S}$ and let $\mathfrak{F}$ be the section filter of $\mathscr{F}$. Then $\mathfrak{F}$ converges to $\bigwedge \mathscr{F}$ (resp. $\bigvee \mathscr{F}$) (R.-M. Hervé 1960 [3]).

11.2.2. The bounded sets of $[\mathscr{S}]$ are precompact (use Proposition 11.2.9).

11.2.3. For any $u \in \mathscr{S}$, the set $\{v \in \mathscr{S} \mid v \leq u\}$ is compact (use Exercise 11.2.2).

11.2.4. Let $[\mathscr{S}]'$ be the dual of $[\mathscr{S}]$ (endowed with the strong topology), let $[\mathscr{S}]''$ be the dual of $[\mathscr{S}]'$ and assume $[\mathscr{S}]$ imbedded in $[\mathscr{S}]''$. Let $\mathscr{S}'$ (resp. $\mathscr{S}''$) be the set of elements of $[\mathscr{S}]'$ (resp. $[\mathscr{S}]''$) which are positive on $\mathscr{S}$ (resp. $\mathscr{S}'$). Then $\mathscr{S} = \mathscr{S}''$. (Since any bounded set of $[\mathscr{S}]$ is precompact (use Exercise 11.2.2) on any equicontinuous set of $[\mathscr{S}]'$, the weak topology $\sigma([\mathscr{S}]', [\mathscr{S}]'')$ and the strong topology coincide (N. Bourbaki [4], Ch. III, § 3, Proposition 5). Hence the equicontinuous sets of $[\mathscr{S}]'$ are relatively compact with respect to the weak topology $\sigma([\mathscr{S}]', [\mathscr{S}]'')$ and thus the topology $\mathscr{T}''$ on $[\mathscr{S}]''$, of uniform convergence on equicontinuous sets of $[\mathscr{S}]'$, is consistent with the duality $\langle [\mathscr{S}]', [\mathscr{S}]'' \rangle$ (N. Bourbaki [4], Ch. IV, § 2, Théorème 2). Since the restriction to $\mathscr{S}$ of $\mathscr{T}''$ coincides with $\mathscr{T}$, $\mathscr{S}$ is complete and therefore closed with respect to $\mathscr{T}''$. Hence $\mathscr{S} = \mathscr{S}^{00}$ in the duality $\langle [\mathscr{S}]', [\mathscr{S}]'' \rangle$. We have, trivially, that $\mathscr{S}^0 = -\mathscr{S}'$, $\mathscr{S}^{00} = \mathscr{S}''$.) (The assertion follows also from the fact that $\mathscr{S}$ is weakly complete C. Constantinescu 1970 [9].)

11.2.5. If $\mathscr{H}$ has the property of nuclearity, then $[\mathscr{S}]$ is nuclear (use Theorem 11.1.2 $a \Rightarrow b$ and A. Pietsch [1] Satz 4.1.5).

11.2.6. Assume that the sheaf of harmonic functions on X has the Doob convergence property. If there exist two measures μ, ν on X with compact, disjoint carriers such that any absorbent set containing the carrier of one of the measures is equal to X, then there exists a continuous linear form l on $[\mathcal{S}]$, strictly positive on $\mathcal{S} \setminus \{0\}$, such that the set

$$\{u \in \mathcal{S} \mid l(u) = 1\}$$

is compact (C. Constantinescu 1965 [2]).

11.2.7. For any filter $\mathfrak{F}$ on $\mathcal{S}$ we set

$$L(\mathfrak{F}) := \bigvee_{\mathcal{F} \in \mathfrak{F}} \wedge \mathcal{F},$$

and for any $f \in \mathscr{C}_+(X_0)$, we denote by $f \cdot \mathfrak{F}$ the image of the filter $\mathfrak{F}$ through the map $u \mapsto f \cdot u \colon \mathcal{S} \to \mathcal{S}$. Then $L(\mathfrak{F})$ is hyperharmonic. If $\mathfrak{U}$ is an ultrafilter on $\mathcal{S}$, then for any $f, g \in \mathscr{C}_+(X_0)$

$$L((f+g) \cdot \mathfrak{U}) = L(f \cdot \mathfrak{U}) + L(g \cdot \mathfrak{U})$$

(C. Constantinescu 1965 [2]).

11.2.8. Let X be the harmonic space of Theorem 2.1.2, let Y be the topological subspace $]-\infty, +\infty]$ of $\overline{\mathbb{R}}$ and let $\mathfrak{M}(Y)$ be the locally convex space of signed measures on Y endowed with the vague topology. For any $\mu \in \mathfrak{M}(Y)$, we denote by u_μ the function on X

$$x \mapsto \mu(]x, \infty]).$$

Show that $u_\mu \in [\mathcal{S}]$ and that the map

$$\mu \mapsto u_\mu \colon \mathfrak{M}(Y) \to [\mathcal{S}]$$

is an isomorphism of locally convex vector spaces. $u_\mu \in \mathcal{S}$ if and only if μ is measure.

11.2.9. If we endow $\mathscr{H}(X)$ with the induced topology $\mathcal{T}$, then $\mathscr{H}_+(X)$ is weakly complete. (Let $\mathscr{C}(X)$ be the locally convex space of continuous real functions on X (endowed with the topology of compact convergence), let $\mathscr{C}(X)'$ (resp. $\mathscr{C}(X)''$) be the strong dual of $\mathscr{C}(X)$ (resp. $\mathscr{C}(X)'$). Then $\mathscr{C}(X)'$ may be identified with the set of measures with compact carriers on X and is therefore, a vector lattice. Show that every positive linear form on $\mathscr{C}(X)'$ belongs to $\mathscr{C}(X)''$. Hence, the cone of positive elements of $\mathscr{C}(X)''$ is complete with respect to the weak topology $\sigma(\mathscr{C}(X)'', \mathscr{C}(X)')$. The assertion now follows from the fact that the weak topology of $\mathscr{H}(X)$ coincides with the restriction to $\mathscr{H}(X)$ of the weak topology $\sigma(\mathscr{C}(X)'', \mathscr{C}(X)')$ and from Exercise 11.1.12.)

11.2.10. Let X be the harmonic space defined on $\{(x, y) \in \mathbb{R}^2 \mid y > 0\}$ with respect to the Laplace equation. For any $n \in \mathbb{N}$, $n \neq 0$ let h_n be the function on X defined by

$$h_n(x, y) = \frac{1}{n}\left(\frac{y}{\left(x - \frac{1}{n}\right)^2 + y^2} - \frac{y}{x^2 + y^2}\right).$$

Show that the set $\{h_n \mid n \in \mathbb{N}, n \neq 0\}$ is a bounded set of $[\mathscr{S}]$ which is not relatively compact.

§ 11.3. Abstract Integral Representation

Throughout this section we shall denote by E a locally convex (separated), real vector space and by C, a closed convex cone in E with vertex 0 such that $C \cap (-C) = \{0\}$ *and such that every bounded subset of C is relatively compact. C is then the set of positive elements of E with respect to an order relation on E, denoted by* $\preccurlyeq$, *which is consistent with the structure of real vector space.* We say that C is *a lattice with respect to its own order* if it is a lattice with respect to this order relation; in this case we denote by $\curlyvee$ and $\curlywedge$, the lattice operations in C. A linear form on E is called *positive* if it is positive on C.

For any subset A of $E \setminus \{0\}$, we set

$$\tilde{A} := \{\alpha x \mid \alpha \in]0, \infty[, x \in A\}.$$

A *ray of* E is a set of the form $\widetilde{\{x\}}$, where $x \in E \setminus \{0\}$. An *extremal ray of* C is a ray R of E such that

$$x, y \in C \setminus 0, \quad x + y \in R \Rightarrow x, y \in R.$$

We denote by C_e the set of points of C which lie on extremal rays of C. A *section in* C is a subset A of C_e such that its intersection with any extremal ray contains at most one point. For any section A in C, we denote by ρ_A, the function $\tilde{A} \to]0, \infty[$ such that for any $x \in \tilde{A}$

$$\frac{1}{\rho_A(x)} x \in A.$$

Obviously, for any $x \in \tilde{A}$ and for any $\alpha \in]0, \infty[$,

$$\rho_A(\alpha x) = \alpha \rho_A(x).$$

Conversely, let B be a subset of C_e such that $B = \tilde{B}$ and let ρ be a map $B \to]0, \infty[$ such that for any $x \in B$ and any $\alpha \in]0, \infty[$,

$$\rho(\alpha x) = \alpha \rho(x).$$

Then the set

$$A := \{x \in B \mid \rho(x) = 1\}$$

is a section in C such that $\tilde{A} = B$ and $\rho_A = \rho$. We say that the section A in C is *continuous* (resp. *measurable*) if ρ_A is continuous (resp. if the restriction of ρ_A to any compact K of A is measurable with respect to any measure on K).

Proposition 11.3.1. *If the origin has a countable fundamental system of neighbourhoods in C, then there exists a measurable section A in C such that $\tilde{A} = C_e$ and such that ρ_A is bounded on every compact set of C_e.*

Let V be a convex circled neighbourhood of the origin of E and let p_V be the gauge of V. Then

$$\{x \in C_e \mid p_V(x) = 1\}$$

is a continuous section in C which will be denoted by A_V. $\tilde{A}_V$ is obviously open in C_e and $\rho_{A_V} = p_V|_{\tilde{A}_V}$.

Let $(V_n)_{n \in \mathbb{N}}$ be a decreasing sequence of convex circled neighbourhoods of the origin of E such that $(V_n \cap C)_{n \in \mathbb{N}}$ is a fundamental system of neighbourhoods of the origin in C. For any $n \in \mathbb{N}$, we set

$$A_n := A_{V_n} \setminus \bigcup_{\substack{m \in \mathbb{N} \\ m < n}} \tilde{A}_{V_m}.$$

By the above considerations, A_n is a continuous section in C. For any $m, n \in \mathbb{N}$, $m \neq n$, we have

$$\tilde{A}_m \cap \tilde{A}_n = \varnothing.$$

Hence $A := \bigcup_{n \in \mathbb{N}} A_n$ is a section in C, which is obviously measurable.

Let x be any point lying on an extremal ray of C. Then there exists $n \in \mathbb{N}$ such that $x \notin V_n$ and we get

$$x \in \bigcup_{\substack{m \in \mathbb{N} \\ m \leq n}} \tilde{A}_m.$$

Hence

$$C_e = \bigcup_{n \in \mathbb{N}} \tilde{A}_n = \widetilde{\bigcup_{n \in \mathbb{N}} A_n} = \tilde{A}.$$

Let K be a compact set of C_e. Since

$$K \subset \bigcup_{n \in \mathbb{N}} \tilde{A}_{V_n}$$

and since $(\tilde{A}_{V_n})_{n \in \mathbb{N}}$ is an increasing sequence of open sets of C_e there exists $n \in \mathbb{N}$ such that

$$K \subset \tilde{A}_{V_n}.$$

We get

$$\sup_K \rho_{\tilde{A}} \leq \sup_K \rho_{V_n} < \infty . \quad \square$$

In order to find an integral representation on sections of C, we have to use measures on C, i.e., measures defined on a Hausdorff topological space. We now define this more general notion of measure.

Let Y be a Hausdorff topological space. A *measure* on Y is a function which associates to every compact set K of Y, a measure, μ_K, on K such that if K, L are compact sets of Y with $K \subset L$, then the restriction of μ_L to K is equal to μ_K. If Y is locally compact, then any Radon measure μ on Y defines, in a natural way, a measure in the above sense: namely, for any compact set K of Y, we take as μ_K the restriction of μ to K. Conversely, any measure μ on Y in the above sense defines a unique Radon measure on Y whose restriction to an arbitrary compact set K of Y is equal to μ_K. This enables us to identify the Radon measures on locally compact spaces with the measures introduced above.

Let $\mathfrak{K}$ be the set of compact sets of Y ordered by the inclusion relation, let $\mathfrak{F}_Y$ be the section filter of $\mathfrak{K}$ and let μ be a measure on Y. A map ψ of Y into a topological space is called *μ-measurable* if its restriction to any compact set K of Y is μ_K-measurable. A numerical function f on Y is called *μ-integrable* if its restriction to any compact set K of Y is μ_K-integrable and if

$$\int f \, d\mu := \lim_{K, \mathfrak{F}_Y} \int f|_K \, d\mu_K$$

exists in $\mathbb{R}$. If f is μ-integrable, then every μ-measurable function g on Y such that $|g| \leq |f|$ is μ-integrable. For any positive numerical function f on Y, we set

$$\overset{\bullet}{\int} f \, d\mu := \sup_{K \in \mathfrak{K}} \overset{*}{\int} f|_K \, d\mu_K .$$

If A is a subset of Y, we say that A is *μ-measurable* (resp. *μ-integrable*) if its characteristic function χ_A is μ-measurable (resp. μ-integrable); if A is μ-integrable, we set

$$\mu(A) := \int \chi_A \, d\mu .$$

Let μ, ν be two measures on Y, and let $\alpha \in \mathbb{R}_+$. By $\mu + \nu$ (resp. $\alpha \mu$) we denote the measure on Y

$$K \mapsto \mu_K + \nu_K \qquad (\text{resp. } K \mapsto \alpha \mu_K).$$

With respect to these two operations, the set of measures on Y is a prevector lattice (see p. 183). For any measures μ, ν on Y and any compact set K of Y, $(\mu \wedge \nu)_K$ is the greatest lower bound of μ_K, ν_K in the set of all measures on K.

Proposition 11.3.2. *Let A be a subset of C and μ be a measure on A. If the restriction to A of any continuous linear form on E is μ-integrable, then there exists a unique point $x_\mu \in C$, called the barycenter of μ, such that*

$$\int x'|_A \, d\mu = x'(x_\mu)$$

for every continuous linear form x' on E. If $\mathfrak{K}$ denotes the set of compact sets of A, then x_μ is the least upper bound in C of the upper directed family $(x_{\mu_K})_{K \in \mathfrak{K}}$.

The unicity is trivial. Let K be a compact set of A and let K' be the closed convex hull of K. Since K' is bounded and closed, it is compact by the hypotheses about C. Hence μ_K has a barycenter $x_{\mu_K} \in C$ (N. Bourbaki [5], Ch. IV, § 7, Corollaire de la Proposition 1). We set

$$B := \{x_{\mu_K} \mid K \in \mathfrak{K}\}.$$

Let x' be a continuous linear form on E. Then the restriction of $|x'|$ to A is μ-integrable and therefore

$$\sup_{x \in B} |x'(x)| = \sup_{K \in \mathfrak{K}} |x'(x_{\mu_K})| \le \int \big| x'|_A \big| \, d\mu < \infty .$$

Hence B is weakly bounded and therefore bounded (N. Bourbaki [4], Ch. IV, § 2, Théorème 3). By the hypotheses about C, $\bar{B}$ is a compact subset of C. On $\bar{B}$, the initial topology and the weak topology coincide and therefore there exists a point, $x_\mu \in \bar{B}$, such that

$$\lim_{K, \mathfrak{F}_A} x_{\mu_K} = x_\mu ,$$

where $\mathfrak{F}_A$ denotes the section filter on $\mathfrak{K}$. All of the assertions of the proposition now follow without difficulty. □

Proposition 11.3.3. *For any compact set K of C such that $0 \notin K$, there exists a continuous linear form on E which is positive on C and strictly positive on K. In particular 0 does not belong to the closed convex hull of K.*

Let $x \in C$, $x \neq 0$. Then $-x \notin C$ and there exists by the Hahn-Banach theorem, a continuous linear form x'_x of E which is positive on C and equal to -2 at x. The set $U_x := \{y \in K \mid x'_x(y) > 1\}$ is a neighbourhood of x in K. But K is compact and thus there exists a finite subset A of

K such that

$$K \subset \bigcup_{x \in A} U_x.$$

The linear form $\sum_{x \in A} x'_x$ has the required properties. $\square$

Proposition 11.3.4. *Let A, B be two sections in C such that A is measurable and $B \subset \tilde{A}$. Let μ be a measure on B such that the restriction to B of any continuous linear form is μ-integrable. Then there exists a unique measure ν on A such that:*

a) for any positive function f on A whose restriction to any compact set of A is a Borel function, we have

$$\int^{\bullet} f \, d\nu = \int^{\bullet} f\left(\frac{1}{\rho_A(x)} x\right) \rho_A(x) \, d\mu(x);$$

b) the restriction to A of every continuous linear form on E is ν-integrable;

c) the barycenters of μ and ν coincide.

The unicity follows immediately from *a)*.

Let L be a compact set of A. For any numerical function f on L, we denote by $\bar{f}$ the function on A which is equal to f on L and equal to 0 on $A \setminus L$. By the preceding proposition, there exists a continuous linear form x'_0 on E, greater than 1 on L. For any $x \in B$ and any numerical function f on L, we have

$$\left|\bar{f}\left(\frac{1}{\rho_A(x)} x\right)\right| \rho_A(x) \leq \left|\bar{f}\left(\frac{1}{\rho_A(x)} x\right)\right| x'_0(x).$$

Since for any $f \in \mathscr{C}(L)$, the function on B

$$x \mapsto \bar{f}\left(\frac{1}{\rho_A(x)} x\right) \rho_A(x)$$

is μ-measurable, it is also μ-integrable and the map

$$f \mapsto \int \bar{f}\left(\frac{1}{\rho_A(x)} x\right) \rho_A(x) \, d\mu(x): \ \mathscr{C}(L) \to \mathbb{R}$$

defines a measure ν_L on L. It is immediate that for every bounded Borel function f on L, we have

$$\int f \, d\nu_L = \int \bar{f}\left(\frac{1}{\rho_A(x)} x\right) \rho_A(x) \, d\mu(x).$$

Hence if L, L' are two compact sets of A with $L \subset L'$, then the restriction of $\nu_{L'}$ to L is equal to ν_L. It follows that the family $(\nu_L)_L$ defines a measure ν on A.

a) For any compact set L of A, we have

$$\overset{*}{\int} f|_L \, d\nu_L = \overset{*}{\int} \overline{f|_L}\left(\frac{1}{\rho_A(x)}x\right) \rho_A(x)\, d\mu(x) \leq \overset{\bullet}{\int} f\left(\frac{1}{\rho_A(x)}x\right) \rho_A(x)\, d\mu(x).$$

Since L is arbitrary, we obtain

$$\overset{\bullet}{\int} f\, d\nu \leq \overset{\bullet}{\int} f\left(\frac{1}{\rho_A(x)}\right) \rho_A(x)\, d\mu(x).$$

In order to prove the converse inequality, we denote by $\mathfrak{K}$, the set of compact subsets K of B such that the restriction of ρ_A to K is continuous. Since ρ_A is measurable,

$$\overset{\bullet}{\int} f\left(\frac{1}{\rho_A(x)}x\right) \rho_A(x)\, d\mu(x) = \sup_{K \in \mathfrak{K}} \overset{*}{\int} f\left(\frac{1}{\rho_A(x)}x\right) \rho_A(x)|_K \, d\mu_K(x).$$

Let $K \in \mathfrak{K}$ and let L be its image with respect to the map

$$x \mapsto \frac{1}{\rho_A(x)}x : K \to A.$$

Since the map is continuous, L is a compact set of A. We have

$$\begin{aligned}\overset{\bullet}{\int} f\, d\nu \geqq \overset{*}{\int} f|_L \, d\nu_L &= \overset{\bullet}{\int} \overline{f|_L}\left(\frac{1}{\rho_A(x)}x\right) \rho_A(x)\, d\mu(x) \\ &= \overset{*}{\int} f\left(\frac{1}{\rho_A(x)}x\right) \rho_A(x)|_K \, d\mu_K(x).\end{aligned}$$

Since K is arbitrary we get

$$\overset{\bullet}{\int} f\, d\nu \geqq \overset{\bullet}{\int} f\left(\frac{1}{\rho_A(x)}x\right) \rho_A(x)\, d\mu(x).$$

b & c). Let x' be a continuous linear form on E. We set

$$f := \sup(x', 0).$$

By *a)* we have

$$\overset{\smile}{\int} f|_A \, d\nu = \overset{\bullet}{\int} f\left(\frac{1}{\rho_A(x)}x\right) \rho_A(x)\, d\mu(x) = \int f|_B \, d\mu < \infty.$$

Hence $f|_A$ is integrable. A similar result may be proved for $\inf(x', 0)$. We deduce that $x'|_A$ is ν-integrable and

$$\int x'|_A \, d\nu = \int x'|_B \, d\mu. \quad \square$$

A *cap* of C is a compact convex set K of C such that $C \setminus K$ is convex. A point x of a cap K is called an *extremal point* if for any $y, z \in K$ and

for any $\alpha \in]0, 1[$ with
$$x = \alpha y + (1-\alpha) z$$
we have $y = z = x$. Every extremal point of a cap which is different from 0, lies on an extremal ray of C (N. Bourbaki [3], Ch. II, § 7, Corollaire 1 de la Proposition 4). The set A of non-zero extremal points of a cap is a section in C for which ρ_A is lower semi-continuous (N. Bourbaki [3], Ch. II, § 7, Proposition 4); in particular A is a measurable section in C.

Corollary 11.3.1. *Let A be a measurable section in C and K be a metrisable cap of C such that every non-zero extremal point of K belongs to $\tilde{A}$. Then any point of $\tilde{K}$ is the barycenter of a measure on A.*

Let B be the set of non-zero extremal points of K and let $x \in \tilde{K}$. By N. Bourbaki [5], Ch. IV, § 7, Théorème 1, B is a G_δ-set and there exists a measure μ' on K such that
$$\mu'(K \setminus B) = 0, \qquad x_{\mu'} = x.$$
For any compact set L of B, we denote by μ_L the restriction of μ' to L. The family $(\mu_L)_L$ defines a measure on B whose barycenter is equal to x. The corollary now follows from the proposition since B is a section. ▯

Theorem 11.3.1. *Let A be a measurable section in C such that $\tilde{A} = C_e$. If C endowed with the weak topology is metrisable and complete, then any point of C is the barycenter of a measure on A.*

By N. Bourbaki [3], Ch. II, § 7, Proposition 5, any point of C belongs to a cap of C which is obviously metrisable. ▯

Remark. This theorem still holds under the following weaker conditions about C: *a)* C is metrisable; *b)* the origin has a countable fundamental system of weak neighbourhoods in C; *c)* every continuous linear form on E is the difference of two continuous linear forms on E which are positive on C.

Proposition 11.3.5. *Let K be a cap of C and let $x, y \in C$. If $x + y \in K$, then $x, y \in K$.*

We may assume $x \neq 0$ and $y \neq 0$. Since K is compact, there exists $\alpha \in]0, 1[$ such that $\frac{1}{\alpha} y \in C \setminus K$. From
$$\alpha\left(\frac{1}{\alpha} y\right) + (1-\alpha)\left(\frac{1}{1-\alpha} x\right) = x + y \in K$$
it follows that $\frac{1}{1-\alpha} x \in K$. Hence $x \in K$. ▯

Proposition 11.3.6. *If the closure of the union of an upper directed family of caps of C is compact, then it is a cap.*

Let $(K_\iota)_{\iota \in I}$ be an upper directed family of caps of C such that the closure K of its union is compact. Since the family $(K_\iota)_{\iota \in I}$ is upper directed, K is convex. Hence, to show that K is a cap, we have only to prove that $C \setminus K$ is convex. Let $x, y \in C \setminus K$ and let $\alpha \in [0, 1]$. Let V be a convex neighbourhood of the origin of E such that

$$(x+V) \cap K = (y+V) \cap K = \emptyset .$$

Let z be a point of $\alpha x + (1-\alpha) y + V$. Then there exists $v \in V$ such that

$$z = \alpha(x+v) + (1-\alpha)(y+v).$$

Since for any $\iota \in I$, $C \setminus K_\iota$ is convex, it follows that $z \in C \setminus K_\iota$. Hence

$$\Big(\bigcup_{\iota \in I} K_\iota\Big) \cap (\alpha x + (1-\alpha) y + V) = \emptyset ,$$

$$\alpha x + (1-\alpha) y \in C \setminus K . \quad \square$$

Corollary 11.3.2. *Let K be a compact section in C. If C is a lattice with respect to its own order, the closed convex hull of $K \cup \{0\}$ is a cap.*

We remark first that the closed convex hull of $K \cup \{0\}$, denoted by L, is bounded, and therefore precompact.

Assume first that K is finite. In order to show that L is a cap, we only have to prove that $C \setminus L$ is convex. Let $x, y \in C \setminus L$ and let $\alpha \in [0, 1]$. Assume that

$$\alpha x + (1-\alpha) y \in L .$$

Then there exists a family $(\alpha_z)_{z \in K}$ of positive real numbers such that

$$\sum_{z \in K} \alpha_z \leq 1 , \qquad \alpha x + (1-\alpha) y = \sum_{z \in K} \alpha_z z .$$

Let $(x_z)_{z \in K}$, $(y_z)_{z \in K}$ be two families in C such that

$$\alpha x = \sum_{z \in K} x_z , \qquad (1-\alpha) y = \sum_{z \in K} y_z ,$$

$$z \in K \Rightarrow \alpha_z z = x_z + y_z$$

(Proposition 8.1.1 *d)*). Let $z \in K$. Since z lies on an extremal ray of C, it follows from the last relation that there exists $\beta_z \in [0, \alpha_z]$ such that

$$x_z = \beta_z z , \qquad y_z = (\alpha_z - \beta_z) z .$$

We get

$$\alpha x = \sum_{z \in K} \beta_z z , \qquad (1-\alpha) y = \sum_{z \in K} (\alpha_z - \beta_z) z .$$

Since x, y do not belong to L, it follows

$$\sum_{z \in K} \beta_z > \alpha, \qquad \sum_{z \in K} (\alpha_z - \beta_z) > 1 - \alpha.$$

This leads to the contradictory relation

$$\sum_{z \in K} \alpha_z > 1.$$

Let $\mathfrak{A}$ be the set of finite subsets of K; for any $A \in \mathfrak{A}$, let A' be the closed convex hull of $A \cup \{0\}$. The family $(A')_{A \in \mathfrak{A}}$ is an upper directed family of caps and the closed convex hull of $K \cup \{0\}$ is the closure of its union. The corollary now follows immediately from the proposition. □

Proposition 11.3.7. *Let K be a compact section in C and let μ, ν be two measures on K such that*

$$x_\mu = x_\nu.$$

If C is a lattice with respect to its own order, then $\mu = \nu$.

Let L be the closed convex hull of K. By Proposition 11.3.3, $0 \notin L$ and by N. Bourbaki [3], Ch. II, § 7, Proposition 5, there exists a closed hyperplane H of E such that: *a)* $0 \notin H$; *b)* H intersects any ray of $\tilde{L}$; *c)* $H \cap \tilde{L}$ is compact. Hence $A := H \cap \tilde{K}$ is a compact section of C such that $K \subset \tilde{A}$. By Proposition 11.3.4, there exists a unique measure μ' (resp. ν') on A such that

$$x_{\mu'} = x_\mu \qquad (\text{resp. } x_{\nu'} = x_\nu).$$

By N. Bourbaki [5], Ch. IV, Corollaire du Théorème 3, $\mu' = \nu'$. Let $f \in \mathscr{C}(K)$. We denote by g, the function on A

$$x \mapsto f\left(\frac{1}{\rho_K(x)} x\right) \rho_K(x).$$

We have for any $x \in K$,

$$f(x) = g\left(\frac{1}{\rho_A(x)} x\right) \rho_A(x).$$

Hence by Proposition 11.3.4,

$$\mu(f) = \mu'(g) = \nu'(g) = \nu(f).$$

Since f is arbitrary, it follows that $\mu = \nu$. □

Theorem 11.3.2. *Let A be a section in C and let $\mathfrak{M}_A$ be the set of measures on A for which the restriction to A of any continuous linear form is integrable.*

a) $\mathfrak{M}_A$ is a convex cone.

b) If μ, ν are measures on A such that $\mu \preccurlyeq \nu$ and $\nu \in \mathfrak{M}_A$, then $\mu \in \mathfrak{M}_A$.

c) If C is a lattice with respect to its own order, then the map

$$\mu \mapsto x_\mu : \mathfrak{M}_A \to \mathscr{S}$$

is an additive and positively homogeneous injection which preserves the lattice operations.

a) is trivial.

b) Let x' be a continuous linear form on E. Then $|x'|_A|$ is ν-integrable and therefore μ-integrable. Hence $\mu \in \mathfrak{M}_A$.

c) Let $\mu, \nu \in \mathfrak{M}_A$. We set

$$\mu' := \mu - \mu \curlywedge \nu, \qquad \nu' := \nu - \mu \curlywedge \nu.$$

$$\mu' \curlywedge \nu' = 0.$$

Let K be a compact set of A. We denote by L, the closed convex hull of $K \cup \{0\}$ and set

$$y = x_{\mu'_K} \curlywedge x_{\nu'_K}.$$

By Corollary 11.3.2, Proposition 11.3.5 and N. Bourbaki [5], Ch. IV, §7, Proposition 1, there exist three measures λ, μ'', ν'' on K such that

$$x_\lambda = y, \qquad x_{\mu''} = x_{\mu'_K} - y, \qquad x_{\nu''} = x_{\nu'_K} - y.$$

Since

$$x_{\lambda + \mu''} = x_{\mu'_K}, \qquad x_{\lambda + \nu''} = x_{\nu'_K}$$

we have the preceding proposition,

$$\mu'_K = \lambda + \mu'', \qquad \nu'_K = \lambda + \nu'',$$

$$\lambda \preccurlyeq \mu'_K \curlywedge \nu'_K = 0, \qquad x_{\mu'_K} \curlywedge x_{\nu'_K} = 0.$$

Since K is arbitrary, we get by Proposition 11.3.2

$$x_{\mu'} \curlywedge x_{\nu'} = 0.$$

Hence

$$x_\mu \curlywedge x_\nu = (x_{\mu'} + x_{\mu \curlywedge \nu}) \curlywedge (x_{\nu'} + x_{\mu \curlywedge \nu}) = x_{\mu'} \curlywedge x_{\nu'} + x_{\mu \curlywedge \nu} = x_{\mu \curlywedge \nu}.$$

Choose $\mu, \nu \in \mathfrak{M}_A$ such that

$$x_\mu = x_\nu.$$

From

$$x_\mu = x_\mu \curlywedge x_\nu = x_{\mu \curlywedge \nu}$$

we get

$$x_{(\mu - \mu \curlywedge \nu)} = 0, \qquad \mu - \mu \curlywedge \nu = 0, \qquad \mu \preccurlyeq \nu.$$

Similarly, we have

$$\nu \preccurlyeq \mu$$

and therefore

$$\mu = \nu.$$

Hence the map of the theorem is an injection. ▯

Corollary 11.3.3. *Let A be a measurable section in C such that $\tilde{A} = C_e$ and let $\mathfrak{M}_A$ be the set of measures on A for which the restriction to A of any continuous linear form on E is integrable. If C is reticulated with respect to its own order and if C, endowed with weak topology, is metrisable and complete, then the map*

$$\mu \mapsto x_\mu : \mathfrak{M}_A \to C$$

is an isomorphism of prevector lattices.

The assertion follows immediately from the theorem and from Corollary 11.3.1. ▯

§ 11.4. Riesz-Martin Kernels

Throughout this section, X will denote a $\mathfrak{P}$-harmonic space.

We endow $[\mathcal{S}]$ with the topology $\mathcal{T}$ defined in §11.2. Then $\mathcal{S}$ is a closed convex cone of $[\mathcal{S}]$ with vertex 0 such that $\mathcal{S} \cap (-\mathcal{S}) = \{0\}$ and such that every bounded subset of $\mathcal{S}$ is relatively compact (Corollary 11.2.3). This enables us to apply all of the notions and results from the preceding section to $[\mathcal{S}]$ and $\mathcal{S}$. We remark further that for any subset $\mathcal{A}$ of $\mathcal{S}$, if we endow $\mathcal{A}$ with the topology $\mathcal{T}$ or with the weak topology, we get the same measures on $\mathcal{A}$.

Proposition 11.4.1. *Let μ be a measure on a subset $\mathcal{A}$ of $\mathcal{S}$. The function u_μ on X*

$$x \mapsto \int^{\bullet} u(x)\, d\mu(u)$$

is hyperharmonic. The following assertions are equivalent:

a) u_μ is superharmonic;

b) the restriction to $\mathcal{A}$ of every continuous linear form on $[\mathcal{S}]$ is μ-integrable.

If those assertions hold, then u_μ is the barycenter of μ.

Let $\mathcal{K}$ be a compact set of $\mathcal{A}$. By Corollary 11.2.3, the closed convex hull of $\mathcal{K}$ is compact. By N. Bourbaki [5], Ch. IV, Corollaire de la Proposition 1, μ has a barycenter $u_{\mu_{\mathcal{K}}}$. Let $x \in X$. For any $(V, f) \in I_x$, the map

$$u \mapsto A_{(V,f)} u(x)$$

is the restriction to $\mathscr{S}$ of a continuous linear form on $[\mathscr{S}]$ (Proposition 11.2.4). We have by Proposition 11.2.2 *b)* and *c)*

$$\begin{aligned} u_{\mu_{\mathscr{K}}}(x) &= \sup_{(V,f)\in I_x} A_{(V,f)} u_{\mu_{\mathscr{K}}}(x) = \sup_{(V,f)\in I_x} \int_{\mathscr{K}} A_{(V,f)} u(x)\, d\mu_{\mathscr{K}}(u) \\ &= \int_{\mathscr{K}} \sup_{(V,f)\in I_x} A_{(V,f)} u(x)\, d\mu_{\mathscr{K}}(u) = \int_{\mathscr{K}} u(x)\, d\mu_{\mathscr{K}}(u). \end{aligned}$$

Let $\mathfrak{K}$ be the set of compact sets of $\mathscr{A}$. Then $(u_{\mu_{\mathscr{K}}})_{\mathscr{K}\in\mathfrak{K}}$ is a specifically upper directed family of superharmonic functions and u_μ is obviously its least upper bound. Hence u_μ is hyperharmonic and for every positive continuous linear form on $[\mathscr{S}]$, we have

$$\dot{\int} l|_{\mathscr{A}}\, d\mu = \sup_{\mathscr{K}\in\mathfrak{K}} \int l|_{\mathscr{K}}\, d\mu_{\mathscr{K}} = \sup_{\mathscr{K}\in\mathfrak{K}} l(u_{\mu_{\mathscr{K}}}).$$

$a \Rightarrow b$ follows from this relation since

$$l(u_{\mu_{\mathscr{K}}}) \leq l(u_\mu) < \infty$$

for any $\mathscr{K}\in\mathfrak{K}$. $b \Rightarrow a$ as well as the last assertion follow from the same relation and from Proposition 11.3.2. □

A non-identically 0, positive superharmonic function on X is called *extremal* if its positive specific minorants are proportional. A positive superharmonic function is extremal if and only if it lies on an extremal ray of the convex cone $\mathscr{S}$. We denote by $\mathscr{S}_e$ the set of extremal superharmonic functions on X.

Proposition 11.4.2. *Let $(X_\iota)_{\iota\in I}$ be the family of connected components of X. For any $\iota\in I$, let $\mathscr{S}_{\iota e}$ be the set of extremal superharmonic functions on X which vanish identically on $X \setminus X_\iota$. Then $(\mathscr{S}_{\iota e})_{\iota\in I}$ is a family of pairwise disjoint, open and closed sets of $\mathscr{S}_e$ whose union is equal to $\mathscr{S}_e$.*

Let $u\in\mathscr{S}_e$. For any $\iota\in I$, we denote by u_ι, the superharmonic function on X which is equal to u on X_ι and equal to 0 elsewhere. Then

$$u = \sum_{\iota\in I} u_\iota .$$

Since u is extremal, any u_ι is proportional to u. Hence there exists $\iota\in I$ such that $u_\iota = u$ and $u_\kappa = 0$ for all $\kappa\in I$, $\kappa \neq \iota$. This shows that $(\mathscr{S}_{\iota e})_{\iota\in I}$ is a family of pairwise disjoint subsets of $\mathscr{S}_e$ whose union is equal to $\mathscr{S}_e$.

Let $\iota\in I$ and $u\in\mathscr{S}_{\iota e}$. There exists $f\in\mathscr{C}_+(X_0)$ and a compact subset K of $X \setminus \operatorname{Supp} f$ such that

$$\sup_K f \cdot u > 1 .$$

Then for any $\kappa \in I$, $\kappa \neq \iota$:

$$(u+\mathscr{V}(f,K)) \cap \mathscr{S}_\kappa = \emptyset .$$

Hence $\mathscr{S}_{\iota e}$ is open and since ι is arbitrary, $\mathscr{S}_{\iota e}$ is also closed. □

Proposition 11.4.3. *For every extremal superharmonic function u on X, $S_0(u)$ is a one point set. If we denote this point by $\pi(u)$, then π is a continuous map from $\mathscr{S}_e$ into X_0. If $\pi(u) \in X$ (resp. $\pi(u) \notin X$), then u is a potential (resp. a positive harmonic function).*

Assume that u is an extremal superharmonic function and that $S_0(u)$ is not equal to a point. Then there exist two functions $f, g \in \mathscr{C}_+(X_0)$ such that $\inf(f,g)=0$ and $f \cdot u$, $g \cdot u$ are not identically 0. Since u is extremal, $f \cdot u$ and $g \cdot u$ are proportional and this contradicts the relation

$$(f \cdot u) \wedge (g \cdot u) = \inf(f,g) \cdot u = 0$$

(Theorem 8.1.3 and Proposition 8.2.1).

Let now $u \in \mathscr{S}_e$ and let U be a neighbourhood of $\pi(u)$. Let $f \in \mathscr{C}_+(X_0)$ such that

$$\operatorname{Supp} f \subset U, \quad f(\pi(u)) = 1,$$

and such that there exists $x \in X \setminus \operatorname{Supp} f$ at which u is strictly positive. Let α be a real number such that $\alpha u(x) > 1$ and let

$$v \in \mathscr{S}_e \cap (u + \mathscr{V}(\alpha f, \{x\})).$$

If $\pi(v) \notin U$, then

$$f \cdot v(x) = 0$$

which contradicts the relation

$$|\alpha f \cdot v(x) - \alpha f \cdot u(x)| = \alpha u(x) \leq 1 .$$

The last assertion is trivial. □

For any extremal superharmonic function u, the point $\pi(u)$ is called the *pole of u*. We denote by $\mathscr{P}_e$ (resp. $\mathscr{H}_e$) the set of extremal superharmonic functions which are potentials (resp. harmonic functions). By $\sim$ we denote the equivalence relation on $\mathscr{S}_e$

$$u \sim v :\Leftrightarrow (\exists \alpha)(\alpha \in \mathbb{R}_+,\ u = \alpha v).$$

The quotient topological space of $\mathscr{S}_e$ (resp. $\mathscr{P}_e$, $\mathscr{H}_e$) with respect to this equivalence relation is called the *Martin* (resp. *Riesz*, *Poisson*) space of X and will be denoted by M_X (resp. R_X, P_X). We also denote by φ the canonical map $\mathscr{S}_e \to M_X$. Since the saturated set for $\sim$ of any open set U of $\mathscr{S}_e$ is the open set

$$\bigcup_{\substack{\alpha \in \mathbb{R}_+ \\ \alpha \neq 0}} \alpha U,$$

it follows that φ is open. We identify R_X and P_X with subsets of M_X. Obviously

$$R_X \cup P_X = M_X, \qquad R_X \cap P_X = \varnothing.$$

By the preceding proposition, there exists a continuous map π' of M_X into X_0 such that

$$\pi = \pi' \circ \varphi.$$

We have

$$R_X = \overset{-1}{\pi'}(X), \qquad P_X = \overset{-1}{\pi'}(X_0 \setminus X).$$

Hence R_X is open and P_X closed. We do not know whether or not R_X is dense in M_X.

Proposition 11.4.4. *Let $(X_\iota)_{\iota \in I}$ be the family of connected components of X. With the notation of Proposition 11.4.2, $(\varphi(\mathcal{S}_{\iota e}))_{\iota \in I}$ is a family of pairwise disjoint open and closed sets of M_X whose union is equal to M_X.*

The proposition follows immediately from Proposition 11.4.2 and from the fact that for any $\iota \in I$,

$$\overset{-1}{\varphi}(\varphi(\mathcal{S}_{\iota e})) = \mathcal{S}_{\iota e}. \quad \square$$

Proposition 11.4.5. *The Martin space of X is a Hausdorff space. If X has a countable base, then the Martin space of X has a countable base; hence if the Martin space of X is completely regular, it is metrisable.*

Let ξ, η be two different points of M_X and let $u \in \overset{-1}{\pi}(\xi)$, $v \in \overset{-1}{\pi}(\eta)$. Assume that there exist no disjoint neighbourhoods of ξ and η. Then for any neighbourhood $\mathcal{V}$ of the origin in $[\mathcal{S}]$, there exists a point $\zeta_{\mathcal{V}}$ of M_X belonging to

$$\pi(\mathcal{S}_e \cap (u + \mathcal{V})) \cap \pi(\mathcal{S}_e \cap (v + \mathcal{V})).$$

Let $u_{\mathcal{V}}$ (resp. $v_{\mathcal{V}}$) be a point of

$$\mathcal{S}_e \cap (u + \mathcal{V}) \cap \overset{-1}{\pi}(\zeta_{\mathcal{V}}) \qquad (\text{resp. } \mathcal{S}_e \cap (u + \mathcal{V}) \cap \overset{-1}{\pi}(\zeta_{\mathcal{V}})).$$

Since $u_{\mathcal{V}}$ and $v_{\mathcal{V}}$ are proportional and converge to u and v respectively as $\mathcal{V}$ runs through the set of neighbourhoods of the origin, it follows that u and v are proportional which is a contradiction.

The last assertion follows immediately from Proposition 11.2.10 and from the fact that every completely regular space which has a countable base is metrisable. $\square$

Remark. There are examples in which X has a countable base but the Martin space of X is not regular (Exercise 11.4.11).

Let $\mathcal{A}$ be a section in $\mathcal{S}$. The subset $\varphi(\mathcal{A})$ of the Martin space M_X will be called *the projection of the section $\mathcal{A}$*. The map $\mathcal{A} \to \varphi(\mathcal{A})$

defined by φ is obviously a one-to-one map. We denote by $\psi_{\mathscr{A}}$, its inverse from $\varphi(\mathscr{A})$ into $\mathscr{A}$. Obviously

$$\overset{-1}{\varphi}(\varphi(\mathscr{A}))=\tilde{\mathscr{A}}$$

and for any $u\in\tilde{\mathscr{A}}$

$$\psi_{\mathscr{A}}(\varphi(u))=\frac{1}{\rho_{\mathscr{A}}(u)}\,u.$$

From this relation it follows that $\rho_{\mathscr{A}}$ is continuous if and only if $\psi_{\mathscr{A}}$ is continuous.

Let A be a subset of the Martin space of X. A *Riesz-Martin kernel on A* is a function

$$k\colon\ X\times A\to[0,\infty]$$

such that for any $\xi\in A$, the function k_ξ on X

$$x\mapsto k(x,\xi)$$

is an extremal superharmonic function on X whose pole is $\pi'(\xi)$ (or equivalently $k_\xi\in\overset{-1}{\pi}(\xi)$). The kernel is called *regular* if it is lower semi-continuous and if its restriction to $\{(x,\xi)\in X\times A\,|\,x\neq\pi'(\xi)\}$ is continuous. The kernel is called *measurable* if its restriction to any compact set of $X\times A$ is universally measurable. The kernel is called *semi-regular* if it is measurable and if, for any compact set K of A, the origin of $\mathscr{S}$ does not belong to the closure of the set $\{k_\xi\,|\,\xi\in K\}$. Every regular kernel is obviously semi-regular. The subset of $\mathscr{S}_e$

$$\{k_\xi\,|\,\xi\in A\}$$

is a section in $\mathscr{S}$ whose projection is equal to A. We call this section *the section generated by the Riesz-Martin kernel k and denote it by $\mathscr{A}_k$. We shall write simply ψ_k and ρ_k instead of $\psi_{\mathscr{A}_k}$ and $\rho_{\mathscr{A}_k}$* respectively. Obviously

$$\psi_k(\xi)=k_\xi$$

for any $\xi\in A$. Conversely, any section $\mathscr{A}$ in $\mathscr{S}$ is generated by the kernel defined on its projection by

$$(x,\xi)\mapsto(\psi_{\mathscr{A}}(\xi))(x)\colon\ X\times\varphi(\mathscr{A})\to[0,\infty].$$

Proposition 11.4.6. *Let k be a Riesz-Martin kernel defined on a subset A of the Martin space of X. The following assertions are equivalent:*

a) k is regular;

b) the restriction of k to

$$\{(x,\xi)\in X\times A\,|\,x\neq\pi'(\xi)\}$$

is continuous;

c) ψ_k *is continuous;*

d) *the section* $\mathscr{A}_k$ *is continuous.*

$a \Rightarrow b$ is trivial.

$b \Rightarrow c$. Let $\xi \in A$, let $f \in \mathscr{C}_+(X_0)$ and let K be a compact subset of $X \setminus \operatorname{Supp} f$. We have to find a neighbourhood V of ξ such that for any $\eta \in V$

$$|f \cdot \psi_k(\eta) - f \cdot \psi_k(\xi)| \leq 1$$

on K. The existence of V is trivial if $\pi'(\xi) \notin \operatorname{Supp} f$, so that we may assume $\pi'(\xi) \in \operatorname{Supp} f$. This implies $\pi'(\xi) \notin K$. Hence for any $x \in K$, there exists a neighbourhood U_x of x and a neighbourhood V_x of ξ such that for any $(y, \eta) \in U_x \times V_x$ we have

$$|k(y, \eta) - k(x, \xi)| \leq \frac{1}{4 \sup_{X_0} f}.$$

Since K is compact there exists a finite subset B of K such that

$$K \subset \bigcup_{x \in B} U_x.$$

We set

$$V := \Big(\bigcap_{x \in B} V_x\Big) \cap \left\{\eta \in A \,\middle|\, |f(\pi'(\eta)) - f(\pi'(\xi))| \leq \frac{1}{2 \sup_{y \in K} k(y, \xi)}\right\}.$$

For any $\eta \in V$ and any $y \in K$ we have

$$\begin{aligned} |(f \cdot \psi(\eta))(y) - (f \cdot \psi(\xi))(y)| &= |f(\pi'(\eta))\, k(y, \eta) - f(\pi'(\xi))\, k(y, \xi)| \\ &\leq \sup_{X_0} f\, |k(y, \eta) - k(y, \xi)| \\ &\quad + |f(\pi'(\eta)) - f(\pi'(\xi))| \sup_{y \in K} k(y, \xi) \leq 1. \end{aligned}$$

$c \Leftrightarrow d$ is trivial.

$c \Rightarrow a$. Let us denote by τ, the map

$$(x, u) \mapsto u(x) \colon X \times \mathscr{S} \to [0, \infty]$$

and by ψ, the map

$$(x, \xi) \mapsto (x, \psi_k(\xi)) \colon X \times A \to X \times \mathscr{S}.$$

Obviously

$$k = \tau \circ \psi.$$

The assertion now follows from Corollary 11.2.1 and Proposition 11.2.5 *b)*. □

An important question is how to find a regular Riesz-Martin kernel on the whole Martin space or at least on the Riesz or Poisson spaces. Such kernels do not always exist even if π' is a one-to-one map of the

Riesz-Martin space onto X_0 (Exercise 11.4.11). We give now some criteria (Proposition 11.4.7, Proposition 11.4.8, Corollary 11.4.1, Corollary 11.4.2) for the existence of regular kernels.

Proposition 11.4.7. *If a subset A of the Martin space of X is paracompact with respect to the induced topology, then there exists a regular Riesz-Martin kernel on it. In particular, this is the case for a compact or a metrisable subset of the Martin space.*

Let $\mathscr{L}$ be the set of positive continuous linear forms on $[\mathscr{S}]$. For any $l \in \mathscr{L}$, the set

$$\mathscr{S}_l := \{u \in \mathscr{S}_e \mid l(u) > 0\}$$

is open in $\mathscr{S}_e$. Hence $(\varphi(\mathscr{S}_l))_{l \in \mathscr{L}}$ is an open covering of A. Since A is paracompact, there exists a continuous partition of unity $(f_l)_{l \in \mathscr{L}}$ subordinate to this covering (N. Bourbaki [1], Corollaire de la Proposition 4). The function on $\tilde{A}$

$$\rho := \sum_{l \in \mathscr{L}} (f_l \circ \varphi)\, l$$

is continuous and strictly positive and for any $\alpha \in]0, \infty[$, $u \in \overset{-1}{\varphi}(A)$, we have

$$\rho(\alpha u) = \alpha \rho(u).$$

Hence

$$\{u \in \overset{-1}{\varphi}(A) \mid \rho(u) = 1\}$$

is continuous section in $\mathscr{S}$ whose projection is equal to A. The assertion now follows from the preceding proposition, $d \Rightarrow a$.

The last assertion follows the fact that compact and metrisable spaces are paracompact. □

Remark. If X has a countable base, it follows from this proposition, Proposition 11.4.6 $a \Rightarrow c$, and Proposition 11.2.10 that the existence of a regular kernel on a subset of the Riesz-Martin space of X is equivalent with the metrisability of this set.

Proposition 11.4.8. *Let A be a subset of M_X for which there exists a set $\mathfrak{W}$, of open sets of X_0, such that:*

a) $\pi'(A) \subset \bigcup_{W \in \mathfrak{W}} W$;

b) for any $W \in \mathfrak{W}$, every extremal superharmonic function of $\overset{-1}{\varphi}(A)$ whose pole lies in W does not vanish identically on $X \setminus W$.

If X is metrisable, then there exists a regular Riesz-Martin kernel on A.

We set

$$Y := \bigcup_{W \in \mathfrak{W}} W.$$

For any $x \in Y$, let W_x be an open neighbourhood of x such that $X \cap \partial W_x$ is compact and such that there exists $W \in \mathfrak{W}$ with the property that $\overline{W}_x \subset W$. Let $(g_x)_{x \in Y}$ be a continuous partition of unity on the topological space Y subordinate to the open covering $(W_x)_{x \in Y}$ of Y (N. Bourbaki [1], § 4, Corollaire de la Proposition 4 et Théorème 4). For any $x \in Y$, we denote by f_x the function on X_0 which is equal to g_x on Y and equal to 0 on $X_0 \setminus Y$. Obviously $f_x \in \mathscr{C}_+(X_0)$.

For any $x \in Y$, we denote by ρ_x, the function

$$u \mapsto \sup_{\partial W_x}(f_x \cdot u) \colon \overset{-1}{\varphi}(A) \to \mathbb{R}_+ .$$

It is obvious that ρ_x is continuous and that for any $u \in \overset{-1}{\varphi}(A)$ and any $\alpha \in]0, \infty[$, we have

$$\rho_x(\alpha u) = \alpha \rho_x(u) .$$

We set

$$\rho := \sum_{x \in Y} \rho_x .$$

For any $u \in \overset{-1}{\varphi}(A)$ and any $\alpha \in]0, \infty[$, we have

$$\rho(\alpha u) = \alpha \rho(u) .$$

Let $u \in \overset{-1}{\varphi}(A)$. There exists $x \in Y$ such that

$$f_x(\pi(u)) > 0 .$$

Hence $\pi(u) \in W_x$ and by hypothesis

$$\sup_{\partial W_x} u > 0 .$$

Therefore

$$\rho(u) \geq \rho_x(u) = f_x(\pi(u)) \sup_{\partial W_x} u > 0 .$$

Since the set

$$Y' = \{y \in Y \mid (\operatorname{Supp} f_y) \cap W_x \neq \emptyset\}$$

is finite and since for any $v \in \overset{-1}{\pi}(W_x)$,

$$\rho(v) = \sum_{y \in Y'} \rho_y(v)$$

it follows that ρ is finite and continuous at u. The set $\{u \in \overset{-1}{\varphi}(A) \mid \rho(u) = 1\}$ is a continuous section in $\mathscr{S}$ whose projection is equal to A. The assertion now follows from Proposition 11.4.6 $d \Rightarrow a$. □

Corollary 11.4.1. *If the complementary set of any absorbent set of X is either empty or not relatively compact, then there exists a continuous Riesz-Martin kernel on R_X.*

We may take as $\mathfrak{W}$ in the proposition, the set of relatively compact sets of X which are different from X. □

Corollary 11.4.2. *If the harmonic space X is elliptic, then there exists a continuous Riesz-Martin kernel on M_X.*

By Proposition 11.4.4, we may assume that X connected. We may then take as $\mathfrak{W}$ in the proposition, the set $\{X_0 \setminus \{x\}, X_0 \setminus \{y\}\}$, where x, y are two different points of X. □

Proposition 11.4.9. *Let k be a Riesz-Martin kernel defined on a subset A of the Martin space of X. The following assertions are equivalent:*

a) k is measurable;

b) ψ_k is measurable;

c) the section $\mathscr{A}_k$ is measurable.

$a \Rightarrow b$. Let K be a compact set of A let μ be a measure on K and let l be a continuous linear form on $[\mathscr{S}]$. We want to show that the map

$$\xi \mapsto l(k_\xi): \ K \to \mathbb{R}$$

is μ-measurable. By Proposition 11.2.4, it is sufficient to prove this assertion for any map of the form

$$\xi \mapsto v(f \cdot k_\xi): \ K \to \mathbb{R},$$

where $f \in \mathscr{C}_+(X_0)$ and v is a measure on X whose carrier, L, is compact and contained in $X \setminus \operatorname{Supp} f$. Now we have

$$f \cdot k_\xi = f(\pi'(\xi))\, k_\xi$$

and therefore, by hypothesis,

$$(x, \xi) \mapsto (f \cdot k_\xi)(x): \ L \times K \to \mathbb{R}$$

is $v \otimes \mu$ measurable. By Fubini's theorem, the above map

$$\xi \mapsto v(f \cdot k_\xi): \ K \to \mathbb{R}$$

is μ-measurable.

Let k' be a regular kernel on K (Proposition 11.4.7). By Proposition 11.4.6, $\psi_{k'}$ is continuous and therefore $\mathscr{A}_{k'}$ is compact. Let l be a continuous linear form on $[\mathscr{S}]$ which is strictly positive on $\mathscr{A}_{k'}$ (Proposition 11.3.3). For any $\xi \in K$, we have

$$\psi_k(\xi) = k_\xi = \frac{l(k_\xi)}{l(k'_\xi)}\, k'_\xi = \frac{l(k_\xi)}{l(k'_\xi)}\, \psi_{k'}(\xi).$$

From this relation and using the first part of the proof, we see that the restriction of ψ_k to K is μ-measurable. Since K and μ are arbitrary, ψ_k is measurable.

$b \Rightarrow c$. Let $\mathscr{K}$ be a compact set of $\tilde{\mathscr{A}}_k$ and let l be a continuous linear form on $[\mathscr{S}]$ which is strictly positive on $\mathscr{K}$ (Proposition 11.3.3). From

$$\psi_k(\varphi(u)) = \frac{1}{\rho_k(u)} u$$

for all $u \in \mathscr{K}$, we get

$$\rho_k(u) = \frac{l(u)}{l(\psi_k(\varphi(u)))},$$

which proves that the restriction of ρ_k to $\mathscr{K}$ is universally measurable. The assertion follows now from the fact that $\mathscr{K}$ is arbitrary.

$c \Rightarrow a$. Assume that the section $\mathscr{A}_k$ is measurable. Let K (resp. L) be a compact set of A (resp. of X). By the preceding proposition there exists a regular Riesz-Martin kernel k' on K. The measurability of k follows from the relation

$$k_\xi(x) = \frac{k'_\xi(x)}{\rho_k(k'_\xi)},$$

which holds for all $(x, \xi) \in L \times K$. □

Corollary 11.4.3. *If X has a countable base, then there exists a semi-regular Riesz-Martin kernel on the Martin space of X.*

By Proposition 11.2.10, $\mathscr{S}$ is metrisable and by Proposition 11.3.1, there exists a measurable section $\mathscr{A}$ in $\mathscr{S}$ such that $\rho_{\mathscr{A}}$ is bounded on every compact set of $\mathscr{S}_e$. Let k be the kernel on M_X

$$(x, \xi) \mapsto (\psi_{\mathscr{A}}(\xi))(x).$$

By the proposition, k is measurable.

Let K be a compact set of M_X and let k' be a regular kernel on K (Proposition 11.4.7). For any $\xi \in K$, we have

$$k_\xi = \frac{1}{\rho_{\mathscr{A}}(k'_\xi)} k'_\xi.$$

Since $\rho_{\mathscr{A}}$ is bounded on $\mathscr{A}_{k'}$ it follows that 0 does not belong to the closure of $\{k_\xi \mid \xi \in K\}$. □

Proposition 11.4.10. *Let X, Y be two Hausdorff topological spaces, let μ be a measure on X and let $\varphi: X \to Y$ be a μ-measurable map such that for any compact set L of Y, the set $\overset{-1}{\varphi}(L)$ is μ-integrable. Then, there exists a unique measure $\varphi(\mu)$ on Y such that for any positive $\varphi(\mu)$-measurable numerical function f on Y, $f \circ \varphi$ is μ-measurable and*

$$\int^{\bullet} f \, d\varphi(\mu) = \int^{\bullet} f \circ \varphi \, d\mu.$$

Let L be a compact set of Y. For any numerical function f on L, we denote by $\bar{f}$, the function on Y which is equal to f on L and equal to 0 on $Y \setminus L$. For any $f \in \mathscr{C}(L)$, the function $\bar{f} \circ \varphi$ is μ-integrable and the map

$$f \mapsto \int \bar{f} \circ \varphi \, d\mu$$

defines a measure ν_L on L. It is immediate that if L and L' are two compact sets of Y with $L \subset L'$, then the restriction of $\nu_{L'}$ to L coincides with ν_L. Hence the family $(\nu_L)_L$ defines a measure on Y.

Let now f be a positive, ν-measurable, numerical function on Y. Let K be a compact set of X and let ε be a strictly positive real number. Since φ is μ-measurable, there exists a compact subset K' of K such that $\varphi|_{K'}$ is continuous and

$$\mu_K(K \setminus K') < \frac{\varepsilon}{2}.$$

Since f is ν-measurable, there exists a compact subset L of the compact set $\varphi(K')$ such that $f|_L$ is continuous and

$$\nu_{\varphi(K')}(\varphi(K') \setminus L) < \frac{\varepsilon}{2}.$$

Then $\overset{-1}{\varphi}(L) \cap K'$ is compact, $f \circ \varphi|_{\overset{-1}{\varphi}(L) \cap K'}$ is continuous and

$$\mu_K(K \setminus \overset{-1}{\varphi}(L) \cap K') < \varepsilon.$$

Hence $f \circ \varphi$ is μ-measurable. Obviously

$$\overset{\bullet}{\int} f \, d\nu \geq \overset{*}{\int} f \circ \varphi|_{K'} \, d\mu_{K'}.$$

Since ε and K are arbitrary,

$$\overset{\bullet}{\int} f \, d\nu \geq \overset{\bullet}{\int} f \circ \varphi \, d\mu.$$

Now let L be a compact set of Y and let ε be a strictly positive real number. There exists a compact subset L' of L such that $f|_{L'}$ is continuous and

$$\mu_L(L \setminus L') < \varepsilon.$$

Obviously

$$\overset{*}{\int} f|_{L'} \, d\nu_{L'} \leq \overset{\bullet}{\int} f \circ \varphi \, d\mu.$$

Since ε and L are arbitrary, we get

$$\overset{\bullet}{\int} f \, d\nu = \overset{\bullet}{\int} f \circ \varphi \, d\mu.$$

The unicity is trivial. □

Let k be a measurable Riesz-Martin kernel on a subset A of the Martin space of X. For any measure μ on A, we denote by k_μ, the function on X

$$x \mapsto \int k(x, \xi)\, d\mu(\xi).$$

Theorem 11.4.1. *Let k be a measurable Riesz-Martin kernel on a subset A of the Martin space of X.*

a) For any measure μ on A, we have

$$k_\mu = u_{\psi_k(\mu)}.$$

b) For any measure μ on $\mathscr{A}_k$ such that for every compact set K of A the set $\psi_k(K)$ is μ-integrable, we have

$$u_\mu = k_{\varphi_k(\mu)},$$

where $\varphi_k := \varphi\lfloor_{\mathscr{A}_k}$.

c) If μ, ν are two measures on A such that k_μ, k_ν are superharmonic and equal, then $\mu = \nu$.

d) Assume that k is semi-regular and let μ be a measure on $\mathscr{A}_k$ such that u_μ is superharmonic. Then, for any compact set K of A, the set $\psi_k(K)$ is μ-integrable.

Let $x \in X$. We denote by f (resp. g), the function on A (resp. $\mathscr{A}_k$)

$$\xi \mapsto k(x, \xi) \quad (\text{resp. } u \mapsto u(x)).$$

We have

$$f \circ \varphi_k = g, \qquad g \circ \psi_k = f.$$

The assertions *a), b)* now follow by applying the preceding proposition to the functions f and g and using Proposition 11.4.9 $a \Rightarrow b$.

c) By *a)* and Proposition 11.4.1, we deduce that the restriction to $\mathscr{A}_k$ of continuous linear forms on $[\mathscr{S}]$ are integrable with respect to either $\psi_k(\mu)$ or $\psi_k(\nu)$ and that their barycenters coincide. By Theorem 11.3.2,

$$\psi_k(\mu) = \psi_k(\nu).$$

Hence

$$\mu = \varphi_k(\psi_k(\mu)) = \varphi_k(\psi_k(\nu)) = \nu.$$

d) By Proposition 11.4.7, there exists a semi-regular Riesz-Martin kernel k' on K. By Proposition 11.4.6, $\psi_{k'}(K)$ is a compact set. Let l be a positive continuous linear form on $[\mathscr{S}]$ which is strictly positive on $\psi_{k'}(K)$ (Proposition 11.3.3). For any $\xi \in K$, we have

$$k_\xi = \frac{l(k_\xi)}{l(k'_\xi)}\, k'_\xi.$$

If

$$\inf_{\xi \in K} l(k_\xi) = 0,$$

then

$$0 \in \overline{\{k_\xi \mid \xi \in K\}}$$

and this contradicts the hypothesis that k is semi-regular. Hence

$$\alpha := \inf_{\xi \in K} l(k_\xi) > 0.$$

The assertion now follows from Proposition 11.4.1 $a \Rightarrow b$. □

Corollary 11.4.4. *Let k be a measurable Riesz-Martin kernel on a subset A of the Martin space of X and let $\mathfrak{M}_k$ be the set of measures μ on A such that k_μ is superharmonic.*

a) $\mathfrak{M}_k$ is a convex cone.

b) If μ, ν are measures on A such that $\mu \preccurlyeq \nu$, $\nu \in \mathfrak{M}_k$ then $\mu \in \mathfrak{M}_k$.

c) The map

$$\mu \mapsto k_\mu : \mathfrak{M}_k \to \mathscr{S}$$

is an additive and positively homogeneous injection which preserves the lattice operations. □

Proposition 11.4.11. *Let k be a measurable Riesz-Martin kernel on a subset A of the Martin space X. Let μ be a measure on A and let ν be a measure on X for which there exists a K_σ-set B of X such that $\nu(X \setminus B) = 0$. Then the function on A*

$$\xi \mapsto \int^{*} k(x, \xi)\, d\nu(x)$$

is μ-measurable and

$$\int^{\bullet} \left(\int^{*} k(x, \xi)\, d\nu(x)\right) d\mu(\xi) = \int^{*} k_\mu\, d\nu.$$

The measurability follows immediately from Fubini's theorem and the hypothesis about ν. Let K be a compact set of A. By Fubini's theorem, we have

$$\int^{*} \left(\int^{*} k(x, \xi)\, d\nu(x)\right) d\mu_K(\xi) = \int^{*} k_{\mu_K}\, d\nu.$$

The proposition now follows from the fact that $(k_{\mu_K})_K$ is an upper directed family of lower semi-continuous functions (Theorem 11.4.1, Proposition 11.4.1). □

Proposition 11.4.12. *Let k be a measurable Riesz-Martin kernel on a subset A of the Martin space of X and let μ be a measure on A.*

a) If k_μ is superharmonic, then for any $f \in \mathscr{C}_+(X_0)$ and any $x \in X$,

$$(f \cdot k_\mu)(x) = \int^{\bullet} f(\pi'(\xi))\, k(x, \xi)\, d\mu(\xi).$$

b) Let U be an open set of X. If k_μ is superharmonic, then it is harmonic on U if and only if $\mu(\overset{-1}{\pi'}(U)\cap A)=0$.

c) If k_μ is a superharmonic function, then it is harmonic (resp. a potential) if and only if $\mu(A\cap R_X)=0$ (resp. $\mu(A\cap P_X)=0$).

d) If k_μ is a potential, then for any positive, bounded, k_μ-measurable function f on X and any $x\in X$, we have

$$(f\cdot k_\mu)(x)=\int f_0(\pi'(x))\,k(x,\xi)\,d\mu(\xi),$$

where f_0 denotes the function on X_0 which is equal to f on X and equal to 0 at the Alexandroff point of X.

e) If X has a countable base, then for any subset B of X and any $x\in X$, the function on A

$$\xi\mapsto \hat{R}^B_{k_\xi}(x)$$

is μ-measurable and

$$\hat{R}^B_{k_\mu}(x)=\overset{\bullet}{\int}\hat{R}^B_{k_\xi}(x)\,d\mu(\xi).$$

a) For any $f\in\mathscr{C}_+(X_0)$, we denote by $l(f)$, the function on X

$$x\mapsto\overset{\bullet}{\int} f(\pi'(\xi))\,k(x,\xi)\,d\mu(\xi).$$

By hypothesis, $l(f)$ is a superharmonic function on X. Let V be an open set of X whose boundary is compact. For any $x\in V$, we have by Proposition 11.4.11,

$$\overset{*}{\int} l(f)\,d\mu_x^V=\overset{\bullet}{\int}\Big(\overset{*}{\int} k(y,\xi)\,d\mu_x^V(y)\Big)\,f(\pi'(\xi))\,d\mu(\xi).$$

If the closure of V in X_0 is contained in $X_0\setminus\operatorname{Supp} f$, then for any $\xi\in A$, we have

$$f(\pi'(\xi))\overset{*}{\int} k(y,\xi)\,d\mu_x^V(y)=f(\pi'(\xi))\,k(x,\xi)$$

and therefore

$$\mu^V(l(f))=l(f)$$

on V. Hence $l(f)$ is harmonic on $X\setminus\operatorname{Supp} f$ and

$$l(f)=\hat{R}^{X\setminus V}_{l(f)}$$

(Proposition 5.3.3, Theorem 1.2.1). We want to show that

$$S_0(l(f))\subset\operatorname{Supp} f.$$

By the above relation, this is obvious if the Alexandroff point of X_0 belongs to $\operatorname{Supp} f$. If this is not the case, let V be an open set of X such that $X \setminus V$ is a compact neighbourhood of Supp$f$. Since X is a $\mathfrak{P}$-harmonic space, $\hat{R}_{L(f)}^{X \setminus V}$ is a potential (Proposition 5.3.4) and this proves the required relation. The assertion now follows from Theorem 8.1.1 and Proposition 8.2.1.

b) Assume first that k_μ is harmonic on U. Let K be a compact set of $\overset{-1}{\pi'}(U) \cap A$ and let $f \in \mathscr{C}_+(X_0)$ with $f \geq 1$ on $\pi'(K)$ and Supp$f \subset U$. By Proposition 8.1.6 and Proposition 8.2.1, $f \cdot k_\mu = 0$. Hence by *a)*, we get for any $x \in X$

$$\int k(x, \xi)\, d\mu_K(\xi) \leq \int f(\pi'(\xi))\, k(x, \xi)\, d\mu(\xi) = 0.$$

Since x is arbitrary, $k_{\mu_K} = 0$ and therefore

$$\mu(K) = \mu_K(K) = 0$$

(Theorem 11.4.1 *c)*). We obtain immediately that $\mu(\overset{-1}{\pi'}(U) \cap A) = 0$.

Assume now that $\mu(\overset{-1}{\pi'}(U)) = 0$. Let $f \in \mathscr{C}_+(X_0)$, $f \leq 1$ on U and $f = 1$ on $X_0 \setminus U$. By *a)*, $f \cdot k_\mu = k_\mu$ and therefore

$$S_0(k_\mu) \subset \operatorname{Supp} f$$

(Proposition 8.1.6 and Proposition 8.2.1). Hence k_μ is harmonic on U since f is arbitrary.

c) The assertion that k_μ is harmonic if and only if $\mu(A \cap R_X) = 0$ follows immediately from *b)*.

Assume now that $\mu(A \cap P_X) = 0$. We set

$$\mathscr{F} := \{ f \in \mathscr{C}_+(X_0) \mid f \leq 1, \operatorname{Supp} f \subset X \}.$$

By *a)* it follows that

$$k_\mu := \bigvee_{f \in \mathscr{F}} f \cdot k_\mu .$$

The family $(f \cdot k_\mu)_{f \in \mathscr{F}}$, being a specifically upper directed family of potentials k_μ, is a potential.

Assume now that k_μ is a potential and let K be a compact subset of $A \cap P_X$. By the above considerations, k_{μ_K} is a harmonic function. Being a speicific minorant of k_μ, k_{μ_K} must vanish identically. Hence

$$\mu(K) = \mu_K(K) = 0$$

(Theorem 11.4.1 *c)*).

d) Choose $x \in X$ such that $k_\mu(x) < \infty$. We denote by τ the function on A

$$\xi \mapsto k_\xi(x).$$

For any compact set K of A, we denote by ν_K, the measure on K $\tau|_K \cdot \mu$. It is immediate that the family $(\nu_K)_K$ defines a measure on A. Since $\nu(A) < \infty$, the measure $\pi'(\nu)$ on X_0 exists (Proposition 11.4.10). By *a)*, for any $f \in \mathscr{K}_+(X)$

$$\mathsf{V}_{k_\mu, x}(f) = (\pi'(\nu))(f_0).$$

Hence $\mathsf{V}_{k_\mu, x}$ is equal to the restriction of $\pi'(\nu)$ to X. The assertion now follows from Proposition 11.4.10,

e) follows immediately from the preceding proposition and from Corollary 7.1.2. □

Proposition 11.4.13. *If X is a $\mathfrak{P}$-Brelot space, then for any $x \in X$, there exists an extremal superharmonic function on X whose pole is x. If any two extremal superharmonic functions on X having the same pole in X are proportional, then $\pi'|_{R_X}$ is a homeomorphism of the Riesz space of X onto X* (R.-M. Hervé 1962 [4]).

We may assume that X is connected. By Proposition 11.2.11, there exists a continuous linear form l on $[\mathscr{S}]$ which is strictly positive on $\mathscr{S} \setminus \{0\}$ and such that $\{u \in \mathscr{S} \mid l(u) = 1\}$ is compact.

Let $x \in X$. Then $\mathscr{K} := \{u \in \mathscr{S} \mid l(u) = 1, S_0(u) = \{x\}\}$ is a compact convex set of $\mathscr{S}$ and by Corollary 11.1.3, $\mathscr{K}$ is non-empty. By the Krein-Milman theorem (N. Bourbaki [3]), it possesses an extremal point u_0. Let $u, v \in \mathscr{S} \setminus \{0\}$

$$u + v := u_0.$$

Then $\frac{1}{l(u)} u$, $\frac{1}{l(v)} v \in \mathscr{K}$ and

$$u_0 = l(u)\left(\frac{1}{l(u)} u\right) + l(v)\left(\frac{1}{l(v)} v\right).$$

Hence

$$\frac{1}{l(u)} u = \frac{1}{l(v)} v = u_0$$

and u_0 is an extremal superharmonic function.

Assume now that any two extremal superharmonic functions on X having the same pole in X are proportional. In order to show that $\pi'|_{R_X}\colon R_X \to X$ is a homeomorphism, we only have to show that it is proper. Let $\mathfrak{U}$ be an ultrafilter on R_X whose image through π' converges to a point $x \in X$. For any $\xi \in R_X$, choose $u_\xi \in \mathscr{S}_e$ such that $l(u_\xi) = 1$. Since $\{u \in \mathscr{S} \mid l(u) = 1\}$ is compact, there exists $u_0 \in \mathscr{S}$, such that

$$\lim_{\xi, \mathfrak{U}} u_\xi = u_0.$$

Obviously

$$S(u_0) \subset \{x\}.$$

Since

$$l(u_0) = 1,$$

u_0 is an extremal superharmonic function whose pole is x. □

Exercises

11.4.1. Let $(X_\iota)_{\iota\in I}$ be a family of open and closed sets of X whose union is equal to X. For any $\iota \in I$, we set

$$\mathcal{S}_\iota := \{u \in \mathcal{S} \mid u|_{X \setminus X_\iota} = 0\}, \qquad [\mathcal{S}_\iota] = \mathcal{S}_\iota - \mathcal{S}_\iota.$$

Then for any $\iota \in I$, $[\mathcal{S}_\iota]$ is a band of the vector lattice $[\mathcal{S}]$ and it is a closed subspace of the locally convex space $[\mathcal{S}]$. The projection $[\mathcal{S}] \to [\mathcal{S}_\iota]$ is continuous and $[\mathcal{S}]$ is isomorphic, as a locally convex space and as a vector lattice, to $\prod_{\iota\in I} [\mathcal{S}_\iota]$. If we denote by $[\mathcal{S}(X_\iota)]$, the corresponding space $[\mathcal{S}]$ constructed on X_ι instead of X, then $[\mathcal{S}(X_\iota)]$ is isomorphic, as a locally convex space and as a vector lattice, to $[\mathcal{S}_\iota]$.

11.4.2. Let $(X_\iota)_{\iota\in I}$ be a family of open and closed sets of X whose union is equal to X. For any $\iota \in I$, we set

$$\mathcal{S}_{\iota,e} := \{u \in \mathcal{S}_e \mid u|_{X \setminus X_\iota} = 0\}, \qquad M_{\iota,X} := \varphi(\mathcal{S}_{\iota,e}).$$

Then $(\mathcal{S}_{\iota,e})_{\iota\in I}$ (resp. $(M_{\iota,X})_{\iota\in I}$) is a family of pairwise disjoint open and closed sets of $\mathcal{S}_e$ (resp. M_X) whose union is equal to $\mathcal{S}_e$ (resp. M_X). For any $\iota \in I$, $\mathcal{S}_{\iota,e}$ (resp. $M_{\iota,X}$) is canonically homeomorphic to $\mathcal{S}_e$ constructed on the harmonic space X_ι (resp. to M_{X_ι}).

11.4.3. Let u be an extremal superharmonic function on X; whose pole is polar. If A is a subset of X, then either

$$\hat{R}_u^A = u$$

or

$$\hat{R}_u^A \wedge u = 0.$$

(Assume $\hat{R}_u^A \neq u$. Then there exists a fine open set B, containing $A \setminus S_0(u)$, such that $R_u^B \neq u$. There exist $v \in \mathcal{S}$ and $\alpha \in [0, 1[$ such that

$$R_u^B = \alpha u + v, \qquad u \wedge v = 0.$$

From

$$\alpha u + v = R_u^B = R_{R_u^B}^B = \alpha R_u^B + R_v^B$$

we obtain

$$\alpha u = \alpha R_u^B.$$

Hence $\alpha=0$ and $R_u^B \wedge u=0$. Assume that $\hat{R}_u^A \wedge u \neq 0$. Then there exists $\beta \in]0, 1]$ such that $\hat{R}_u^A \wedge u = \beta u$ and we obtain $\beta u \leq R_u^B$, $\beta u \preccurlyeq R_u^B$ (Exercise 8.3.4).)

11.4.4. Let u be an extremal superharmonic function on X whose pole is polar. If A, B are two subsets of X such that

$$\hat{R}_u^A \neq u \quad \text{and} \quad \hat{R}_u^B \neq u,$$

then

$$\hat{R}_u^{A \cup B} \neq u.$$

11.4.5. Let F be a closed set of X and let K be a compact set of M_X. If X has a countable base, then the set A of points $\xi \in K$ such that

$$u \in \overset{-1}{\varphi}(\xi) \Rightarrow \hat{R}_u^F \neq u$$

is a K_σ-set. (Let k be a regular Riesz-Martin kernel on K. For every $x \in X$, the map

$$\xi \mapsto \hat{R}_{k_\xi}^F(x)\colon\ K \to [0, \infty]$$

is lower semi-continuous. Let B be a countable dense subset of $X \setminus F$. Then

$$A := \{\xi \in K \mid (\exists\, x)(x \in B,\ \hat{R}_{k_\xi}^F(x) < k_\xi(x))\}.)$$

11.4.6. Let X be a $\mathfrak{P}$-harmonic space such that for any potential p on X and any closed set F of X which is not thin at any of its points, $R_p^F|_{X \setminus F}$ is the greatest harmonic minorant of $p|_{X \setminus F}$. Then any two potentials p, q on X, for which $S(p)$ and $S(q)$ reduce to the same point, are proportional (R.-M. Hervé 1962 [4]). Any potential p on X is continuous if $p|_{S(p)}$ is continuous, but this condition does not imply the above proportionality (see Exercise 11.4.9).

11.4.7. Let X be a $\mathfrak{P}$-harmonic space, x be a point of X and $\mathfrak{V}$ be a fundamental system of neighbourhoods of x for which there exists a real number α such that for any positive harmonic function h on $X \setminus \{x\}$ and any $V \in \mathfrak{V}$, we have

$$\sup_{\partial V} h \leq \alpha \inf_{\partial V} h.$$

Then any two potentials p, q on X for which $S(p)=S(q)=\{x\}$ are proportional (N. Boboc-P. Mustaţă 1968 [5], Théorème 4.3).

11.4.8. Let X be the harmonic space defined in Exercise 3.2.7. For any point $x \in X$, any two potentials p, q on X for which $S(p)=S(q)=\{x\}$ are proportional (M. Bôcher 1903 [1], R.-M. Hervé 1965 [7], N. Boboc-P. Mustaţă 1967 [2], 1968 [5], Théorème 4.4).

11.4.9. Let $(\alpha_n)_{n\in\mathbb{N}}$, $(\beta_n)_{n\in\mathbb{N}}$ be two sequences of strictly positive real numbers and let X be the topological subspace of $\mathbb{R}^2$

$$\{(x,y)\in\mathbb{R}^2 \mid x>0,\ y=0\} \cup \{(x,y)\in\mathbb{R}^2 \mid x>0,\ y=1\}$$
$$\cup \Big(\bigcup_{\substack{n\in\mathbb{N}\\ n\neq 0}} \{(x,y)\in\mathbb{R}^2 \mid x=n,\ 0<y<1\}\Big).$$

For any open set U of X, we denote by $\mathscr{H}(U)$, the set of real continuous functions h on U such that:

α) h is locally linear in x and y on the set

$$\{(x,y)\in U \mid (x,y)\notin\mathbb{N}\times\{0,1\}\};$$

β) for any $n\in\mathbb{N}$ with $(n,1)\in U$, we have

$$\left(\frac{\partial h}{\partial x}\right)_r(n,1)+\left(\frac{\partial h}{\partial x}\right)_l(n,1)+\alpha_n\left(\frac{\partial h}{\partial y}\right)_l(n,1)=0$$

(here the suffixes l and r mean the left and the right derivative respectively);

γ) for any $n\in\mathbb{N}$ with $(n,0)\in U$, we have

$$\left(\frac{\partial h}{\partial x}\right)_r(n,0)+\left(\frac{\partial h}{\partial x}\right)_l(n,0)+\beta_n\left(\frac{\partial h}{\partial y}\right)_r(n,0)=0.$$

a) $\mathscr{H}$ is a harmonic sheaf on X and X endowed with $\mathscr{H}$ is a Brelot space.

b) The constant functions and the function $(x,y)\mapsto x$ on X are harmonic functions; hence X is a $\mathfrak{P}$-Brelot space.

c) For any $\alpha\in\mathbb{R}_+$ and any lower bounded hyperharmonic function u defined on $U:=\{(x,y)\in X \mid x>\alpha\}$, we have

$$\inf_U u \geq \liminf_{\substack{x\to\alpha\\ x>\alpha}} u(x,y).$$

d) Let h be a harmonic function on X such that

$$\lim_{x\to 0} h(x,y)=0.$$

Then h is completely determined by its values at the points $(1,0)$ and $(1,1)$. If $h(1,0)<h(1,1)$, then $h(n,0)<h(n,1)$ for every $n\in\mathbb{N}$, the function $x\mapsto h(x,0)$ is concave on $]0,\infty[$ and the function $x\mapsto h(x,1)$ is convex on $]0,\infty[$.

e) If the sequence $(\alpha_n)_{n\in\mathbb{N}}$ converges to ∞, then for any $\alpha\in\mathbb{R}_+$ and any bounded harmonic function h on $\{(x,y)\in X \mid x>\alpha\}$, the limit of $h(x,y)$, for x converging to ∞, exists. Moreover for any positive harmonic function h on X for which

$$\lim_{x\to 0} h(x, y) = 0,$$

we have $h(n, 0) \le h(n, 1)$ for every $n \in \mathbb{N}$; hence the harmonic function on X, $(x, y) \mapsto x$, is an extremal positive harmonic function.

f) Let h be a continuous real function defined on X such that for any $n \in \mathbb{N}$, $n \ge 1$,

$$2h(n, 0) - h(n-1, 0) - h(n+1, 0) > 0$$

$$2h(n, 1) - h(n-1, 1) - h(n+1, 1) < 0$$

$$h(n, 0) < h(n, 1), \qquad \lim_{x\to 0} h(x, y) = 0$$

and such that h is locally linear in x and y on $\{(x, y) \in X \mid (x, y) \notin \mathbb{N} \times \{0, 1\}\}$. Then we may choose sequences $(\alpha_n)_{n\in\mathbb{N}}$, $(\beta_n)_{n\in\mathbb{N}}$ such that h is harmonic. Every harmonic function h' on X for which

$$\lim_{x\to 0} h'(x, y) = 0$$

is a linear combination of h and the function $(x, y) \mapsto x$. If

$$\lim_{n\to\infty} \frac{h(n+1, 1)}{h(n, 1)} = \infty,$$

then

$$\lim_{n\to\infty} \alpha_n = \infty.$$

g) Let h be a function as in *f)* such that h is positive and

$$\lim_{n\to\infty} \frac{h(n+1, 1)}{h(n, 1)} = \infty, \qquad \lim_{n\to\infty} \frac{h(n, 0)}{n} = 0.$$

Assume that the sequences $(\alpha_n)_{n\in\mathbb{N}}$, $(\beta_n)_{n\in\mathbb{N}}$ are chosen so that h is harmonic on X. Let X' be the topological space obtained from X by adding a point $x_0 \notin X$ for which the family

$$(\{x_0\} \cup \{(x, y) \in X \mid x > \alpha\})_{\alpha\in\mathbb{R}_+}$$

is a fundamental system of neighbourhoods. If, for any open set U of X', we denote by $\mathscr{H}'(U)$ the set of real continuous functions h on U such that $h|_{U\cap X} \in \mathscr{H}(U \cap X)$, then $\mathscr{H}'$ is a harmonic sheaf on X', X' endowed with $\mathscr{H}'$ is a $\mathfrak{P}$-Brelot space and $\{x_0\}$ is a polar set. Let h' (resp. h'_0) be the function on X' which is equal to h (resp. to $(x, y) \mapsto x$) on X and equal to

$$\liminf_{x\to\infty} h(x, y) \qquad (\text{resp. } \infty)$$

at x_0. Then h' and h'_0 are non-proportional potentials on X' which are harmonic outside $\{x_0\}$ and are extremal positive superharmonic functions. Every potential on X which is harmonic outside $\{x_0\}$ is a linear combination of h' and h'_0. It is possible to construct h' such that $h'(x_0)$ is either finite or infinite.

11.4.10. Show that the Martin space of the example of Theorem 2.1.2 is the subspace $]-\infty, +\infty]$ of $\overline{\mathbb{R}}$ and that $\pi'(x)=x$ for all $x\in R_X$, $\pi'(\infty)=$ the Alexandroff point of X. Hence π' is a one to one map of M_X onto X_0, but π' is not a homeomorphism.

11.4.11. We denote by X, the subspace of $\mathbb{R}^4$

$$\{(x_1, x_2, x_3, x_4)\in\mathbb{R}^4 \mid x_4=0\}$$
$$\cup \bigcup_{\substack{n\in\mathbb{N}\\ n\neq 0}} \left\{(x_1, x_2, x_3, x_4)\in\mathbb{R}^4 \mid x_1=\frac{1}{n},\ x_2=x_3=0, 0<x_4\leq\frac{1}{n}\right\}.$$

For any open set U of X, we denote by $\mathscr{H}(U)$, the set of continuous real functions h on U such that:

a) the function

$$(x_1, x_2, x_3)\mapsto h(x_1, x_2, x_3, 0)\colon \{(x_1, x_2, x_3)\in\mathbb{R}^3 \mid (x_1, x_2, x_3, 0)\in U\}\to\mathbb{R}$$

is a solution for the Laplace equation;

b) for any $n\in\mathbb{N}$, $n\neq 0$, the function

$$x_4\mapsto h\left(\frac{1}{n}, 0, 0, x_4\right)\colon \left\{x_4\in\left]0, \frac{1}{n}\right[\,\middle|\, \left(\frac{1}{n}, 0, 0, x_4\right)\in U\right\}\to\mathbb{R}$$

is locally linear;

c) for any $n\in\mathbb{N}$, $n\neq 0$, such that $\left(\frac{1}{n}, 0, 0, \frac{1}{n}\right)\in U$, the function h is constant on a neighbourhood of $\left(\frac{1}{n}, 0, 0, \frac{1}{n}\right)$. $\mathscr{H}$ is a harmonic sheaf on X having the Doob convergence property and X, endowed with $\mathscr{H}$, is a $\mathfrak{P}$-Bauer space such that π' is a one-to-one map of the Martin space of X onto X_0. The map

$$M_X\setminus\{\overset{-1}{\pi'}(0,0,0,0)\}\to X\setminus\{(0,0,0,0)\},$$

induced by π', is a homeomorphism and thus there exists a regular Riesz-Martin kernel on $M_X\setminus\{\overset{-1}{\pi'}(0,0,0,0)\}$. The point x_0 of the Martin space of X for which $\pi'(x_0)=(0,0,0,0)$ has no fundamental system of closed neighbourhoods. Hence the Martin space and the Riesz space of X are not regular topological spaces and no regular Riesz-Martin kernel exists on it.

11.4.12. Prove Proposition 11.4.13 with the hypothesis that X is a Brelot space replaced by the hypothesis of Exercise 11.2.6 (use Exercise 11.2.6 and Proposition 11.1.1).

11.4.13. If X is a connected Brelot space, then P_X is a G_δ-set of a compact metrisable space (use Exercise 6.2.9).

11.4.14. Let f be a real function on X whose infimum on every compact set of X is strictly positive. For any positive harmonic function h on X, we set

$$\rho_f(h) := \sup \sum_{\iota \in I} f(x_\iota)\, h_\iota(x_\iota),$$

where I is an arbitrary finite set, $(x_\iota)_{\iota \in I}$ is an arbitrary family in X and $(h_\iota)_{\iota \in I}$ is an arbitrary family of positive harmonic functions on X whose sum is equal to h. Then

$$\{h \in \mathscr{H}_+(X) \mid \rho_f(h) \leq 1\}$$

is a cap. If X is σ-compact, then for any compact set $\mathscr{K}$ of $\mathscr{H}_e$ $(= \mathscr{S}_e \cap \mathscr{H}(X))$, there exists a strictly positive continuous real function f on X such that for any measure μ on $\mathscr{K}$ with $\mu(\mathscr{K}) \leq 1$ we have

$$\rho_f(u_\mu) \leq 1.$$

§ 11.5. Integral Representation of Positive Superharmonic Functions

Throughout this section X will denote a $\mathfrak{P}$-harmonic space with a countable base for which the sheaf of harmonic functions on X has the property of nuclearity.

Proposition 11.5.1. *Every weakly Cauchy filter on $\mathscr{S}$ is a Cauchy filter on $\mathscr{S}$. In particular, $\mathscr{S}$ is weakly complete and the weak topology on $\mathscr{S}$ coincides with the topology $\mathscr{T}$.*

Let $\mathfrak{F}$ be a weakly Cauchy filter on $\mathscr{S}$. Let $f \in \mathscr{C}_+(X_0)$ and let K be a compact subset of $X \setminus \operatorname{Supp} f$. Let L be a compact neighbourhood of K contained in $X \setminus \operatorname{Supp} f$. By Theorem 11.1.2, there exists a measure μ on X whose carrier is compact and contained in $X \setminus \operatorname{Supp} f$ and such that for any positive harmonic function h on $X \setminus \operatorname{Supp} f$, we have

$$\sup_L h \leq \mu(h).$$

Since $\mathfrak{F}$ is a weakly Cauchy filter, there exists $\mathscr{F} \in \mathfrak{F}$ such that

$$\mathscr{F} - \mathscr{F} \subset \mathscr{V}(f, \mu) := \{u \in [\mathscr{S}] \mid |\mu(f \cdot u)| \leq 1\}.$$

Let $u \in \mathscr{F}$. From

$$\mathscr{F} \subset u + \mathscr{V}(f, \mu),$$

it follows that

$$\mu(f \cdot v) \leq \mu(f \cdot u) + 1$$

for every $v \in \mathscr{F}$. Hence

$$\sup_{v \in \mathscr{F}} \sup_L (f \cdot v) \leq \mu(f \cdot u) + 1 < \infty.$$

By Theorem 11.1.1, $\{f \cdot v|_K \,|\, v \in \mathscr{F}\}$ is an equicontinuous set of real functions on K. Since

$$\lim_{v, \mathfrak{F}} f \cdot v(x)$$

exists for all $x \in K$, it follows that it converges uniformly on K. Hence there exists $\mathscr{F}' \in \mathfrak{F}$ such that

$$\sup_K |f \cdot v - f \cdot w| \leq 1$$

for all $v, w \in \mathscr{F}'$. This means

$$\mathscr{F}' - \mathscr{F}' \subset \mathscr{V}(f, K).$$

$\mathfrak{F}$ is therefore a Cauchy filter on $\mathscr{S}$. □

Remark. C. Constantinescu 1965 [2], using a slightly different hypothesis, proved the weak completeness of $\mathscr{S}$. G. Mokobodzki 1964 [unpublished] proved the same property for another topology under the hypothesis that $\mathscr{H}$ has the Doob convergence property.

In fact any normal complete cone of a Schwartz space is weakly complete and so $\mathscr{S}$ is weakly complete without the hypotheses that X has a countable base and that the sheaf of harmonic functions on X has the property of nuclearity (C. Constantinescu 1970 [9]).

Theorem 11.5.1. *There exists a semi-regular Riesz-Martin kernel on the Martin space of X. If k is such a kernel, then for any positive superharmonic function u on X, there exists a unique measure μ on the Martin space of X such that*

$$u = k_\mu.$$

The first assertion is exactly Corollary 11.4.3. The preceding proposition and Proposition 11.2.10 show that $\mathscr{S}$ is weakly complete and weakly metrisable. By Corollary 11.3.3, there exists a measure ν on the section $\mathscr{A}_k$ generated by k such that u is its barycenter. By Proposition 11.4.1,

$$u = u_\nu$$

and by Theorem 11.4.1 *d)* and *b)*, there exists a measure μ on the Martin space of X such that

$$u = k_\mu.$$

The unicity follows from the same theorem *c)*. □

Remark. This theorem was proved by M. Brelot 1959 [14] for a $\mathfrak{P}$-Brelot space with a countable base and satisfying a supplementary condition. This additional condition was later removed in a proof given by R.-M. Hervé 1960 [3].

Corollary 11.5.1. *Let k be a semi-regular Riesz-Martin kernel on the Martin space M_X of X and let $\mathfrak{M}_k$ be the set of measures μ on M_X such that k_μ is superharmonic.*

a) $\mathfrak{M}_k$ is a prevector lattice and for any two measures μ, ν on M_X such that $u \preccurlyeq \nu$, $\nu \in \mathfrak{M}_k$, we have $\mu \in \mathfrak{M}_k$.

b) The map

$$\mu \mapsto k_\mu : \mathfrak{M}_k \to \mathscr{S}$$

is an isomorphism of prevector lattices.

The corollary follows from the theorem with the aid of Corollary 11.4.4. □

Corollary 11.5.2. *Let F be a closed set of X_0. If there exists a positive superharmonic function u on X such that*

$$\emptyset \neq S(u) \subset F,$$

then there exists an extremal superharmonic function whose pole lies in F.

By Theorem 11.5.1, there exists a semi-regular Riesz-Martin kernel on M_X and a measure μ on M_X such that

$$u = k_\mu .$$

By Proposition 11.4.12 *b), c)*, we have

$$\mu(\overset{-1}{\pi'}(X_0 \setminus F)) = 0 .$$

Hence

$$\mu(\overset{-1}{\pi'}(F)) \neq 0, \qquad \overset{-1}{\pi'}(F) \neq \emptyset . \quad \square$$

Corollary 11.5.3. *If there exists a non-identically 0 positive harmonic function on X, then there exists an extremal superharmonic function which is harmonic.* □

Corollary 11.5.4. *Let K be a compact, non-polar set of X. Then there exists an extremal superharmonic function whose pole lies in K.*

Let p be a strictly positive potential on X. Since K is not polar

$$S_0(\hat{R}_p^K) \neq \emptyset . \quad \square$$

Theorem 11.5.2. *Assume that X is a $\mathfrak{P}$-Brelot space with a countable base and that any two extremal superharmonic functions on X having the* *same pole in X are proportional. Then there exists a lower semi-continuous, positive numerical function g on X^2 such that:*

a) the restriction of g to $\{(x,y)\in X^2 \mid x\neq y\}$ is finite and continuous;

b) for any $y\in X$, the function $x\mapsto g(x,y)$ is an extremal superharmonic function having y as a pole;

c) for any potential p on X, there exists a unique measure μ on X such that for any $x\in X$

$$p(x)=\int^{*} g(x,y)\,d\mu(y)$$

(R.-M. Hervé 1962 [4]).

By Proposition 11.4.13, $\pi'|_{R_X}$ is a homeomorphism of R_X onto X. By Corollary 11.4.2, there exists a continuous Riesz-Martin kernel k on M_X. For any $(x,y)\in R_X^2$, we set

$$g(x,y):=k\big(x, \overset{-1}{\pi'}|_{R_X}(y)\big).$$

a) and *b)* are obvious, *c)* follows from Theorem 11.5.1 and Proposition 11.4.12 *c)*. □

Remark 1. In classical potential theory, *c)* was proved by F. Riesz 1925 [1].

Remark 2. The property of proportionality from the theorem is not always fulfilled (Exercise 9.2.6, Exercise 11.4.9). Sufficient conditions for this property to hold are given in Exercises 11.4.6 and 11.4.7.

Exercises

11.5.1. Let A be a subset of X. If $\hat{R}_u^A\neq u$ for any extremal superharmonic function on X, then A is semi-polar. (Assume the contrary and let p be a non-identically 0 potential on X such that

$$\hat{R}_p^A=p$$

(Exercise 7.2.19).) Let k be a semi-regular Riesz-Martin kernel on M_X and let μ be a measure on M_X such that $p=k_\mu$ (Theorem 11.5.1). Let ν be a measure on X such that $\nu\in\Lambda$, p is ν-integrable and every fine open set is ν-integrable and of positive ν-measure (Exercise 8.3.9). We have

$$\int p\,d\nu=\int \hat{R}_p^A\,d\nu=\int p\,d\nu^A.$$

By Proposition 11.4.11, we obtain

$$((\int k_\xi\,d\nu)\,d\mu(\xi)=\int(\int k_\xi\,d\nu^A)\,d\mu(\xi).$$

Hence, there exists $\xi \in A$ such that

$$\int \hat{R}^A_{k_\xi}\, dv = \int k_\xi\, dv^A = \int k_\xi\, dv < \infty, \qquad k_\xi = \hat{R}^A_{k_\xi}.)$$

11.5.2. If the axiom of polarity holds, then a subset A of X is polar if and only if $\hat{R}^A_u \neq u$ for any extremal superharmonic function u on X (use preceding exercise).

11.5.3. Every finite potential on X is the sum of a family $(p_\iota)_{\iota \in I}$ of potentials on X such that for any $\iota \in I$, $p_\iota|_{S(p_\iota)}$ is continuous and there exists a compact set $K \subset R_X$ with $\pi'(K) = S(p_\iota)$. (By Theorem 8.2.2, we may assume that the restriction of the given potential p to $S(p)$ is continuous. By Theorem 11.5.1 there exists a semi-regular Riesz-Martin kernel k on M_X and a measure μ on M_X such that $p = k_\mu$. By Proposition 11.4.12 *d)* $\mu(M_X \setminus \overset{-1}{\pi'}(S(p))) = 0$. For any compact set $L \subset \overset{-1}{\pi'}(S(p))$, $k_{\mu_L}|_{S(k_{\mu_L})}$ is continuous and thus there exists a compact set $K \subset R_X$ with $\pi'(K) = S(k_{\mu_L})$.)

11.5.4. If every pair of extremal superharmonic functions on X having the same pole in X are proportional, then the axiom of polarity implies the axiom of domination (R.-M. Hervé 1962 [4]). (Let p be a potential on X such that $p|_{S(p)}$ is finite, continuous and such that there exists a compact set $K \subset R_X$ with $\pi'(K) = S(p)$. By the preceding exercise, it is sufficient to show that $\hat{R}^{S(p)}_p = p$ (Theorem 8.3.2 $c \Rightarrow a$). Let k be a semi-regular Riesz-Martin kernel on M_X and let μ be a measure on K with $k_\mu = p$ (Theorem 11.5.1). By Proposition 11.4.12 *d)*, $\mu(M_X \setminus K) = 0$. By Exercise 11.4.5, there exists an increasing sequence $(K_n)_{n \in \mathbb{N}}$ of compact subsets of K such that

$$\{\xi \in K \mid R^{S(p)}_{k_\xi} \neq k_\xi\} = \bigcup_{n \in \mathbb{N}} K_n.$$

By Exercise 11.5.2, $\pi'(K_n)$ is polar and therefore $p_{\pi'(K_n)} = 0$ (Proposition 8.2.3). By Proposition 11.4.12 *d)*, $\mu(K_n) = 0$. The assertion now follows from Proposition 11.4.12 *d)*.)

11.5.5. Let U be an open set of X and let A be the set of points $x \in \partial U$ such that there exists an extremal superharmonic function u on X having x as pole and $u \neq \hat{R}^{X \setminus U}_u$. If A is polar, then for any bounded potential p on X, $\bar{H}^U_p$ is the greatest harmonic minorant of $p|_U$. (We may assume p harmonic on U. Let k be a semi-regular Riesz-Martin kernel on M_X and μ be a measure on M_X such that $p = k_\mu$. By Proposition 8.2.3, $p_A = 0$. By Proposition 11.4.12 *d)*, $\mu(\overset{-1}{\pi'}(A)) = 0$. The assertion now follows from Proposition 11.4.12 *e)*.)

11.5.6. Let X be a Brelot space and assume that every potential p on X is continuous if its restriction to $S(p)$ is finite and continuous.

Then the set of points $x \in X$ such that there exists two non-proportional, extremal superharmonic functions on X, having x as pole, is polar (G. Mokobodzki 1968 [2]).

11.5.7. Let k be a semi-regular Riesz-Martin kernel on R_X. Then for any kernel V on X, there exists a unique measure μ on R_X such that for all $f \in \mathscr{L}^1_+(\mathsf{V})$, the function $f \circ (\pi'|_{R_X})$ is μ-measurable and for all $x \in X$

$$\mathsf{V} f(x) = \overset{\bullet}{\int} f(\pi'(\xi))\, k(x, \xi)\, d\mu(\xi).$$

11.5.8. Prove Theorem 11.5.2 without the hypothesis that X has a countable base. Let $\mathfrak{M}_g$ be the set of measures μ on X such that the function p_μ

$$x \mapsto \overset{*}{\int} g(x, y)\, d\mu(y)$$

is a potential. Then $\mathfrak{M}_g$ is a prevector lattice and

$$\mu \mapsto p_\mu \colon \mathfrak{M}_g \to \mathscr{P}$$

is an isomorphism of prevector lattices; if we endow $\mathfrak{M}_g$ with the vague topology $\mu \mapsto p_\mu$ is also a homeomorphism.

References

Albinus, G., Boboc, N., Mustaţă, P.:

[1] to appear.

Bauer, H.:

[1] Minimalstellen von Funktionen und Extremalpunkte I. Arch. Math. **9**, 389–393 (1958).

[2] Minimalstellen von Funktionen und Extremalpunkte II. Arch. Math. **11**, 200–205 (1960).

[3] Une axiomatique du problème de Dirichlet pour certaines équations elliptiques et paraboliques. C. R. Acad. Sci. Paris **250**, 2672–2674 (1960).

[4] Šilovscher Rand und Dirichletsches Problem. Ann. Inst. Fourier **11**, 89–136 (1961).

[5] Axiomatische Behandlung des Dirichletschen Problems für elliptische und parabolische Differentialgleichungen. Math. Ann. **146**, 1–59 (1962).

[6] Weiterführung einer axiomatischen Potentialtheorie ohne Kern (Existenz von Potentialen). Z. Wahrscheinlichkeitstheorie und Verw. Gebiete **1**, 197–229 (1963).

[7] Propriétés fines des fonctions hyperharmoniques dans une théorie axiomatique du potential. Ann. Inst. Fourier **15**, 1, 137–154 (1965).

[8] Zum Cauchyschen und Dirichletschen Problem bei elliptischen und parabolischen Differentialgleichungen. Math. Ann. **164**, 142–153 (1966).

[9] Harmonische Räume und ihre Potentialtheorie (Lecture Notes in Math. 22). Berlin-Heidelberg-New York: Springer 1966.

[10] Harmonic spaces and associated Markov processes. In: Potential theory, 23–67 (CIME, 1° Ciclo, Stresa 2–10 Luglio 1969). Roma: Edizioni Cremonese 1970.

Berg, C.:

[1] Quelques propriétés de la topologie fine dans la théorie du potentiel et des processus standard. Bull. Sci. Math. **95**, 27–31 (1971).

Boboc, N., Constantinescu, C., Cornea, A.:

[1] Teoria potenţialului pe suprafeţe riemanniene. Bucureşti: Editura Academiei P. P. R. (Seminar S. Stoilow 1959–1960) 1962.

[2] On the Dirichlet problem in the axiomatic theory of harmonic functions. Nagoya Math. J. **23**, 73–96 (1963).

[3] Axiomatic theory of harmonic functions. Non-negative superharmonic functions. Ann. Inst. Fourier **15**, 1, 283–312 (1965).

[4] Axiomatic theory of harmonic functions. Balayage. Ann. Inst. Fourier **15**, 2, 37–70 (1965).

[5] Semigroups of transitions on harmonic spaces. Rev. Roumaine Math. Pures Appl. **12**, 763–805 (1967).

Boboc, N., Cornea, A.:

[1] Behaviour of harmonic functions at a nonregular boundary point. Bull. Math. Soc. Sci. Math. R. S. Roumanie **10**, 63–74 (1966).

[2] Convex cones of lower semicontinuous functions on compact spaces. Rev. Roumaine Math. Pures Appl. **12**, 471–525 (1967).
[3] Comportement des balayées des mesures ponctuelles. Comportement des solutions du problème de Dirichlet aux points irréguliers. C. R. Acad. Sci. Paris Sér. A **264**, 995–997 (1967).
[4] Nonnegative hyperharmonic functions. Balayage and natural order. Rev. Roumaine Math. Pures Appl. **13**, 609–618 (1968).
[5] Espaces harmoniques. Axiome *D* et théorème de convergence. Rev. Roumaine Math. Pures Appl. **13**, 933–947 (1968).
[6] Cônes convexes ordonnés. *H*-cônes et adjoints de *H*-cônes. C. R. Acad. Sci. Paris Sér. A **270**, 596–599 (1970).

Boboc, N., Mustaţă, P.:
[1] Remarks on the existence of solutions of Dirichlet problem for strongly elliptic linear operators of second order. Bull. Math. Soc. Sci. Math. R. S. Roumanie **10**, 75–85 (1966).
[2] Considérations sur les espaces harmoniques associés aux opérateurs linéaires elliptiques à coefficients continus. Bull. Math. Soc. Sci. Math. R. S. Roumanie **11**, 11–19 (1967).
[3] Sur un problème concernant les domaines d'unicité pour le problème de Dirichlet associé à un opérateur elliptique. Atti Accad. Naz. Lincei Rend. Cl. Sci. Fis. Mat. Natur. **VIII**, Ser. 42, 181–186 (1967).
[4] Sur l'équicontinuité de certaines familles de fonctions continues. C. R. Acad. Sci. Paris Sér. A **266**, 802–805 (1968).
[5] Espaces harmoniques associés aux opérateurs différentiels linéaires du second ordre de type elliptique (Lecture Notes in Math. 68). Berlin-Heidelberg-New York: Springer 1968.

Boboc, N., Radu, N.:
[1] Une classe de fonctions définies sur des variétés topologiques triangulables. Mesures associées. Fonctions de Green associées. Rev. Math. Pures Appl. **3**, 309–323 (1958).

Bôcher, M.:
[1] Singular points of fractions which satisfy partial differential equations of the elliptic type. Bull. Amer. Math. Soc. **9**, 455–465 (1903).

Bony, J.-M.:
[1] Détermination des axiomatiques de théorie du potentiel dont les fonctions harmoniques sont différentiables. Ann. Inst. Fourier **17**, 1, 353–382 (1967).
[2] Principe du maximum dans les espaces de Sobolev. C. R. Acad. Sci. Paris Sér. A **265**, 333–336 (1967).
[3] Problème de Dirichlet et inégalité de Harnack pour une classe d'opérateurs elliptiques dégénérés du second ordre. C. R. Acad. Sci. Paris Sér. A **266**, 830–833 (1968).
[4] Principe du maximum et inégalité de Harnack pour les opérateurs elliptiques dégénérés. Sém. Théorie Potentiel. **12**, 10.01–10.20 (1969).
[5] Principe du maximum, inégalité de Harnack et unicité du problème de Cauchy pour les opérateurs elliptiques dégénérés. Ann. Inst. Fourier **19**, 1, 277–304 (1969).
[6] Opérateurs elliptiques dégénérés associés aux axiomatiques de la théorie du potentiel. In: Potential theory 69–119 (CIME, 1° ciclo, Stresa 2–10 Luglio 1969). Roma: Edizioni Cremonese 1970.

Bouligand, G.:
[1] Sur le problème de Dirichlet. Ann. Soc. Polonaise **4**, 59–112 (1926).

Bourbaki, N.:
[1] Topologie générale. Ch. 9, 2° éd. Paris: Hermann 1958.
[2] Topologie générale. Ch. 10, 2e éd. Paris: Hermann 1961.
[3] Espaces vectoriels topologiques. Ch. 1–2, 2e éd. Paris: Hermann 1966.
[4] Espaces vectoriels topologiques. Ch. 3–5. Paris: Hermann 1955.
[5] Intégration. Ch. 1–4, 2e éd. Paris: Hermann 1965.
[6] Intégration. Ch. 7–8. Paris: Hermann 1963.
[7] Théories spectrales. Ch. 1–2. Paris: Hermann 1967.

Brelot, M.:
[1] Sur le potentiel et les suites de fonctions sous-harmoniques. C. R. Acad. Sci. Paris **207**, 836–838 (1938).
[2] Sur la théorie moderne du potentiel. C. R. Acad. Sci. Paris **209**, 828–830 (1939).
[3] Familles de Perron et problème de Dirichlet. Acta Sci. Math. (Szeged) **9**, 133–153 (1938–40).
[4] Sur la théorie autonome des fonctions sousharmoniques. Bull. Sci. Math. France **65**, 78–98 (1941).
[5] Sur le rôle du point à l'infini dans la thèorie des fonctions harmoniques. Ann. Sci. École Norm. Sup. **61**, 301–332 (1944).
[6] Sur les ensembles effilés. Bull. Sci. Math. **68**, 12–36 (1944).
[7] Sur l'approximation et la convergence dans la théorie des fonctions harmoniques ou holomorphes. Bull. Soc. Math. France **73**, 55–70 (1945).
[8] Minorant sous-harmoniques extrémales et capacités. J. Math. Pures Appl. **24**, 1–32 (1945).
[9] Quelques propriétés et applications du balayage. C. R. Acad. Sci. Paris **227**, 19–21 (1948).
[10] Extension axiomatique des fonctions sous-harmoniques I. C. R. Acad. Sci. Paris **245**, 1688–1690 (1957).
[11] Extension axiomatique des fonctions sous-harmoniques II. C. R. Acad. Sci. Paris **246**, 2334–2337 (1958).
[12] La convergence des fonctions surharmoniques et des potentiels généralisés. C. R. Acad. Sci. Paris **246**, 2709–2712 (1958).
[13] Une axiomatique générale du problème de Dirichlet dans les espaces localement compacts. Sém. Théorie du Potentiel. **1**, 6.01–6.16 (1958).
[14] Axiomatique des fonctions harmoniques et surharmoniques dans un espace localement compact. Sém. Théorie du Potentiel. **2**, 1.01–1.40 (1959).
[15] Lectures on potentiel theory. Part IV. Bombay: Tata Institute of Fundamental Research 1960. Reissued 1967.
[16] Sur un théorème du prolongement fonctionnel de Keldych concernant le problème de Dirichlet. J. Analyse Math. **8**, 273–288 (1961).
[17] Étude comparée de quelques axiomatiques des fonctions harmoniques et surharmoniques. Sém. Théorie du Potentiel. **6**, 1.13–1.26 (1962).
[18] Quelques propriétés et applications nouvelles de l'effilement. Sém. Théorie du Potentiel. **6**, 1.27–1.40 (1962).
[19] Capacité et balayage pour ensembles décroissants. C. R. Acad. Sci. Paris **260**, 2683–2685 (1965).
[20] Einige neuere Fortschritte in der axiomatischen Theorie der harmonischen Funktionen. In: Vorträge 3. Tagg Prob. Meth. Math. Phys. Karl-Marx-Stadt 1966, H. 1, p. 28–35. Technische Hochschule Karl-Marx-Stadt 1966.
[21] Capacity and balayage for decreasing sets. In: Proceedings of the Fifth Berkeley Symposium on Mathematical Statistics and Probability 1965–66, p. 279–293. Berkeley: University of California Press 1967.

[22] Recherches axiomatiques sur un théorème de Choquet concernant l'effilement. Nagoya Math. J. **30**, 33–46 (1967).

Brelot, M., Choquet, G.:
[1] Espaces et lignes de Green. Ann. Inst. Fourier **3**, 199–263 (1951).

Brelot, M., Hervé, R.-M.:
[1] Introduction de l'effilement dans une théorie axiomatique du potentiel. C. R. Acad. Sci. Paris **247**, 1956–1959 (1958).

Cartan, H.:
[1] Capacité extérieure et suites convergentes de potentiels. C. R. Acad. Sci. Paris **214**, 944–946 (1942).
[2] Théorie du potentiel Newtonien: énergie, capacité, suites de potentiels. Bull. Soc. Math. France **73**, 74–106 (1945).

Choquet, G.:
[1] Capacitabilité. Théorèmes fondamentaux. C. R. Acad. Sci. Paris **234**, 784 (1952).
[2] Sur les points d'effilement d'un ensemble. Application à l'étude de la capacité. Ann. Inst. Fourier **9**, 91–101 (1959).
[3] Démonstration non probabiliste d'un théorème de Getoor. Ann. Inst. Fourier **15**, 2, 409–414 (1965).
[4] Sur un théorème de Keldych concernant le problème de Dirichlet. Ann. Inst. Fourier **18**, 1, 309–315 (1968).

Collin, B.:
[1] Remarques axiomatiques sur les points-frontières irréguliers dans le problème de Dirichlet. Ann. Inst. Fourier **14**, 2, 485–492 (1964).

Constantinescu, C.:
[1] An example of harmonic space. Rev. Roumaine Math. Pures Appl. **10**, 267–270 (1965).
[2] A topology on the cone of non-negative superharmonic functions. Rev. Roumaine Math. Pures Appl. **10**, 1331–1348 (1965).
[3] Harmonic spaces. Absorbent sets and balayage. Rev. Roumaine Math. Pures Appl. **11**, 887–910 (1966).
[4] Familles continues de mesures et équicontinuité. C. R. Acad. Sci. Paris **262**, 1309–1312 (1966).
[5] Equicontinuity on harmonic spaces. Nagoya Math. J. **29**, 1–6 (1967).
[6] Some properties of the balayage of measures on a harmonic space. Ann. Inst. Fourier **17**, 1, 273–293 (1967).
[7] Kernels and nuclei on harmonic spaces. Rev. Roumaine Math. Pures Appl. **13**, 35–57 (1968).
[8] Markov processes on harmonic spaces. Rev. Roumaine Math. Pures Appl. **13**, 627–654 (1968).
[9] L'ensemble des fonctions surharmoniques positives sur un espace harmonique est faiblement complet. C. R. Acad. Sci. Paris Sér. A **271**, 549–551 (1970).

Constantinescu, C., Cornea, A.:
[1] On the axiomatic of harmonic functions I. Ann. Inst. Fourier **13**, 2, 373–388 (1963).
[2] On the axiomatic of harmonic functions II. Ann. Inst. Fourier **13**, 2, 389–394 (1963).
[3] Examples in the theory of harmonic spaces. In: Seminar über Potentialtheorie, p. 161–171. Berlin-Heidelberg-New York: Springer (Lecture Notes in Math. 69) 1968.
[4] Nuclearity on harmonic spaces. Tôhoku Math. J. **21**, 558–572 (1969).

Cornea, A.:
[1] Sur la démonbrabilité à l'infini d'un espace harmonique de Brelot. C. R. Acad. Sci. Paris Sér. A **264**, 190–191 (1967).
[2] Weakly compact sets in vector lattices and convergence theorems in harmonic spaces. In: Seminar über Potentialtheorie, p. 173–180. Berlin-Heidelberg-New York: Springer (Lecture Notes in Math. 69) 1968.

Doob, J. L.:
[1] Probability methods applied to the first boundary value problem. In: Proc. 3rd Berkeley Sym. on Math. Stat. and Prob., 1954–55, p. 49–80. Berkeley: University of California Press 1956.
[2] Applications to analysis of a topological definition of smallness of a set. Bull. Amer. Math. Soc. **72**, 579–600 (1966).

Evans, G. C.:
[1] Applications of Poincaré's sweeping out process. Proc. Nat. Acad. Sci. U.S.A. **19**, 457–461 (1933).

Frostman, O.:
[1] Potentiel d'équilibre et capacité des ensembles avec quelques applications à la théorie des fonctions. Medd. Lunds Univ. Mat. Sem. **3**, 1–118 (1935).
[2] Les points irréguliers dans la théorie du potentiel et le critère de Wiener. Medd. Lunds Univ. Mat. Sem. **4**, 1–10 (1939).
[3] Sur le balayage des mesures. Acta Sci. Math. (Szeged) **3**, 43–51 (1938–40).

Fuglede, B.:
[1] Esquisse d'une théorie axiomatique de l'effilement et de la capacité. C. R. Acad. Sci. Paris **261**, 3272–3274 (1965).
[2] Connexion en topologie fine et balayage des mesures. Ann. Inst. Fourier **21**, 3, 227–244 (1971).

Grothendieck, A.:
[1] Critères de compacité dans les espaces fonctionnels généraux. Amer. J. of Math. **74**, 168–186 (1952).
[2] Sur les applications linéaires faiblement compactes d'espaces du type $C(K)$. Canad. J. Math. **5**, 129–173 (1953).

Guber, S.:
[1] On the potentiel theory of linear homogeneous parabolic partial differential equations of second order. In: Symposium on Probability Methods in Analysis (Loutraki 1966), p. 112–117 (Lecture Notes in Math. 31). Berlin-Heidelberg-New York: Springer 1967.

Hansen, W.:
[1] Konstruktion von Halbgruppen und Markoffschen Prozessen. Invent. Math. **3**, 179–214 (1967).

Harnack, A.:
[1] Existenzbeweise zur Theorie des Potentials in der Ebene und im Raume. Ber. Verhandl. Königl. Sächs. Ges. Wiss. Leipzig **1886**, 144–169 (1886).

Herglotz, G.:
[1] Über Potenzreihen mit positivem reellem Teil im Einheitskreis. Ber. Verhandl. Königl. Sächs. Ges. Wiss. Leipzig Math.-Phys. Kl. **63**, 501–511 (1911).

Hervé, R.-M.:
[1] Développements sur une théorie axiomatique des fonctions surharmoniques. C. R. Acad. Sci. Paris **248**, 179–181 (1959).

[2] Recherches sur la théorie axiomatique des fonctions surharmoniques. Sém. Théorie du Potentiel. **3**, 11.01–11.06 (1960).

[3] Topologie de l'ensemble des fonctions surharmoniques ≥ 0 et représentation intégrale. C. R. Acad. Sci. Paris **250**, 2834–2836 (1960).

[4] Recherches axiomatiques sur la théorie des fonctions surharmoniques et du potentiel. Ann. Inst. Fourier **12**, 415–571 (1962).

[5] Un principe du maximum pour les sous-solutions locales d'une équation uniformément elliptique de la forma $Lu=\sum_i \frac{\partial}{\partial x_i}\left(\sum_j a_{ij}\frac{\partial u}{\partial x_j}\right)=0$. Ann. Inst. Fourier **14**, 2, 493–508 (1964).

[6] Quelques propriétés des fonctions surharmoniques associés à un opérateur uniformément elliptique de la forme $Lu=-\sum_i \frac{\partial}{\partial x_i}\left(\sum_j a_{ij}\frac{\partial u}{\partial x_j}\right)=0$. Sém. Théorie du Potentiel. **9**, 7.01–7.11 (1965).

[7] Quelques propriétés des fonctions surharmoniques associées à une équation uniformément elliptique de la forme $Lu=-\sum_i \frac{\partial}{\partial x_i}\left(\sum_j a_{ij}\frac{\partial u}{\partial x_j}\right)=0$. Ann. Inst. Fourier **15**, 2, 215–224 (1965).

Hervé, R.-M., Hervé, M.:

[1] Les fonctions surharmoniques associées à un opérateur elliptique du second ordre à coefficients discontinues. Ann. Inst. Fourier **19**, 1, 305–359 (1969).

Hinrichsen, D.:

[1] Randintegrale und nukleare Funktionsräume. Ann. Inst. Fourier **17**, 1, 225–271 (1967).

Hörmander, L.:

[1] Hypoelliptic second order differential equations. Acta Math. Uppsala **119**, 147–171 (1967).

Hunt, G.A.:

[1] Markoff processes and potentials II. Illinois J. Math. **1**, 316–369 (1957).

Ionescu Tulcea, A.:

[1] On equicontinuity of harmonic functions in axiomatic potential theory. Illinois J. Math. **11**, 529–534 (1967).

Keldych, M.V.:

[1] Sur la résolubilité et la stabilité du problème de Dirichlet. Dokl. Akad. Nauk SSSR **18**, 315–318 (1938).

[2] On the resolutivity and the stability of Dirichlet problem [Russian]. Uspehi Mat. Nauk **8**, 172–231 (1941).

Kellogg, O.D.:

[1] Unicité des fonctions harmoniques. C.R. Acad. Sci. Paris **187** II, 526–527 (1928).

[2] Fondations of potential theory. Berlin: Springer (Die Grundlehren der Math. Wissenschaften 31) 1929 (reissued 1967).

Köhn, J.:

[1] Harmonische Räume mit einer Basis semiregulärer Mengen. In: Seminar über Potentialtheorie, p. 1–12. Berlin-Heidelberg-New York: Springer (Lecture Notes in Math. 69) 1968.

Köhn, J., Sieveking, M.:

[1] Reguläre und extremale Randpunkte in der Potentialtheorie. Rev. Roumaine Math. Pures Appl. **12**, 1489–1502 (1967).

Kori, T.J.:

[1] Axiomatic theory of non-negative fullsuperharmonic functions. J. Math. Soc. Japan **23**, 481–526 (1971).

Král, J., Lukeš, J.:
[1] Harmonic continuation (to appear).

Král, J., Lukeš, J., Netuka, I.:
[1] Elliptic points in one-dimensional harmonic spaces. Comment. Math. Univ. Carolinae **12**, 3, 453–483 (1971).

Lebesgue, H.:
[1] Sur le problème de Dirichlet. C.R. Acad. Sci. Paris **154**, 335–337 (1912).
[2] Sur des cas d'impossibilité du problème de Dirichlet. C.R. Soc. Math. France **1913**, 17 (1913).
[3] Conditions de régularité conditions d'irrégularité, condition d'impossibilité dans le problème de Dirichlet. C.R. Acad. Sci. Paris **178**, 349–354 (1924).

Loeb, P., Walsh, B.:
[1] The equivalence of Harnack's principle and Harnack inequality in the axiomatic potential theory. Ann. Inst. Fourier **15**, 2, 597–600 (1965).
[2] Nuclearity in axiomatic potential theory. Bull. Amer. Math. Soc. **72**, 685–689 (1966).

Maeda, F.-Y.:
[1] Harmonic and fullharmonic structures on a differentiable manifold. J. Sci. Hiroshima Univ. Ser. A-I **31**, 271–312 (1970).

Maria, A.J.:
[1] The potential of a positive mass and the weight function of Wiener. Proc. Nat. Acad. Sci. U.S.A. **20**, 485–489 (1934).

Martin, R.S.:
[1] Minimal positive harmonic functions. Trans. Amer. Math. Soc. **49**, 137–172 (1941).

Meyer, P.-A.:
[1] Brelot's axiomatic theory of the Dirichlet problem and Hunt's theory. Ann. Inst. Fourier **13**, 2, 357–372 (1963).
[2] Probability and Potentials. Waltham-Massachusets-Toronto-London: Blaisdell Publishing Company 1966.

Mokobodzki, G.:
[1] Espcés de Riesz complètement réticulés et ensembles équicontinus de fonctions harmoniques. Sém. Choquet. **5**, 601–608 (1968).
[2] Rareté de l'ensemble des pôles de non unicité en théorie axiomatique de Brelot. Sém. Théorie du Potentiel. **11**, 17.01 (1968).

Mokobodzki, G., Sibony, D.:
[1] Principe du minimum et maximalité en théorie du potentiel. Esquisse d'une théorie globale. C.R. Acad. Sci. Paris Sér. A **263**, 126–129 (1966).
[2] Principe du minimum et maximalité en théorie du potentiel. Ann. Inst. Fourier **17**, 1, 401–442 (1967).
[3] Théorie globale du potentiel. C.R. Acad. Sci. Paris Sér. A **264**, 238–241 (1967).
[4] Sur une propriété caractéristique des cônes de potentiels. C.R. Acad. Sci. Paris Sér. A **266**, 215–218 (1968).

Mustaţă, P.:
[1] (to appear)

Myrberg, P.J.:
[1] Über die Existenz der Green'schen Funktionen auf einer gegebenen Riemannschen Fläche. Acta Math. **61**, 39–79 (1933).

Osgood, W.F.:
[1] On the existence of the Green's function for the most general simply connected region. Trans. Amer. Math. Soc. **1**, 310–314 (1900).

Perron, O.:
[1] Eine neue Behandlung der ersten Randwertaufgabe für $\Delta u=0$. Math. Z. **18**, 42–54 (1923).

Pietsch, A.:
[1] Nukleare lokalkonvexe Räume. Berlin: Akademie Verlag 1965. (Second edition 1969.)

Poincaré, H.:
[1] Sur les equations aux dérivés partielles de la physique mathématiques. Amer. J. Math. **12**, 211–294 (1890).

Poisson, S.D.:
[1] Suite du mémoire sur les intégrales définies et sur la sommation des séries. J. éc. royal polyt. **12**, 404–509 (1823).

Remark, R.:
[1] Über potentialkonvexe Funktionen. Math. Z. **20**, 126–130 (1924).

Riesz, F.:
[1] Über die subharmonischen Funktionen und ihre Rolle in der Funktionentheorie und in der Potentialtheorie. Acta Sci. Math. (Szeged) **2**, 2, 87–100 (1925).
[2] Sur les fonctions subharmoniques et leur rapport à la théorie du potentiel. Acta Math. **54**, 321–360 (1930).

Schaefer, H.:
[1] Topological vector spaces. New York: Macmillan; London: Collier-Macmillan 1966 (republished Berlin-Heidelberg-New York: Springer 1971).

Schläfli, L.:
[1] Über die partielle Differentialgleichung $\frac{dw}{dt}=\frac{d^2w}{dx^2}$. J. Reine Angew. Math. **72**, 263–284 (1870).

Schwarz, H.A.:
[1] Über die Integration der partiellen Differentialgleichungen $\frac{d^2u}{dx^2}+\frac{d^2u}{dy^2}=0$ für die Fläche eines Kreises. Vierteljschr. Naturforsch. Ges. Zürich **15**, 113–128 (1870).
[2] Zur Integration der partiellen Differentialgleichungen. J. Reine Angew. Math. **74**, 218–253 (1872).

Sternberg, W.:
[1] Über die Gleichung der Wärmeleitung. Math. Ann. **101**, 394–398 (1929).

Tautz, G.:
[1] Zur Theorie der elliptischen Differentialgleichungen II. Math. Ann. **118**, 733–770 (1941–43).
[2] Zur Theorie der ersten Randwertaufgabe. Math. Nachr. **2**, 279–303 (1949).
[3] Zum Umkehrungsproblem bei elliptischen Differentialgleichungen I, II, Bemerkungen. Arch. Math. **3**, 232–238, 239–250, 361–365 (1952).

Taylor, J.C.:
[1] Strict potentials and Hunt processes. Invent. Math. **16**, 249–259 (1972).
[2] Potential kernels of Hunt processes (to appear).

Vallée Poussin, Ch. de la:
[1] Sur quelques extentions de la méthode du balayage de Poincaré et sur le problème de Dirichlet. C. R. Acad. Sci. Paris **192**, 651–653 (1931).
[2] Extension de la méthode du balayage de Poincaré et problème de Dirichlet. Ann. Inst. H. Poincaré **2**, 169–232 (1932).

Vasilesco, F.:
[1] Sur les singularités des fonctions harmoniques. J. Math. Pures Appl. **9**, 81–111 (1930).
[2] Sur la continuité du potentiel à travers les masses et la démonstration d'un lemme de Kellogg. C. R. Acad. Sci. Paris **200**, 1173–1174 (1935).

Wallin, H.:
[1] Continuous functions and potential theory. Ark. Math. **5**, 55–84 (1963).

Wiener, N.:
[1] Certain notions in potential theory. J. Math. Massachussetts **3**, 24–51 (1924).
[2] Note on a paper of O. Perron. J. Math. Massachussetts **4**, 21–32 (1925).

Zaremba, S.:
[1] Sur le principe de Dirichlet. Acta Math. **34**, 293–316 (1911).

Bibliography

We list here several papers which were not quoted in this book but may present interest for the researcher in this field.

Anandam, V.:

[a] Espaces harmoniques sans potentiel positiv. Ann. Inst. Fourier (to appear).

Avanissian, V.:

[a] Formes linéaires dépendant harmoniquement d'un paramètre. J. Analyse Math. **20**, 397–405 (1967).

Bauer, H.:

[a] Recent developments in axiomatic potential theory. In: Symposium on Probability Methods in Analysis (Loutraki 1966), p. 20–27. Berlin-Heidelberg-New York: Springer (Lecture Notes in Math. 31) 1967.

Bertin, E. M. J.:

[a] Limites projectives d'espaces harmoniques. C. R. Acad. Sci. Paris Sér. A **268**, 869–871 (1969).

[b] Fonctions harmoniques non-réelles. Indag. Math. **32**, 45–52 (1970).

[c] Espaces harmoniques généralisés. Indag. Math. **33**, 10–25 (1971).

Bliedtner, J.:

[a] Groupes harmoniques et noyaux de convolution de Hunt. C. R. Acad. Sci. Paris Sér. A **266**, 529–531 (1968).

[b] Harmonische Gruppen und Huntsche Faltungskerne. In: Seminar über Potentialtheorie, p. 69–102. Berlin-Heidelberg-New York: Springer (Lecture Notes in Math. 69) 1968.

[c] On the analytic structure of harmonic groups. Manuscripta Math. **1**, 289–292 (1969).

Bliedtner, J., Janssen, K.:

[a] Bezugsmasse und dominante Masse in harmonischen Räumen. Rev. Roumaine Math. Pures Appl. **17** (1972) (to appear).

[b] Integraldarstellung positiver Lösungen von linearen elliptischen und parabolischen Differentialgleichungen mit konstanten Koeffizienten. Rev. Roumaine Math. Pures Appl. **17** (1972) (to appear).

[c] Harnacksche Kegel und Metrik in harmonischen Räumen. Math. Ann. (to appear).

Boboc, N., Bucur, Gh.:

[a] Cônes convexes ordonnés. Rev. Roumaine Math. Pures Appl. **14**, 283–309 (1969).

Boboc, N., Bucur, Gh., Cornea, A.:

[a] Cones of potentials on topological spaces. Rev. Roumaine Math. Pures Appl. **18**, (1973) (to appear).

Boboc, N., Cornea, A.:
[a] Balayage des mesures par rapport à un cône de fonctions inférieurement semi-continues sur un espace localement compact. Sém. Théorie Potentiel. **11**, 22.01–22.11 (1967).
[b] Cônes convexes ordonnés, H-cônes et biadjoints des H-cônes. C. R. Acad. Sci. Paris Sér. A **270**, 1679–1682 (1970).
[c] Cônes convexes ordonnés. Représentations integrales. C. R. Acad. Sci. Paris Sér. A **271**, 880–883 (1970).

Boboc, N., Mustaţă, P.:
[a] Sur les domaines d'unicité dans les espaces harmoniques. In: Elliptische Differentialgleichungen, Bd. 2, p. 97–107, Kolloquium Berlin 1969. Berlin: Akademie Verlag 1971.
[b] Considérations axiomatiques sur les fonctions polysurharmoniques. Rev. Roumaine Math. Pures Appl. **16**, 1167–1184 (1971).
[c] Fonctions polysurharmoniques associées aux opérateurs différentiels linéaires du type elliptique. Ultrapotentiels. Rev. Roumaine Math. Pures Appl. **18** (1973) (to appear).

Brelot, M.:
[a] La théorie moderne du potentiel. Ann. Inst. Fourier **4**, 113–140 (1954).
[b] Introduction axiomatique de l'effilement. Ann. Mat. Pura Appl. **57**, 77–95 (1962).
[c] Intégrabilité uniforme. Quelques applications à la théorie du potentiel. Sém. Théorie Potentiel. **6**, 101–112 (1962).
[d] Etude comparée des deux types d'effilement. Ann. Inst. Fourier **15**, 1, 155–168 (1965).
[e] Axiomatique des fonctions harmoniques. Montréal: Les Presses de l'Université de Montréal 1966.
[f] Recherches sur la topologie fine et ses applications. Ann. Inst. Fourier **17**, 2, 395–423 (1967).
[g] Historical introduction. In: Potential theory, 1–21 (CIME, 1° ciclo Stresa 2–10 Luglio 1969). Roma: Edizione Cremonese 1970.
[h] On topologies and boundaries in potential theory. Berlin-Heidelberg-New York: Springer (Lecture Notes in Math. 175) 1971.

Bullen, P. S.:
[a] A general Perron integral. Canad. J. Math. **17**, 17–30 (1965).
[b] A general Perron integral II. Canad. J. Math. **19**, 457–473 (1967).
[c] On a theorem of Privaloff. Canad. Math. Bull. **10**, 353–359 (1967).

Constantinescu, C.:
[a] An axiomatic theory for the non-linear Dirichlet problem. Rev. Roumaine Math. Pures Appl. **10**, 755–764 (1965).
[b] Die heutige Lage der Theorie der harmonischen Räume. Rev. Roumaine Math. Pures Appl. **11**, 1041–1056 (1966).
[c] Harmonic spaces and their connections with the semi-elliptic differential equations and with the Markov processes. In: Elliptische Differentialgleichungen, Bd. 1, p. 19–30, Kolloquium Berlin 1969. Berlin: Akademie Verlag 1970.

Constantinescu, C., Cornea, A.:
[a] Compactifications of harmonic spaces. Nagoya Math. J. **25**, 1–57 (1965).

Curtis, jr., Ph. C.:
[a] On a theorem of Keldysh and Wiener. In: Abstract Spaces Approx. Proc. Conf. Math. Res. Inst. Oberwolfach 1968, 351–356 (1969).

Feyel, D., Pradelle, A. de la:
[a] Principes du minimum et maximalité dans les préfaisceaux. Esquisse d'une théorie locale. C. R. Acad. Sci. Paris Sér. A **272**, 19–22 (1971).
[b] Quelques propriétés de la réduite dans les préfaisceaux maximaux. C. R. Acad. Sci. Paris Sér. A **274**, 1285–1288 (1972).

Gowrisankaran, K.:

[a] Extreme harmonic functions and boundary value problems. Ann. Inst. Fourier **13**, 2, 307–356 (1963).

[b] Extreme harmonic functions and boundary value problems II. Math. Z. **94**, 256–270 (1966).

[c] Fatou-Naim-Doob limit theorems in the axiomatic system of Brelot. Ann. Inst. Fourier **16**, 2, 455–467 (1966).

[d] Multiply harmonic functions. Nagoya Math. J. **28**, 27–48 (1966).

[e] Iterated fine limits and iterated non-tangential limits. Trans. Amer. Math. Soc. (to appear).

Hansen, W.:

[a] Potentialtheorie harmonischer Kerne. In: Seminar über Potentialtheorie, p. 103–159. Berlin-Heidelberg-New York: Springer (Lecture Notes in Math. 69) 1968.

[b] Fegen und Dünnheit mit Anwendungen auf die Laplace- und Wärmeleitungsgleichung. Ann. Inst. Fourier **21**, 1, 79–121 (1971).

[c] Potentialtheorie semi-regulärer harmonischer Kerne. Z. Wahrscheinlichkeitstheorie und Verw. Gebiete **18**, 298–304 (1971).

[d] Cohomology in harmonic spaces. In: Seminar on potential theory II, p. 63–101. Berlin-Heidelberg-New York: Springer (Lecture Notes in Math. 226) 1971.

[e] Abbildungen harmonischer Räume mit Anwendung auf die Laplace- und Wärmeleitungsgleichung. Ann. Inst. Fourier **21**, 3, 203–216 (1971).

[f] Hunt's theorem and axiomatic potential theory. Invent. Math. **14**, 242–252 (1971).

Helms, L.L.:

[a] Introduction to potential theory. New York-London-Sydney-Toronto: Wiley 1969 (or Einführung in Potentialtheorie; to appear).

Hervé, R.-M., Hervé, M.:

[a] Les fonctions surharmoniques dans l'axiomatique de Monsieur Brelot associées à un opérateur elliptique dégénéré. Ann. Inst. Fourier (to appear).

Ikegami, T.:

[a] A note on axiomatic Dirichlet problem. Osaka Math. J. **6**, 39–47 (1969).

Janssen, K.:

[a] Martin boundary and $\mathscr{H}^p$-theory of harmonic spaces. In: Seminar on potential theory II, p. 102–151. Berlin-Heidelberg-New-York: Springer (Lecture Notes in Math. 226) 1971.

Köhn, J.:

[a] Die Harnacksche Metrik in der Theorie der harmonischen Funktionen. Math. Z. **91**, 50–64 (1966).

Köhn, J., Sieveking, M.:

[a] Zum Cauchyschen und Dirichletschen Problem. Math. Ann. **177**, 133–142 (1968).

Loeb, P.A.:

[a] An axiomatic treatment of pairs of elliptic differential equations. Ann. Inst. Fourier **16**, 2, 167–208 (1966).

[b] A criterion for the proportionality of potentials with polar point support. Rev. Roumaine Math. Pures Appl. **13**, 1121–1125 (1968).

Loeb, P.A., Walsh, B.:

[a] A maximal regular boundary for solutions of elliptic differential equations. Ann. Inst. Fourier **18**, 1, 283–308 (1968).

[b] An axiomatic approach to the boundary theories of Wiener and Royden. Bull. Amer. Math. Soc. **74**, 1004–1007 (1968).

Lumer, L.:
[a] $\mathscr{H}^p$ spaces of harmonic functions. Ann. Inst. Fourier **17**, 2, 425–469 (1967).

Maeda, F.-Y.:
[a] Axiomatic treatment of full-superharmonic functions. J. Sci.Hiroshima Univ. Ser. A-I Math. **30**, 197–215 (1966).
[b] Comparison of the classes of Wiener functions. J. Sci. Hiroshima Univ. Ser. A-I **33**, 231–235 (1969).

Maurin, K.:
[a] General eigenfunction expansions on harmonic spaces. Existence of reproducing kernels. Bull. Acad. Polon. Sci. Sér. Sci. Math. Astronom. Phys. **15**, 503–507 (1967).

Meghea, C.:
[a] Compactification des espaces harmoniques (Lecture Notes in Math. 222). Berlin-Heidelberg-New York: Springer 1971.

Mokobodzki, G.:
[a] Représentation intégrale des fonctions surharmoniques au moyen des réduites. Ann. Inst. Fourier **15**, 1, 103–112 (1965).
[b] Cônes normaux et espaces nucléaires. Cônes semi-complets. Sém. Choquet. **7**, B 6.01–B 6.14 (1968).
[c] Cônes de potentiels et noyaux subordonnés. In: Potential theory 207–248 (CIME, 1° ciclo Stresa 2–10 Luglio 1969). Roma: Edizione Cremonese 1970.
[d] Eléments extrémaux pour le balayage. Sém. Théorie du Potentiel. **13**, 5.01–5.14 (1970).
[e] Structure des cônes de potentiels. In: Sém. Bourbaki 1969/70, 239–252 (Lecture Notes in Math. 180). Berlin-Heidelberg-New York: Springer 1971.
[f] Noyaux absolument mesurables et opérateurs nucléaires. C. R. Acad. Sci. Paris Sér A **270**, 1673–1675 (1970).

Mokobodzki, G., Sibony, D.:
[a] Familles additives de cônes convexes et noyaux subordonnés. Ann. Inst. Fourier **18**, 2, 205–220 (1969).

Monna, A. F.:
[a] Remarques sur la théorie axiomatique des fonctions harmoniques et la théorie des faisceaux. Nieuw Arch. Wisk. **14**, 214–221 (1966).

Ow, W. H.:
[a] Wiener's compactifications and Φ-bounded harmonic functions in the classification of harmonic spaces. Pacific J. Math. **38**, 759–769 (1971).

Plessis, N. du:
[a] An introduction to potential theory. Edinburgh: Oliver & Boyd 1970.

Pokryvaĭlo, V. D.:
[a] On the balayage operator (russian) Vestnik Leningrad Univ. 13 (Mat. Meh. Astron. Nr. 3) 72–80 (1971).

Pradelle, A. de la:
[a] Approximation et caractère de quasi-analyticité dans la théorie axiomatique des fonctions harmoniques. Ann. Inst. Fourier **17**, 1, 383–399 (1967).
[b] Remarque sur la valeur d'un potentiel à support ponctuel polaire en son pôle en théorie axiomatique. Ann. Inst. Fourier **19**, 1, 275–276 (1969).

[c] A propos du mémoire de G.F. Vincent-Smith sur l'approximation des fonctions harmoniques. Ann. Inst. Fourier **19**, 2, 355–370 (1970).

Schneider, F.:

[a] Dualität und Martinrand. In: Elliptische Differentialgleichungen, Bd. 1, p. 119–136, Kolloquium Berlin 1969. Berlin: Akademie Verlag 1970.

Sibony, D.:

[a] Cônes de fonctions et potentiels. Cours de 3ème Cycle à la Faculté des Sciences de Paris (1967–68) et à l'Université McGill de Montréal (été 68).

[b] Allure à la frontière minimale d'une classe de transformations. Théorème de Doob généralisé. Ann. Inst. Fourier **18**, 2, 91–120 (1969).

Sieveking, M.:

[a] Integraldarstellung superharmonischer Funktionen mit Anwendung auf parabolische Differentialgleichungen. In: Seminar über Potentialtheorie, p. 13–68. Berlin-Heidelberg-New York: Springer (Lecture Notes in Math. 69) 1968.

Smyrnelis, E.:

[a] Allure des fonctions harmoniques au voisinage d'un point-frontière irrégulier. C. R. Acad. Sci. Paris Sér. A **267**, 157–159 (1968).

Taylor, J.C.:

[a] The Martin boundaries of equivalent sheaves. Ann. Inst. Fourier **20**, 1, 433–456 (1970).

[b] Balayage de fonctions excessives. Sém. Théorie du Potentiel. **14**, 2.01–2.11 (1971).

[c] The Martin boundary and adjoint harmonic functions. Contrib. Theory. Topol. Struct. Proc. Sympos. Berlin 1967, 221–233 (1969).

[d] The Martin compactification in axiomatic potential theory. Proc. Prague Symposium General Topology (to appear).

[e] On the existence of submarkovian resolvents. Invent. Math. (to appear).

[f] Duality and the Martin compactification. Ann. Inst. Fourier (1972) (to appear).

[g] On the existence of resolvents. In: Sém. Probabilité (to appear).

Vincent-Smith, G.F.:

[a] Uniform approximation of harmonic functions. Ann. Inst. Fourier **19**, 2, 339–353 (1970).

Walsh, B.:

[a] Flux in axiomatic potential theory I: Cohomology. Invent. Math. **8**, 175–221 (1969).

[b] Flux in axiomatic potential theory II: Duality. Ann. Inst. Fourier **19**, 2, 371–417 (1970).

[c] Spaces of continuous functions characterized by kernel-like conditions. Math. Z. **109**, 71–86 (1969).

[d] Perturbation of harmonic structures and an index-zero theorem. Ann. Inst. Fourier **20**, 1, 317–359 (1970).

[e] Operator theory of degenerate elliptic-parabolic equations. Math. J. Indiana Univ. **20**, 959–964 (1971).

Index

The notions defined in "Terminology and Notation" are not listed here

Notation

The notations introduced in "Terminology and Notation" are not listed here.

Die Grundlehren der mathematischen Wissenschaften in Einzeldarstellungen mit besonderer Berücksichtigung der Anwendungsgebiete

Eine Auswahl

---

74. Boerner: Darstellungen von Gruppen
76. Tricomi: Vorlesungen über Orthogonalreihen
77. Behnke/Sommer: Theorie der analytischen Funktionen einer komplexen Veränderlichen
78. Lorenzen: Einführung in die operative Logik und Mathematik
86. Richter: Wahrscheinlichkeitstheorie
87. van der Waerden: Mathematische Statistik
94. Funk: Variationsrechnung und ihre Anwendung in Physik und Technik
97. Greub: Linear Algebra
99. Cassels: An Introduction to the Geometry of Numbers
104. Chung: Markov Chains with Stationary Transition Probabilities
107. Köthe: Topologische lineare Räume
114. MacLane: Homology
116. Hörmander: Linear Partial Differential Operators
117. O'Meara: Introduction to Quadratic Forms
120. Collatz: Funktionalanalysis und numerische Mathematik
121./122. Dynkin: Markov Processes
123. Yosida: Functional Analysis
124. Morgenstern: Einführung in die Wahrscheinlichkeitsrechnung und mathematische Statistik
125. Itô/McKean jr.: Diffusion Processes and their sample Paths
126. Lehto/Virtanen: Quasikonforme Abbildungen
127. Hermes: Enumerability, Decidability, Computability
128. Braun/Koecher: Jordan-Algebren
129. Nikodým: The Mathematical Apparatus for Quantum Theories
130. Morrey jr.: Multiple Integrals in the Calculus of Variations
131. Hirzebruch: Topological Methods in Algebraic Geometry
132. Kato: Perturbation Theory for Linear Operators
133. Haupt/Künneth: Geometrische Ordnungen
134. Huppert: Endliche Gruppen I
135. Handbook for Automatic Computation. Vol. 1/Part a: Rutishauser: Description of ALGOL 60
136. Greub: Multilinear Algebra
137. Handbook for Automatic Computation. Vol. 1/Part b: Grau/Hill/Langmaack: Translation of ALGOL 60
138. Hahn: Stability of Motion
139. Mathematische Hilfsmittel des Ingenieurs. 1. Teil
140. Mathematische Hilfsmittel des Ingenieurs. 2. Teil
141. Mathematische Hilfsmittel des Ingenieurs. 3. Teil
142. Mathematische Hilfsmittel des Ingenieurs. 4. Teil
143. Schur/Grunsky: Vorlesungen über Invariantentheorie
144. Weil: Basic Number Theory
145. Butzer/Berens: Semi-Groups of Operators and Approximation
146. Treves: Locally Convex Spaces and Linear Partial Differential Equations
147. Lamotke: Semisimpliziale algebraische Topologie.

148. Chandrasekharan: Introduction to Analytic Number Theory
149. Sario/Oikawa: Capacity Functions
150. Iosifescu/Theodorescu: Random Processes and Learning
151. Mandl: Analytical Treatment of One-dimensional Markov Processes
152. Hewitt/Ross: Abstract Harmonic Analysis. Vol. 2: Structure and Analysis for Compact Groups. Analysis on Locally Compact Abelian Groups
153. Federer: Geometric Measure Theory
154. Singer: Bases in Banach Spaces I
155. Müller: Foundations of the Mathematical Theory of Electromagnetic Waves
156. van der Waerden: Mathematical Statistics
157. Prohorov/Rozanov: Probability Theory
159. Köthe: Topological Vector Spaces I
160. Agrest/Maksimov: Theory of Incomplete Cylindrical Functions and Their Applications
161. Bhatia/Szegö: Stability Theory of Dynamical Systems
162. Nevanlinna: Analytic Functions
163. Stoer/Witzgall: Convexity and Optimization in Finite Dimensions I
164. Sario/Nakai: Classification Theory of Riemann Surfaces
165. Mitrinović/Vasić: Analytic Inequalities
166. Grothendieck/Dieudonné: Eléments de Géométrie Algébrique I
167. Chandrasekharan: Arithmetical Functions
168. Palamodov: Linear Differential Operators with Constant Coefficients
170. Lions: Optimal Control Systems Governed by Partial Differential Equations
171. Singer: Best Approximation in Normed Linear Spaces by Elements of Linear Subspaces
172. Bühlmann: Mathematical Methods in Risk Theory
173. F. Maeda/S. Maeda: Theory of Symmetric Lattices
174. Stiefel/Scheifele: Linear and Regular Celestial Mechanics. Perturbed Two-body Motion—Numerical Methods—Canonical Theory
175. Larsen: An Introduction of the Theory of Multipliers
176. Grauert/Remmert: Analytische Stellenalgebren
177. Flügge: Practical Quantum Mechanics I
178. Flügge: Practical Quantum Mechanics II
179. Giraud: Cohomologie non abélienne
180. Landkoff: Foundations of Modern Potential Theory
181. Lions/Magenes: Non-Homogeneous Boundary Value Problems and Applications I
182. Lions/Magenes: Non-Homogeneous Boundary Value Problems and Applications II
183. Lions/Magenes: Non-Homogeneous Boundary Value Problems and Applications III. In preparation
184. Rosenblatt: Markov Processes. Structure and Asymptotic Behavior
185. Rubinowicz: Sommerfeldsche Polynommethode
186. Wilkinson/Reinsch: Handbook for Automatic Computation II, Linear Algebra
187. Siegel/Moser: Lectures on Celestial Mechanics
188. Warner: Harmonic Analysis on Semi-Simple Lie Groups I
189. Warner: Harmonic Analysis on Semi-Simple Lie Groups II
190. Faith: Algebra: Rings, Modules, and Categories I. In preparation
191. Faith: Algebra: Rings, Modules, and Categories II. In preparation
192. Maltsev: Algebraic Systems. In preparation
193. Pólya/Szegö: Problems and Theorems in Analysis. Vol. 1
194. Igusa: Theta Functions
195. Berberian: Baer $*$-Rings
196. Athreya/Ney: Branching Processes